NATO ASI Series

Advanced Science Institutes Series

A series presenting the results of activities sponsored by the NATO Science Committee, which aims at the dissemination of advanced scientific and technological knowledge, with a view to strengthening links between scientific communities.

The Series is published by an international board of publishers in conjunction with the NATO Scientific Affairs Division

A Life Sciences B Physics	Plenum Publishing Corporation London and New York
C Mathematical and Physical Sciences D Behavioural and Social Sciences E Applied Sciences	Kluwer Academic Publishers Dordrecht, Boston and London
F Computer and Systems Sciences G Ecological Sciences H Cell Biology I Global Environmental Change	Springer-Verlag Berlin Heidelberg New York London Paris Tokyo Hong Kong Barcelona Budapest

PARTNERSHIP SUB-SERIES

1. Disarmament Technologies	Kluwer Academic Publishers
2. Environment	Springer-Verlag
3. High Technology	Kluwer Academic Publishers
4. Science and Technology Policy	Kluwer Academic Publishers
5. Computer Networking	Kluwer Academic Publishers

The Partnership Sub-Series incorporates activities undertaken in collaboration with NATO's Cooperation Partners, the countries of the CIS and Central and Eastern Europe, in Priority Areas of concern to those countries.

NATO-PCO DATABASE

The electronic index to the NATO ASI Series provides full bibliographical references (with keywords and/or abstracts) to about 50000 contributions from international scientists published in all sections of the NATO ASI Series. Access to the NATO-PCO DATABASE compiled by the NATO Publication Coordination Office is possible in two ways:

- via online FILE 128 (NATO-PCO DATABASE) hosted by ESRIN, Via Galileo Galilei, I-00044 Frascati, Italy.

- via CD-ROM "NATO Science & Technology Disk" with user-friendly retrieval software in English, French and German (© WTV GmbH and DATAWARE Technologies Inc. 1992).

The CD-ROM can be ordered through any member of the Board of Publishers or through NATO-PCO, Overijse, Belgium.

2. Environment – Vol. 6

Springer
Berlin
Heidelberg
New York
Barcelona
Budapest
Hong Kong
London
Milan
Paris
Santa Clara
Singapore
Tokyo

Sustainable Development of the Lake Baikal Region

A Model Territory for the World

Edited by

Valentin A. Koptyug

Siberian Branch of the Russian Academy of Sciences
Lavrentiev Avenue 17
630090 Novosibirsk, Russia

Martin Uppenbrink

Federal Agency for Nature Conservation
Konstantinstraße 110
D-53179 Bonn, Germany

Springer

Published in cooperation with NATO Scientific Affairs Division

Proceedings of the NATO Advanced Research Workshop "Sustainable Development of the Lake Baikal Region as a Model Territory for the World", held in Ulan-Ude, Republic of Buryatia, September 12–17, 1994

Cataloging-in-Publication Data applied for

Die Deutsche Bibliothek - CIP-Einheitsaufnahme

Sustainable development of the Lake Baikal region : a model territory for the world ; [proceedings of the NATO Advanced Research Workshop "Sustainable Development of the Lake Baikal as a Model Territory for the World", held in Ulan-Ude, Republic of Buryatia, September 12 - 17, 1994] / ed. by Valentin A. Koptyug ; Martin Uppenbrink. Publ. in cooperation with NATO Scientific Affairs Division. - Berlin ; Heidelberg ; New York ; Barcelona ; Budapest ; Hong Kong ; London ; Milan ; Paris ; Santa Clara ; Singapore ; Tokyo : Springer, 1996
(NATO ASI series : Sub-ser. 2, Environment ; Vol. 6)
ISBN 3-540-60560-6
NE: Koptjug, Valentin A. [Hrsg.]; Advanced Research Workshop Sustainable Development of the Lake Baikal as a Model Territory for the World <1994, Ulan-Udė>; NATO: NATO ASI series / 02

ISBN 3-540-60560-6 Springer-Verlag Berlin Heidelberg New York

Printed in Germany

Typesetting: Camera-ready by authors/editors
SPIN: 10492681 31/3137 – 5 4 3 2 1 0 – Printed on acid-free paper

Preface

The Advanced Research Workshop (ARW) "Sustainable Development of the Lake Baikal Region as a Model Territory for the World" was held from 12-17 September, 1994 in Ulan-Ude, capital of the Republic of Buryatia of the Russian Federation.

The opening session of the workshop was chaired by Mr. Leonid Potapov, President of the Republic of Buryatia, and attracted an audience of over 400 people, which indicated the interest aroused in Ulan-Ude by the NATO Advanced Research Workshop.

The Russian authorities have proposed that the Lake Baikal Region be considered a model territory for working out the concept of sustainable development as a follow-up to the United Nations Conference on Environment and Development held in Rio de Janeiro in 1992. Lake Baikal is the oldest, largest and deepest lake in the world - the object of great scientific interest and of strong spiritual and emotional significance to all Russians, and particularly to the people of Siberia. In spite of the pollution threatening the lake, it is not yet considered to be irreversible. It is for this reason that the lake appears to be a legitimate candidate for the system of world model territories for sustainable development. The workshop was devoted to the discussion of the scientific basis of this project.

The principal aims of the workshop were to consolidate existing data on the current state of the environment and economy of the region, to develop a system of indicators for sustainable development, to make recommendations on additional components to the existing monitoring system, and to consider a legal framework and mechanism for its implentation.

The recommendations of the workshop covered the following issues: environmental strategies for industrial development and energy production; best available technologies and best environmental practice; agriculture, forestry, nature conservation and landscape planning; eco-tourism; legislative and administrative support for sustainable development; environment monitoring and control; indigenous population; education and human health.

Participants agreed that the Lake Baikal Region required an overall regional strategy for the most important forms of land use, and also that the number and size of protected areas should be enlarged. Due to the extent of the Lake Baikal Region it was recommended that the entire process, from planning to implementation, should be tested as pilot projects in selected areas. The participants further recommended that UNESCO should be asked to recognise Lake Baikal as a World Natural Heritage Site.

August 1995 The Editors

CONTENTS

II. Main Problems of the Lake Baikal Region Development and the Role of Science and Culture for Their Solution

Introductory Address

L.V. Potapov
President of the Republic of Buryatia, Ulan-Ude, Russia

Dear guests,

Dear participants of the International Symposium,

Two years have passed since the International Conference on Environment Protection and Development was held in Rio de Janeiro and which was an important milestone in the history of mankind.This meeting of heads of countries and governments that took place under the aegis of the United Nations stated that nature protection is an indispensable component of sustainable development. So it is only natural that we are now considering this particular meeting as the next stage of implementation of the decisions taken in Rio de Janeiro. Of course, the practical implementation of those recommendations has to be tried on some concrete object. And we are grateful for the fact that the Lake Baikal Region has been recognised as a potential model territory for sustainable development. We believe that the recommendations of the Symposium that is being inaugurated today will allow us, the Republic of Buryatia, and neighbouring regions of the Lake Baikal Region to find the right balance between the solution of our socio-economic problems and environmental protection. It is especially important for us who live in the Republic of Buryatia because all our economic activities take place in the water-catchment area of Lake Baikal.

A month ago, an international meeting of astronauts and cosmonauts was held here. Right now, an international exhibition "Sibecologia '94" is taking place in Irkutsk. And this meeting today will be an important stage in the solution of the urgent problems Buryatia is facing, as well as the Irkutsk and Chita regions.

We always stress that Lake Baikal is the largest fresh-water reservoir of the planet; meanwhile, the shortage of fresh water is increasing everywhere. Besides, the Lake Baikal Region is also a unique natural complex. And we must preserve its uniqueness. This is Russia's and Buryatia's responsibility. It is also the responsibility of all mankind.

We wish all the participants of the Symposium good health, a pleasant trip to Lake Baikal and successful work in this field.

NATO ASI Series, Partnership Sub-Series, 2. Environment – Vol. 6
Sustainable Development of the Lake Baikal Region
Edited by Valentin A. Koptyug and Martin Uppenbrink

We are grateful to everybody who is taking part in this Symposium, especially to our guests from abroad. We are inclined to discuss our problems concerning transition to market economy. And we think it right to discuss the questions included in the agenda in spite of the difficulties of the socio-economic situation in Russia.

I wish you success in your work!

Welcome Address

Ch.-D. Budayev
Humbo-Lama, Chairman of the Central Buddhist Board of the
Russian Federation

On behalf of the Central Buddhist Board of the Russian Federation I greet you, the participants of the International Symposium, on the sacred Lake Baikal land. At the present time it is obvious that the environmental situation has become a general global problem. The solving of this problem requires mutual efforts of the whole human community. I am sure it is necessary to achieve the unification and harmonization of East and West. Traditionally, the West is distinguished by developed technology and the East has achieved much in developing moral sciences and the science of the mind. These two directions of perfecting human nature ought to be united. Without such a combination it is hard to imagine that the sustainable development of the society will ever be possible.

I do hope this high forum will further promote the ways of harmonic development of progress and national traditions, economic success and preservation of nature, problems of the moral revival of society.

Today, with the growth of material consumption taking such a dangerous lead over the development of the culture of the spirit, when in modern society the problem of choice is becoming more and more acute, the Buddhist spiritual potential is attracting the attention of scientists throughout the world. Buddhism encourages the moral perfecting of human nature and ecological responsibility, personal self-restraint and social solidarity.

The teachings of Buddha brings prosperity and tranquillity to us all. I wish you peace and perfection irrespective of your religious beliefs.

NATO ASI Series, Partnership Sub-Series, 2. Environment – Vol. 6
Sustainable Development of the Lake Baikal Region
Edited by Valentin A. Koptyug and Martin Uppenbrink

I. Sustainable Development: General Aspects, Requirements and Indicators

Main Factors Necessitating Mankind's Transition to Sustainable Development

V.A. Koptyug
Siberian Branch of the Russian Academy of Sciences, Novosibirsk, Russia

The UN Conference on Environment and Development, held in June, 1992 in Rio de Janeiro on the level of Heads of States or Governments, declared a new paradigm of socio-economic development for the 21st century, the concept of sustainable development. The paper discusses the basic reasons for unsustainability of the present development of the world, and also the main criteria and requirements of the future model.

The title of this workshop contains the words "sustainable development". What does this term mean? In a broad sense this implies the acknowledgement that the previous development of our civilization has been unstable, disturbing the equilibrium existing in nature so drastically that it has become dangerous for all life on the planet Earth. At the end of the 20th century, mankind discovered some changes of global scale in the environment:

– depletion of the ozone layer, with increasing negative impact of the shortwave UV radiation on living organisms;

– increasing concentration of carbon dioxide and other "greenhouse gases" in the atmosphere, leading, in the opinion of many experts, to the threat of global warming of the climate;

– large-scale deforestation with diminishing consumption of atmospheric carbon dioxide;

– fast degradation of the land mainly devoted to food production;

– a significant decrease in the biodiversity of the planet.

These and some other global changes in the environment are caused by depredatory exploitation of natural resources and large-scale anthropogenic pollution of the environment by toxic substances. Both types of impact lead to the rapid degradation of nature.

Anthropogenic impact became commensurable in scale with natural processes and, in some cases, even exceeded them. As an example, the emission of heavy metals into the atmosphere from anthropogenic and natural sources may be mentioned.

Why has the anthropogenic impact on nature been increasing so fast over the last decades? The main cause is the fast development of the world's industrial and

NATO ASI Series, Partnership Sub-Series, 2. Environment – Vol. 6
Sustainable Development of the Lake Baikal Region
Edited by Valentin A. Koptyug and Martin Uppenbrink

agricultural production accompanied by increasing total consumption at the expense and to the detriment of nature. In order to reduce the anthropogenic impact on nature, we must strengthen the policy of resources and energy saving per unit of each product. This is an absolutely necessary but insufficient measure. Why?

Let us assume hypothetically that current resources and energy consumption will be halved over a rather short period, disregarding at the moment the cost and time of the realization of required measures. Very important for the evaluation of the situation is also a social factor, namely the enormous inequality, or, to be more precise, enormous economic disparity between and within various countries in consumption scales. Today 20% of the richest part of the world population earn 82.7% of global income, whereas 20% of the poorest population earns only 1.4%. This disparity increased from 30:1 in 1960 to 59:1in 1989. A similar but much faster increase in this disparity was observed in Russia in the course of the so-called reforms.

The consumption of world resources by industry and of goods by the population of the developed countries is much greater than in the developing countries. For example, developed, transitional and developing countries consume today aluminum in the ratio 68:16:15 and copper in the ratio 64:18:18, respectively, but the population of developed countries comprises only 20 to 25% of the world's population. Therefore, it is more informative to compare consumption of developed and developing countries per capita. For example, the per-capita consumption in the United States exceeds that in India by a factor of 34 for aluminum, 45 for copper, 58 for phosphate rock, 12 for coal, 43 for petroleum, 184 for natural gas, 10.5 for beef, etc. However, the ratio of the population sizes between the United States and India is about 1:6!

Still, the poorest people would also like to have adequate rights to adequate living standards, including a corresponding level of consumption. If we attempt to increase the standard of living of 80% of the population earning 17.3% of the world income to the level of 20% with 82.7% of the world income, it will be necessary, in the framework of existing technologies of production, to use 20 times more natural resources than today.

Then we should also take into account the growth of the world's population, which, over the last decades, has taken on the character of an explosion, with a much more rapid increase in the population of the developing countries. It is expected that the population of the world will double in the next 30 to 40 years.

Is it possible to diminish resource and energy consumption per unit of the various products 40 times during this perod through the corresponding changing technologies in order to protect our civilization from the disaster caused by the depletion of non-renewable resources and rapid degradation of the environment? We would have to be very optimistic in order to give a positive answer.

It would mean that on the threshold of the new millennium mankind must globally change the paradigm of development of our civilization.

Maurice Strong, Secretary-General of the UN Conference on Environment and Development held in Rio de Janeiro in June, 1992 at the level of the heads of states and governments, said in his opening statement at the Conference:

> "The same processes of economic growth which have produced such unprecedented levels of wealth and power for the rich minority have also given rise to the risks and imbalances that now threaten the future of rich and poor alike. This growth model, and the patterns in production and consumption which have accompanied it, is not sustainable for the rich, nor can it be replicated by the poor. To continue along this pathway could lead to the end of our civilization."

The UN Conference on Environment and Development declared the concept of sustainable development as the basis of a new paradigm for the future development of civilization and accepted the agenda of actions for the forthcoming 21st century (Agenda 21).

I would like to mention some general principles constituting the foundation of the proposed concept and partly formulated in the Rio Declaration on Environment and Development:

1. The orientation of the development process only to conventional economic indicators cannot be tolerated any longer. Principle 4 of the Rio Declaration proclaims: "In order to achieve sustainable development, environmental protection shall constitute an integral part of the development process and cannot be considered in isolation from it." Thus, we must find an acceptable balance between economic development and conservation of nature. It is necessary in calculating the GNP of individual countries to take into account not only any increase in man-made capital, but also the corresponding decrease in "natural capital". Therefore, the economic approach should be replaced by income calculation methods within the framework of ecological economics.

2. The pursuit of profit can no longer be considered as the main driving force of development. The market system will become more and more regulated by parliaments and governments under the pressure of global changes in the environment and by society. Moral components of development, taking the culture and traditions of a specific nation into account, will play an increasingly vital role. The role of non-governmental entities, including religious communities, will also increase.

3. The disparity of living standards for the rich and the poor must be reduced both within each country and in the world as a whole. Principle 5 of the Rio Declaration says: "All states and all people shall cooperate in the essential task of eradicating poverty as an indispensable requirement for sustainable development in order to

decrease disparities in standards of living and better meet the needs of the majority of the people of the world."

4. Very important is also Principle 3 of the Rio Declaration: "The right to development must be fulfilled so as to equitably meet developmental needs of present and future generations."

This last principle, incidentally, implies among other things the necessity to replace gradually non-renewabl raw materials in industrial production by renewable ones. This is a challenge for scientists.

Science, considering the globality and multidisciplinarity of existing problems, should play the role of primary importance in the transition of civilization to sustainable development.

All this does not mean that the transition to sustainable development will be mainly troublesome for developing countries. Albert Gore, Vice-President of the United States, writes in his book "Earth in the Balance: Ecology and the Human Spirit" (1993): "Many of the impediments to progress lie in the industrial world. Indeed, one of the biggest obstacles to a Global Marshall Plan (a plan of moving to sustainable development —V.K.) is the requirement that the advanced economies must undergo a profound transformation themselves . . . that will be in some ways more wrenching than that of the Third World, simply because powerful established patterns will be disrupted."

Why does Albert Gore believe this? It was demonstrated above that the developed countries are using the major part of the world's resources and generate the major part of harmful emissions and waste. Thus, they are mostly responsible for the emergence of many global problems at the end of the 20th century and should be the first to change their pattern of consumption.

According to Agenda 21 (item 8.7), affirmed by the Heads of States and Governments in Rio de Janeiro: "Governments should adopt a national strategy for sustainable development based on, inter alia, the implementation of decisions taken at the Conference ... Its (the national strategy—V.K.) goals should be to ensure socially responsible economic development while protecting the resource base and the environment for the benefit of future generations. It should be developed through the widest possible participation. It should be based on a thorough assessment of the current situation and initiatives."

In February, 1994, the President of the Russian Federation signed the decree on elaboration of such a strategy for Russia. This will have been done by the end of this year. At the same time in several regions of Russia the elaboration of regional programs of sustainable development was started as a result of the initiatives of local authorities and scientists. Certainly, these programs later must be interfaced with the federal strategy and program.

This workshop is devoted to the discussion of the problem of sustainable development of the Lake Baikal Region. This is a very special case, taking into account a lot of specific features of the Lake and its surroundings.

The scientists of Siberia believe that the Lake Baikal basin ought to have the status of a World Natural Heritage Site. This idea will be discussed in the course of the workshop.

The majority of the people support this proposal. But at the same time the rest are cautious because they fear possible limitations of economic development and excessive internationalization of the region. Apprehension about this last point is partly caused by provocative publications in western newspapers and journals: articles such as “Let’s Buy Siberia” by Walter Russell Mead (CQ, March, 1994).

I hope that the detailed and comprehensive discussion of main regional problems in the course of the workshop will help us to finalize the concept of sustainable development of this unique region.

Environment as a NATO Priority

L. Veiga da Cunha
Director, Priority Area on Environmental Security
NATO Scientific Affairs Division
B-1110 Brussels, Belgium

Water has always been at the centre of many myths, beliefs, magic and conflicts. Like the sun and earth, water is a cosmic element. According to Greek philosophy, water, air, earth and fire were the four elements of which all substances in the world were made; in today´s terms that means water, air, soil and energy.

Sacred and trivial, dominating and submissive, omnipresent and elusive, tender and wild, water is all of these things at once. These different viewpoints reflect the unique role of water in our lives and also justify our inability to comprehend all the intricacies of the water cycle. This has been the case throughout historical times, and it must be said that today we still have a surprising lack of knowledge concerning the storing, transporting and conserving of our water resources.

Lake Baikal is the oldest, largest, deepest lake in the world—the object of great scientific interest, and of strong spiritual and emotional significance to all Russians, and particulary to the people of Siberia. In spite of the pollution threatening the lake, it is not yet considered to be irreversible. This is the reason why the lake appears to be a legitimate candidate for the system of world model territories for sustainable development. The present NATO workshop is expected to contribute towards giving scientific support to this intention.

The importance of the great lakes of the world as reserves of water resources and as support for freshwater ecosystems is unquestionable, and Lake Baikal is a paradigmatic case. Such lakes are in fact a fundamental element of our environment and form part of our world's natural heritage.

Water of rivers and lakes has for centuries been affected by numerous problems. In the future, the quality of water and the imbalance of freshwater ecosystems are expected to change more quickly and more drastically than before. With the changes in lake ecosystems, landscape will also change through the processes of soil erosion, sediment deposition and the circulation of nutrients. Landscape is indeed the result of a delicate balance between water, air and land.

Scientific research has recognised this complex interdependence of water, soil and air, and new technologies have certainly contributed to some progress, but, unfortunately, it has not been possible so far to establish general laws because of the many regional differences that affect the behaviour of freshwater ecosystems. This regional diversity requires from scientists and managers that they consider

NATO ASI Series, Partnership Sub-Series, 2. Environment – Vol. 6
Sustainable Development of the Lake Baikal Region
Edited by Valentin A. Koptyug and Martin Uppenbrink

the development of water and soil resources from an integrated perspective. Water-resources management has to be a process by which the integrated management of water and soil is achieved in a given river basin, taking into consideration all the pertinent economic, social and environmental factors.

Within the framework of the NATO Scientific Affairs Division, a strong emphasis has been and is being put on the environment and environmental security, in such a way that it can be said that the environment and environmental protection are clearly NATO priorities. This NATO concern with environmental issues has developed through the two committees whose activities make up the Third Dimension of NATO, the Science Committee and the Committee on the Challenges of Modern Society (CCMS).

The mechanisms of these two committees are complementary. The Science Committee activities are in most cases designed "bottom-up", as a result of applications submitted spontaneously by individual scientists or groups of scientists, and funded by NATO on a competitive basis, as a foundation would do. The action of CCMS is developed through structured pilot studies designed "topdown" at the initiative of groups of countries and funded directly by the nations.

The Science Committee has, during its 35-year existence, supported activities covering a wide spectrum of scientific interests, making use of mechanisms such as science fellowships, advanced research workshops, advanced study institutes (post-doctoral level schools) and collaborative research grants, having involved hundreds of thousands of scientists. Many of these activities deal with environmental problems. In the 25 years of activity of the Committee on the Challenges of Modern Society, about 70 pilot studies have been developed through international co-operation.

Apart from the NATO Science Committee activities resulting from unsolicited applications, this committee has also approved the so-called special programmes, where activities usually result from applications solicited by NATO. Several of these special programmes have been developed over the years in the area of the environment. The most recent of these programmes still to be active deals with the Science of Global Environmental Change and is aimed at describing and understanding the complex, interactive physical, chemical and biological processes that regulate the total earth system.

Following the recent political changes in Central Europe and the former Soviet Union, the NATO Science Committee has launched a series of activities developed within the Workplan for Dialogue, Partnership and Cooperation approved annually by the North Atlantic Cooperation Council which comprises ministers of the 16 NATO countries and the 22 co-operation partner countries, i.e., the countries in Central Europe and the new countries in the territories of the former USSR. This workplan covers activities in the political, military, economic, scientific and environmental fields. The activities carried out by the NATO Scientific and Environmental Division have been developed within five major priority areas following the preferences expresses by the co-operation partners.

The first priority area covers the scientific problems related to disarmament technologies, including the disposal of nuclear and chemical weapons, defence industry conversion and the safety of nuclear technologies.

The second area covers scientific problems related to the environment, including regional environmental problems (Lake Baikal falls into this category), the reclamation of contaminated military sites and natural and man-made disasters.

The third priority area covers scientific problems related to high technology, including information science, materials science, biotechnology and energy.

The fourth area concerns problems related to science and technology policy, including the redeployment of defence-industry scientists, intellectual property rights and career mobility.

Finally, the fifth area covers problems related to advanced communication technologies and computer networking.

The fact that the current Advanced Research Workshop on Lake Baikal is part of the Priority Area on Environment justifies my adding some brief comments on this area.

During its short life, the Priority Area on Environment has already sponsored numerous advanced research workshops, advanced study institutes, linkage grants, collaborative research grants and expert visits, to which it is obviously not possible to make individual reference here. However, in order to give a flavour of the wide spectrum of activities supported, I will refer to the titles of only a limited number of sponsored meetings and collaborative projects.

Meetings, i.e., advanced research workshops and advanced study institutes, have dealt with a variety of topics such as "Critical Scientific Issues of the Aral Sea Basin: State of Knowledge of Future Research Needs" held in Tashkent; "Integrated Rehabilitation of Environmentally Devastated Areas in River Basins in Eastern Europe" held in Moscow; "Disturbance and Recovery of Arctic Terrestrial Ecosystems" held in Rovaniemi, Finland; and "Transboundary Water Resources Management: Technical and Institutional Issues" held in Skopelos, Greece.

Collaborative Projects, i.e., linkage grants and collaborative research grants, have also covered a variety of topics. Examples are: "Development and Application of Ultrasensitive Method for Plutonium Determination in Environment", a co-operation between Germany and Russsia; "Ecological Restoration and Management of Military Training Areas", a co-operation between the United

Kingdom and Czech Republic; "Environmental Aspects of Agricultural Soil-Water Management", a co-operation between USA and Albania; "Alaska-Russia, Past and Present Tectonic Connections", a co-operation between USA and Russia.

The extensiveness of the list does not allow me to give you any details on the activities referred to above, but I would like to make additional reference to three other activities organised in Siberia. The first is the Advanced Research Workshop held last week in Barnaul on "Long-term Consequences of Nuclear Tests for the Environment and Population Health (Semipalatinsk/Altai Case Study)". This workshop, organised as a follow-up to another NATO workshop on the same topic held in January this year in Vienna at the International Atomic Energy Agency, was a great success. These were in fact historic meetings, where for the first time the United States, Russia, Western European countries and China provided information related to their nuclear tests and the assessment of their consequences for the environment and human health.

The other two activities relating to Siberia in fact also deal with Lake Baikal. The first is a linkage grant on the "Anthropogenic Impact and Aquatic Ecosystem Health of Lake Baikal", a co-operation between Italy and Russia, and the second is a collaborative research grant on "Seismicity, Active Faults, Stresses and Deformations in the Baikal Rift", a co-operation between France and Russia.

I hope that I have been able to give you an initial idea of the scientific and environmental activities of NATO. Further details can of course be shared with you in informal discussions during the course of this workshop and thereafter. I also hope this workshop may be just the start of a fruitful co-operation dealing with Lake Baikal, lasting many years. The local organisers and the scientists involved in this first initiative deserve our thanks and congratulations. I am sure that the conclusions and recommendations of this workshop will provide an agenda of action for the scientists and decision-makers dealing with the sustainable development of freshwater ecosystems in general, and in particular with the sensitive ecological system Lake Baikal.

At the threshold of a new millennium, humankind has a very short time—years or decades, certainly not centuries—to restore the degraded freshwater ecosystems and to prevent the degradation of new ones. Adequate management of natural and ecological resources and sustainable development of these resources are indeed great challenges facing us which will require a clear political will to invest in the future for the welfare of coming generations.

The Global Ecological Challenge: Two Years After the Rio Summit

Martin Uppenbrink
Federal Agency
for Nature Conservation
Konstantinstr. 110
D-53179 Bonn
Germany

The paper deals with the description of the progress that was made at the Rio summit, particularly the UN's foundation of a Commission for Sustainable Development (CSD). Furthermore the implementation of the Rio Conventions and the Rio Declaration are discussed. Finally, the paper points out the challenge facing the international community to formulate a new philosophy of the relationship between society and environment.

Introduction

First of all, I would like to express my thanks to the Government of the Republic of Buryatia, the host of the workshop, and to the Russian Federation, with which my country, Germany, maintains an important bilateral environmental agreement. The heads of the delegation have just now reached St. Petersburg. The Federal Minister for the Environment, Prof. Klaus Töpfer, and the Secretary of State, Mr. Stroctmann, have a strong personal relationship to your country, to the NATO and the Head of the Science Unit, Mr. Luis Veiga da Cunha and, last but not least, to the academician, Prof. Koptyug.

Please allow me to make a review of the subjects of environment and nature. I will talk about it in a global sense, but sometimes I will be looking at it on the European level. Moreover, we also have to see the small-scale level.

1. United Nations Conference on Environment and Development: The Rio Conference

The World Meeting of the four-year process in Rio de Janeiro in June, 1992 was of the greatest importance for the environment, nature, mankind and development. 178 presidents and heads of government from all over the world attended the meeting. On several issues an international legal agreement was signed: a Convention on Climate and a Convention on Biodiversity. Moreover, Agenda 21, a statement of environmental principles and a special declaration on protection of the forests, was passed. This makes clear that this summit was not merely a political show. The Rio Conference was initiated by the Brundtland Commission on Environment and Development. The Brundtland Report had already brought the term *sustainability* onto the political scene. It was the attempt to bring together the surplus of the North with the needs of the South while thinking about future generations. With its far-reaching resolutions,

NATO ASI Series, Partnership Sub-Series, 2. Environment – Vol. 6
Sustainable Development of the Lake Baikal Region
Edited by Valentin A. Koptyug and Martin Uppenbrink

the Conference was a meeting that initiated a new thinking process. The meeting tried to overcome the terminology separating the First, Second and Third Worlds. Rio made clear that there is at least for the North, but not only for it, a new principle of responsibility of shared resources, of shared means and of the shared future. It was the beginning of thinking globally and acting responsibly at home, i.e., in the region, on a national level and locally. It was the first time that "sustainable development" was tackled, defined and was found to be worth fighting for. It was a meeting where politicians decided to be aware, to save the life of mankind and of our planet.

Rio drew the following conclusions on the subject of sustainable development of natural resources:

1. The rate of exploitation of renewable resources or better natural goods or natural elements is not allowed to exceed the rate of regeneration. This corresponds with the demand for maintenance of ecological productivity, which means (at least) the conservation of ecological capital.

2. Immissions can not only be based on the resistance of the function of production, but must also reflect the sensitive function of regulation, in particular by animals and plants.

It was in a way the break with Communism that led worldwide to initiatives for establishing and fighting for peace for the future within a non-military context. Development and the environment as the overriding issues had an important impact on the new partnership for peace. One could even say that what the process of Helsinki was politically for the European region, was taken over globally, as far as the necessity for peace-keeping is concerned, by the Rio Conference.

2. Agenda 21

This document was passed in Rio and makes the attempt to integrate the complicated relationship between environment and development into a kind of notebook, a task book, a work guideline. It deals with analyses and instructions for actions in 40 chapters; the text is easy to read, and the paragraphs are very short. Within this document many transsectorial subjects are dealt with: combating poverty, changing consumer patterns, demographic dynamics, sustainable-development protection and promotion of human health. Also, sectorial issues were considered: protection of the atmosphere, integrated approach to the planning and management of land resources, combating deforestation, combating desertification and conserving the biological diversity. For the first time, Agenda 21 points out the role of important groups: global action for women towards sustainable and equitable development, recognizing and strengthening the role of leaders and their communities and strengthening the role of NGO's. It does not ignore the role of workers and their labor unions as well as business and industry nor science and engineering, questions of financing and international legal instruments and mechanisms. Agenda 21 has thus been accepted on the political scene.

3. Commission on Sustainable Development (CSD)

One of the reasons for the establishment of this Commission is the fact that it was already decided in Rio to implement a Commission on Sustainable Development. This Commission is chaired by the German Federal Minister, Dr. Töpfer. It consists of 53 members, who are to have a political and not a civil-servant mandate. They define their function as the steering of the process of the implementation of Agenda 21. They want to achieve this:

– by surveying,

– by controlling this process,

– by coordinating actions,

– by steering the political direction of the implementation and

– by lending the process a special dynamic impetus.

In connection with this Commission, but as an independent body, the General Secretary of the United Nations, Boutros Ghali, created a High-level Advisory Board on Sustainable Development, and I do not hesitate to mention this important board here in the presence of the academician Prof. Koptyug, a member of that famous advisory group to the Secretary General of the United Nations.

The CSD did something rather new in the context of the United Nations work: they drew up a working program for themselves which will encompass several years. On the basis of this working program, the 1994 meeting covered the following issues:

a) Central Issues

1. Critical elements of sustainability (especially the change of consumer and production patterns; environment and trade)

2. Institutions

3. International law instruments

4. Role of important groups

5. Funding of the implementation of Agenda 21

6. Technology transfer and cooperation, capacity building, education and sciences.

b) Sectorial Issues

1. Help to human settlement

2. Drinking water resources

3. Chemical substances and security

4. Waste

For the 1994 and 1995 sessions of the CSD, it is planned to tackle the following cross- sectorial issues: finances, cooperation in technology and technology transfer, capacity building and the sectorial issues of: protection and use of land resources, mountains, forests, the combating of desertification, sustainable agriculture, biodiversity and biotechnology.

The way international organizations react to the desires of the Commission shows that it seems to be possible to bring together the family of the United Nations, which means that UNEP, UNDP, UNESCO, FAO and other UN organizations are cooperating in questions of sustainable development.

4. The Rio Conventions

The World Conventions on Climate and on Biodiversity signed in Rio were ratified in record time. Conferences of the contracting states have now to be held. According to international law they are test treaties.

a) Convention on Climate

The Convention provides for the reduction of substances which are a danger to the climate (CO_2, carbon dioxide; CFC's, chlorofluorocarbons; methane). As such, it is a framework convention with a technical and substantial approach. In conclusion, the question for the environmental-technical sectors is how to cut back substances that are suspected of damaging the atmosphere, because otherwise more chemical reactions in the atmosphere will have negative impacts on our planet. As of now it is still a moot question if the reduction of CO_2 emissions will be directly listed in the protocol and whether this protocol can be passed at the first conference of the contracting states. 35 small insular states recently presented a draft with the following central demand: reduction by 20 % of CO_2 emissions in the industrialized countries by the year 2005 (compared with the level of 1990). Up to now the Commission has only a general—in the framework of international law non-binding—statement of the industrialized countries to stabilize CO_2 emissions at the level of 1990 without an exact date of fulfillment. It will be crucial whether the idea of stabilization and reduction can be formulated by the European countries and governments with high expectations for environmental policy by the spring of 1995, either separately or jointly in the European Union. In this connection the question should be mentioned whether a CO_2 energy tax can be an incentive in the reduction of CO_2. At the moment there is no consensus on this.

b) Convention on Biodiversity

This Convention constituted an advancement in the conservation of species and their habitats in 1992. The most important aims of this ambitious Convention—designed as a shelter for other conservation-oriented treaties—are the following: the conservation of biodiversity, the sustainable use of biodiversity components and the fair distribution of the expected advantages by using genetic resources. The fair distribution of the advantages is to be realized by providing suitable access to genetic resources and the transmission of technologies. All this is to be made possible by a proper financing.

For the first time in history, this agreement has established a wide international basis for the conservation as well as for the sustainable use of natural resources and especially for the natural heritage of the earth. The Convention already went into effect at the end of last year, and all members of the European Union except Belgium and Ireland have joined it. Through this agreement it has nearly been possible to realize a worldwide protection of all animal and plant species and their habitats. The Convention brings the already extant international regulations for protection of species and habitats to completion.

The first conference of the parties of the Convention on Biological Diversity will be held in Nassau, Bahamas, from 28 November to 9 December 1994. On the basis of two intergovernmental preparatory conferences, the member states will have to tackle complicated institutional matters:

- Questions related to the financial mechanism for the convention
- Consideration of a possible mechanism for safety in biotechnology, including an approach to determining the need and modalities for a protocol on biosafety
- Safety and technical matters
- Preparations for the scientific body of this important convention.

The increasing rate of diminishing of biological diversity, the loss of or the damage to species and their habitats, leads to irreparable infringement on nature and is threatening the life basis of mankind. With the Convention on Biological Diversity the world leaders expressed their wish to stop this development and to guarantee the protection of biological diversity. The most important theme of the first meeting of the parties of this convention will be the creation of a permanent financial mechanism. This mechanism has to give direct help to developing countries. The member states of the European Union feel that the Global Environmental Facility (GEF) should assume this role. The Convention on Biological Diversity is directed towards the global protection of animal and plant species and their threatened habitats and to preserve genetic potentials. The developing countries have the biggest portion of biological diversity. Therefore they should get direct aid for a stronger cooperation, improved technology transfer and funds for the implementation of the Convention.

But this Convention is also important for Europe. The preservation of species can only be realized through the protection of the biotopes. With the ratification by the member states, this Convention will be legally binding for the European Union. This Convention will encourage the EU member states to realize an effective and careful protection of plants and animals and their habitats. At the 4th CSCE Conference in Helsinki it was resolved to protect the natural and semi-natural ecosystems in Central and Eastern Europe. They are to be integrated into the development of a European

Biotope Network, referred to as "Natura 2000" in a statement of intent by the European Union in Brussels.

5. Rio Declaration on Desertification and on Global Development of Forests

a) Convention on Desertification

More than 3 billion hectares of land or 25% of the land surface of the earth are endangered by desertification. For a long time desertification was looked upon as a special problem of Africa. In 1970, the world was alarmed by the dreadful drought in West Africa, where more than 250,000 people died. Since 1977, Africa is trying to solve the problems of accelerating erosion, desertification and droughts without effect. Now, after the Rio Conference, it is evident that this is a global environmental problem. The pressure to negotiate urgently for a Global Convention came from Rio. The Convention was signed in Paris in the presence of many heads of state and government on October 14, 1994. This agreement is intended to support in particular the poorest people in Africa and secure their basis of existence. The efforts of the affected countries and of the donor countries must be enhanced by improved coordination to protect the vegetation, the fertility of soils and water reserves. Because of the geographical proximity and the historical relationship between Africa and Europe, we have a special obligation to fullfill from the viewpoint of Europe. By the completion of the ratified Convention on Desertification we have to think about the waves of migrations out of Africa that will affect Europe. Therefore it is not surprising that the European Union has for a long time been financing about half of the worldwide development funding. The contribution to Africa is even higher. 7.5 billion US $ were paid by the European Union up to 1992. Since then, an additional 434 million US $ have been promised by all programs of the Commission in Brussels for desertification programs, development plans in forestry and agriculture. But there is still one drawback: the international community, especially in the North, still has problems recognizing the global dimension of this problem.

The industrialized countries are still denying the relationship between global-climate change and desertification. If such a correlation were accepted, it would be a boon to the developing countries. They could then refer to the responsibility of the North and demand more action to resolve the problems.

b) Declaration on the Development of Forests

The North joined the Rio Conference with an ambitious idea for the protection of forests. In Rio, the South showed that it is not willing to give up its sovereignty regarding its natural resources, i.e., forests, mainly the tropical rain forest. It does not want to bring all or a part of its forests under a worldwide Convention on Forests. Nevertheless, a political agreement was realized about the use, protection and sustainable development of the forests. The Declaration on the Development of Forests is not binding by international law. Nevertheless, it can be seen as an advancement because negotiations about a possible Convention on Forests or a

Forest Protocol are going to be held. Actually it is being discussed at many levels, but there have been no negotiations yet.

Before we discuss global problems we should look at our own. In Germany, for example, forests are important for the protection of abiotic and biotic resources. In former times a forest ecosystem covered nearly all regions. Today the forests are displaced to locations unsuitable for agriculture. Forests are to some extent isolated and are very seldom in a natural state, with no impact by man. Therefore nature conservationists are demanding more protected forest areas and forest ecosystems in Germany. The forest cover is about 30% of the total area of Germany.

Besides historical and present damage, silvicultural management itself causes impacts on the ecosystem forest. The impact is not a quantitative change of the forest area but a qualitative deterioration of biological diversity. Because of this, an accelerating decrease and endangering of typical species and habitats in the German forests may be observed.

Nature conservation demands the integration of nature-conservation principles in the whole area of forest, based on the principles of a sustainable forestry that protects the biological diversity in the forest. Therefore it is necessary to integrate forestry and conservation aims especially in economically used forests. To point out once more, the principle of sustainability, the magic formula of Rio, i.e., to reach a sustainable development—such was the idea of the developing countries, but of others, too—is primarily geared to utilization and secondarily to the protection of natural resources. The idea of sustainability occurs for the first time in the forest regulation of the monastery of Mauermünster in Alsace in the year 1144. Much later the idea of forest regulation came into effect, and it was not permitted to cut down more trees than were being reproduced. As Fritz Vorholz mentioned in his article "Die Last der Hedonisten"—The Burden of the Hedonists— in *Die Zeit*, the economic principle of living not from capital, but from interest was not applied to the relations of man towards natural resources outside the forests. The forestry term of sustainability has to be extended to meet the demand of nature conservation for forest management. Not only the preservation of the supply of timber and the capacity of timber production can be of primary interest; equally importantly, the functions of the ecosystem forest are to be secured with all its biological and ecological components.

6. Global Environment Facility (GEF)

In the global context, the conservation of environment and nature was content with weak institutions and disastrously low financial support up to the end of 1980. The Conference on Environment in Stockholm was about the protection of man's environment. Questions of nature, natural resources and conservation of nature were not discussed, even though the UNESCO had done good preparatory work. The United Nations Program on Environment (UNEP) was established in Nairobi, with all its weaknesses, which it has thus far not been possible to overcome, thus effecting a change in the damage to nature. Dr. Tolba, the former executive director of the UNEP, introduced an initiative of the North: nature conservation in the Third World should be supported not only with loans but also by cancellation of debts. This Global Environment Facility is a common act of UNEP, UNDP and the World Bank. Actually the dominant institution of this common body is still the World Bank. But the influence of environmental and development organizations as well as the influence of the Third World has grown. The new loans of the North (2 billion US $ for the next year) are indications of a good development. Naturally the North is still far from supplying the funds actually needed. But the North has successfully begun to set priorities. These priorities are the reduction of CFC's, steps to prevent climate changes, progress in the water sector and the protection of biological diversity. The GEF is a first step in the right direction. The North recognizes its responsibility as well as its impact and the damage it causes and is reacting with support and assistance. The South is beginning to accept the priorities and is not simply demanding general support. The conventions are beginning to develop a real impact on the member states because at last international laws are being tied to "financial mechanisms". At last there are international agreements showing good results; the Montreal Protocol is a good example. This development is a hopeful sign. Furthermore, "environment funds" should integrate others, especially "green" international agreements, into a system of financial support. The financial shortcomings of conventions such as RAMSAR, CITES, BONNER and BERNER, as well, are obvious. It is noteworthy that former conventions were established without funds while the more technical protocol (Montreal) on the reduction of the CFC's was directly supported by funding.

7. The Dilemma of Nature

This gives me the opportunity to express my concern about the fact that nature is not given the proper significance by politics nor by the society. The literature showing the lack of acceptance of nature conservation in politics and society is legion. One must not fail to recognize that the broad early nature-conservation movement, including citizens coming both from the right and left, has not reached a breakthrough for nature and nature conservation. Nevertheless, especially in Central Europe it has been possible to establish a citizens' movement which was primarily concerned about chemical substances and excessive air-traffic noise. Therefore the movement was more oriented to technological environmental protection and was a force behind the repression of chemical substances. The term *environment* became a general term. Only with the "ecological turnabout"—which never happened politically—the phenomenon nature was accepted by technological environmental protection. The term ecology—taken from natural sciences—became a catchword of the 80's. I do not want to criticize the terminology and which term covers which field and I don't have the intention of bringing it to the constitutional level. I only want to point out that at least after the Rio Conference and the implementation of the Convention on Biological Diversity, questions about nature and natural resources are now more present in our minds, without directly acknowledging nature "rights of its own".

At least the geneticist, biochemist and biologist tell us that the genetic resources of nature—and man as a part of it—are endlessly complicated and will never be fully understood. Mankind has always had to live with complexity and complex problems. Even if man has found wrong solutions, one has to understand that man had to analyse and solve a lot of things at the same time. This also has to be applied to the ecological demands of our time.

In this context, the social question which has occupied mankind for the last hundred years must also be seen. It crops up in the serious problems of child labor, low wages and the social insecurity of employees. At least in Western Europe many questions have been answered, albeit with some weaknesses; not so in the socialist countries and thus far not in the USA and Japan, either. The "social safety net" tries to help as many people as possible.

If we transfer these issues and solutions to the current ecological questions for the existence of humanity, we realize that there is no possibility to deduce environment and natural conservation from the social regulations of the constitutions. They have a dramatic value in their own right which calls for a secure ecological net of their own. Like the government or employers and employees in a regulated, oppositional relationship, the actors in the fields of nature and environment, "Man and the Biosphere", should also reach solutions through regulated relations and procedures. A triangular relationship can be developed between the government, the users of nature and the conservationists. I don't mean institutions that cannot fight because of their sense for harmony. There is no time left for harmony. The fight for natural resources is already raging.

Summary

1. Nature and the environment are worldwide—certain improvements— in worse condition than at the beginning of the environmental movement and the environmental policies of the 70's.

2. Conservation trends and agreements regarding nature—following Rachel Carson's "Silent spring" of 1962—have remained very weak.

3. They were displaced or dominated by the following phenomena for more than 20 years: the protection of man and especially human health. This was and is a necessary viewpoint; the same is true for the measures taken.

4. Nevertheless: nature was not preceived as capital and thus not defined as such and protected. The interests of nature conservation were not recognized; nature was nearly insignificant.

5. With the Rio Conference and the following events (three important Conventions, with Agenda 21 and especially the Global Environmental Facility), a global view of problems is discernible for the first time. Also, solutions—albeit still at a low level—can be seen. Instead of conventions, regulations will be established: permanently, with a financial base and flexible structures.

6. With the concept of sustainable development a serious debate concern the use and protection of natural resources. Sustainable development should not be a weak term.

We have to overcome Max Weber's view, who said, "Man is the economic animal who destroys nature."

7. The worldwide Convention on Biological Diversity must provide the shelter for the rational balance of demands on natural resources. Diversity ensures survival.

8. The world is familiar with the magic triangle (negative: poverty—population—nature/environment; positive: the social net in Western Europe). For nature, an ecological net must be developed which the goverment, the users and protectors should take part in organizing.

9. No society can exist without a minimum of ethnic consensus. "If the people decide one day 'enough is enough', spirit and culture, consideration and compassion will become reality again" (Marion, Countess Dönhoff). And I add: the capital of nature will be conserved and not wasted, as it is now.

10. Courage should not grow weary (to cite Mayer-Tasch).

The Buddhist Perspectives of Sustainable Ecological Development

C.-D.Budayev
Central Board of Buddhists of the RF, Ulan-Ude, Russia

V.V.Mantatov
Baikal Foundation, Ulan-Ude, Russia

The paper summarises the moral values accepted by Buddhism and complying with the ideas of the sustainable development of human society.

The world we live in is extremely unstable. How can we achieve stability in life? The Buddhist answer to this question opens new perspectives for the sustainable development of the world community and its different regions.

In our opinion, the global perspective of humankind is an ecological civilisation of noosphere orientation. And sustainable development is a gradual, continuous process of transition to the new civilisation, to the harmonic society in which the priorities of development will be spiritual and ecological ones.

The Buddhist philosophers suggest a scientifically founded hierarchy of the new civilisation's values as an alternative to the traditional social idea in the West. Instead of the idea of man's domination over nature and man's uniqueness, Buddhism puts forward the idea of the denial of the individual reality of "the self", of the intrinsic self-importance of every living creature; instead of the idea of nature conquest, the "akhimsa" principle, that is, not to harm any living being; instead of the ideology of political violence, the concept of a natural non-violent way of development, the principle of tolerance; instead of estranged values, genuine happiness.

The ideal of Buddhism is man content with life in the human community and living in harmony with nature.

Buddhism encourages personal restriction and social solidarity, justice and equality, love and compassion, purity of thoughts and deeds. This is a powerful spiritual instrument for ridding oneself of egoism and consumer ideology, of conventional economic ideas and technocratic utopias.

To the Buddhist philosophers, the universe comprises a spiritual and an informational reality and is governed by single laws. Life as a whole is the expression of the spiritual unity of the universe, and the spiritual development of the individual is simply a desire to understand this wholeness. The fundamental principle of Buddhist spirituality is that every living being is capable of achieving a state of Buddha, that is, the cosmic reality of the mind. According to this principle, man is no longer different from all other living creatures.

NATO ASI Series, Partnership Sub-Series, 2. Environment – Vol. 6
Sustainable Development of the Lake Baikal Region
Edited by Valentin A. Koptyug and Martin Uppenbrink

All living beings in the universe are related to each other. “Remember,” Buddha used to say, “everything in nature is your father and mother.” In every flower, in every living thing the secret of man’s vitality is hidden. If nature is ill, man is ill, too. If it is unclean around man, his consciousness becomes unclean, too. We perceive the surrounding world through all our sense organs, and thus it becomes our “self.”

If we destroy nature, we destroy ourselves. To kill an animal is to kill a part of one’s soul. In one of the Buddhist treatises it is said that when birds die, they send their curses. Thus, the requirement not to harm any living being has a deep humanistic meaning and is full of concern for man. In harming nature, man behaves suicidally.

The living ethics of Buddhism originates from the experience of the indissoluble interconnection and interdependency among the living beings on earth and in the cosmos. If everything is interconnected and interrelated, then any egoism towards nature, including human egoism, is unnatural and is destructive to man above all. Selfishness contradicts the natural fundamentals of life itself. There is no fatal struggle for survival between different individuals of the same species. Love and compassion, solidarity and mutual help are the bases of life. According to the famous scientist Lavlok, all organisms live in harmony and are involved in the cyclical interdependency which in the long run allows all beings to exist.

The universal integrity of existence, the primary unity of man and nature is a scientific fact. But how must we restore this broken unity, how can we achieve inner harmony with nature? There are two ways. The first is to learn to apprehend the truth. Knowledge is defined as the apprehension of the true nature of consciousness, its absolute change (alaya-vidjnyan). This is a co-participation in the great void, the cosmic mind which is at the basis of any creative work and life. This high knowledge in turn gives birth to the feeling of compassion for every living creature, without exception. The second way is that of good deeds. The Buddhist philosophers distinguish, as a rule, three levels of moral perfection. The first level of ethics contains the rules of behaviour which improve people’s karma. This is mainly an ordinary man’s ethics, including prohibitions. At the second level people practise and develop such positive qualities as to allow them to get rid of their own ego, “selfness”. These qualities are high morality, tolerance, diligence and contemplation. This is the ethics of compassion, when another person’s reality is of greater value than the reality of one’s own “selfness”. This is the ethics of bodhisatvas. And finally, the third level, the level of pradjnya-paramita, at which one realises the illusoriness of one’s own individual selfness as well as the “ego” of another being, and thus the possibility of human selfishness disappears, and one realises spiritual unity at the cosmic level. This is the ethics of Buddha.

In connection with the topic of our report, we would like to pay special attention to the scientific and philosophical essence of Buddhism. Buddhism is a science of the mind of cosmic philosophy.

The leading idea of the Buddhist outlook is that of the unity of the microcosmos, man, and the macrocosmos, the universe. "Everything is in one thing and one thing is in everything, everything is everywhere, in the smallest thing," says Buddha. Just as a microparticle includes in itself the whole metagalaxy, a person himself contains the whole world. Man's consciousness is connected with the cosmic information sphere. Just as tiny perturbations at the microlevel effect the behaviour of the system at the macrolevel, small human deeds possess a cosmic character. Everything in the world occurs because we exist. We exist because the universe is this way. Between the microcosmos and macrocosmos, between local and global processes, there is a maximum co-ordination of evolutionary rhythms. "Perfecting himself, a man improves the universe," writes Agvan-Lobsan Tentsin-Gjyatso, Dalai-lama XV.

Buddhism acknowledges a creative participation of man in the evolution of the universe, a cosmic dimension of human existence. "In this body, mortal and less than two metres tall," says Buddha, "there exists the world, its origin and its cessation."

In the centre of the Buddhist cosmos is man and his consciousness. The everlasting importance of Buddhism is that it gives proof of man's special role and predestination, as well as his mind in the evolutionary process. The mind guides the world. The mind is the basis of life. From the Buddhist point of view, the sustainable development of the world is a "cosmogenesis", a process governed by reason. The main problem is what spiritual values a man is going to profess, what perspectives and ideals he will follow in his activity. Everything depends on the frame of reference and the perspective he has chosen: "Everything that exists is caused and created by the thought." Thus begins Dhammapada. The Buddhist law of karma makes us responsible for our thoughts, as our actions are only the realisation of thoughts. It means that we ought to foresee and calculate all possible consequences of our decisions.

We are happy to state that today we have here the best intellectual representatives of our planet who have come to help us find a correct choice and to define possible ways of further development. It is just the beginning, and there will be different opinions expressed. However, it is a boundless event from the point of view of the general feeling of responsibility for Lake Baikal, and the choice of the Lake Baikal Region as a world model territory of sustainable development.

Such a choice, in our opinion, is quite justified. First of all, it is the existence of the unique natural phenomenon in the region which has all the characteristics of a World Natural Heritage site. Lake Baikal can be considered as a strategic potential of the planet. Secondly, the Lake Baikal Region possesses colossal ecological opportunities. Thirdly, our region is characterised by a good demographic situation from the ecological point of view. Here, the population density is half the permissible world limits. Fourthly, there is a large scientific and intellectual potential which is of great importance, too. And finally, our region has rich ecological and cultural traditions.

In a definite sense, the Republic of Buryatia can be considered as a micromodel of ecological civilisation going back to the cultures of Central Asia. According to scientific findings, our ancestors lived in harmony with nature, which was not just a material object of conquest, but rather an equal spiritual partner in an endless dialogue. Every event happening in the surrounding world was accepted as a reaction of natural spirits to man's actions. Therefore, in the case of an imbalance with the outside world, people did their best to understand the spiritual structure of the world. Our ancestors understood the interrelation between the degradation of the habitat and the unhealthy symptoms of the human consciousness. They tried to find spiritual contact with nature and treated it like a human being.

Unfortunately, today we frequently encounter cases of barbaric, destructive and consumer attitudes of people towards nature. "The heart burns at what is taking place on the sacred land of Lake Baikal," writes the Buryat author Konstantin Karnyshev. "Let us look around and be frightened by all we have done." In fact, there is a cosmic law of retribution for unrighteous actions. Therefore every action ought to be considered from the point of view of the final results. This idea was clearly understood by the ancient Buddhists. The revival of the ecological intuition of Buddhism and the ecological ethics of the past centuries are the most important factors for the sustainable development of our region.

According to Buddhist teaching, the basis for the sustainable development of society is a spiritual evolution of the individual. This is a moral perfection of a person which requires gradual transition from the state of inner imbalance with himself and with the surrounding world (in the language of Buddhism such a state is called dukha: discontent, suffering) to the state of harmony with the world, with himself, that is, to nirvana. When we feel spiritual harmony we can live in stable relations with people. The stability of social relations could affect the environment, and so on. Development is a matter of man, his psychic energy, his conscience and responsibility. "The most important thing," said Dalai-lama XV in his Nobel speech (1989), "is that everybody must realise his own responsibility for another person and for nature."

Our position is this: sustainable development will become possible only when the idea of ecological development acquires a sacred status in people's minds. In this respect it is hard to overestimate the role of Buddhism for the sustainable development of our republic.

Buryatia is a traditional centre of the Buddhist religion in Russia. The victory of democracy gave new impetus to the development of open Buddhist beliefs. 22

Buddhist temples, dozens of Buddhist communities, centres and schools have already been opened. Buryatia is also a centre of Buddhist scientific ideas. It is necessary to use the scientific and spiritual potential of Buddhism to help introduce the ideas of sustainable development.

Buddhist religious leaders and scientists can greatly contribute to the development of the ecological and ethical code of the Republic of Buryatia.

In addition, in order to strengthen the co-operation and interaction of science, religion and the public in global and regional problems of ecological development, we consider it necessary to organise a Lake Baikal Club (similar to the Club of Rome). Perhaps we are the ones who will open the way leading to solidarity at a new level and to new possibilities of sustainable development.

We realise that the transition to the model of sustainable development under the conditions of the economic crisis will be difficult and complicated. But as Buddha said, "Wrong and bad actions are easy to make, and what is good and useful is extremely difficult to fulfil." Our belief in sustainable development is based not on the power of money, dollars and rubles, but on the new consciousness, spirituality and ethics. Sustainable development largely depends more on the perfection of human qualities than on economic growth. The way of sustainable development is a path of the spiritual awakening of civil society. As long as this idea is understood, the perspective of ecological development will become increasingly real. This is our approach to the problem of sustainable development.

Interaction Between Economics and the Environment from the Point of View of Sustainable Development

Robert Costanza, Alexey Voinov
Maryland International Institute for Ecological Economics
Center for Environmental and Estuarine Studies
University of Maryland
Box 38, Solomons, MD 20688-0038, USA

Sustainable development is a long-term goal for which there should be broad and growing consensus. Establishment of this goal is fundamentally a social decision about the desirability of a survivable ecological economic system. It entails maintenance of (1) a sustainable scale of the economy relative to its ecological life-support system; (2) a fair distribution of resources and opportunities, not only between the current generation of humans, but also between present and future generations and between humans and other species; and (3) an efficient allocation of resources that adequately accounts for natural capital. Conventional economics (including environmental economics—defined as the application of conventional economics to problems of the environment) has concentrated on the third of these problems (efficient allocation) and therefore has not fully addressed the issue of sustainable development. Ecological economics is a transdisciplinary effort to extend the conventional approach by integrating the natural sciences (especially ecology) and the social sciences (especially economics but also sociology, psychology, political science and others) in order to address the interrelated issues of sustainable scale, fair distribution and efficient allocation in an integrated way.

Operationalizing sustainable development thus implies operationalizing this broader academic agenda and developing policies and instruments based on this systems approach. This paper describes: (1) definitions of sustainability; (2) the limits of the conventional approach in dealing with sustainability and the extension of the conventional view through the transdisciplinary field of ecological economics; (3) the research agenda to develop ecological economics and the political and institutional adjustments that are needed with special focus to developing economies; and (5) an attempt to apply ecological economics to the particular situation of Lake Baikal region.

Defining Sustainability

First, we should note that "sustainability" is currently translated into Russian as "stability." However, sustainability is not stability and sustainable development can be hardly identified with only stable development. Sustainable development may be viewed as a long-term goal for which there should be broad and growing

NATO ASI Series, Partnership Sub-Series, 2. Environment – Vol. 6
Sustainable Development of the Lake Baikal Region
Edited by Valentin A. Koptyug and Martin Uppenbrink

consensus. Establishment of this goal is fundamentally a social decision about the desirability of a survivable ecological economic system.

In a way it is a problem of attempting to exert control over the system in such a way that it complies with the chosen goal. On the other hand, it also assumes that the goals and priorities may be and should be changed if the desired behavior cannot be achieved within the existing system properties and if the existing control mechanisms are not sufficient to attain the goal.

Stability has always been a favorite topic in systems ecology. Ecological stability has been discussed extensively in the literature. Some general biological and rigorous mathematical definitions have been suggested.

According to the classic definition, we assume that there is a certain state or trajectory that the stable system returns to or remains close to after being disturbed within certain limits (Fig. 1a). On the contrary, an unstable system actually drifts away from its trajectory if disturbed. The property of stability was mostly analyzed within the framework of ecological systems, described in terms of such variables as population numbers, biomasses, densities, and concentrations. A system was considered as operating on its own, and stability was sought as its intrinsic feature. In this case the dynamics of the system were considered self-contained with the goal of the system imbedded into it. Human social and economic activities were considered external perturbations, and the major issue was whether the system was stable enough to accept and dampen out these perturbations, while maintaining its ecologically natural origin. Stability is similarly treated in some of the economic applications. Again it is a property of the system, and various social or ecological priorities or factors act as forcing functions or perturbations.

The notion of sustainability comes into play when we realize that we cannot delimit the system in such a way that all the essential links, inputs and outputs can be taken into account. "Pristine" ecosystems that can be separated from the human world are impossible to find (Botkin, 1990). The "natural" ecological system operates in a world of human social relations, which are difficult to bound, predict, and understand. (We realize that actually it is just the other way round. However, for particular ecosystems the effect of humans in many cases turns out to be crucial and at the level of those systems, it is the way we have put it above.) Most important is that the social values and goal functions are altered constantly, as a result of changing perceptions and priorities within societies. Therefore, in the case of analyzing sustainability the goal is imposed on the system externally. The sought state or trajectory of the system is to be changed, and the question is whether the system has internal resources to respond appropriately or whether there exist additional external controls that can bring the system to the newly set dynamics (Fig. 1b).

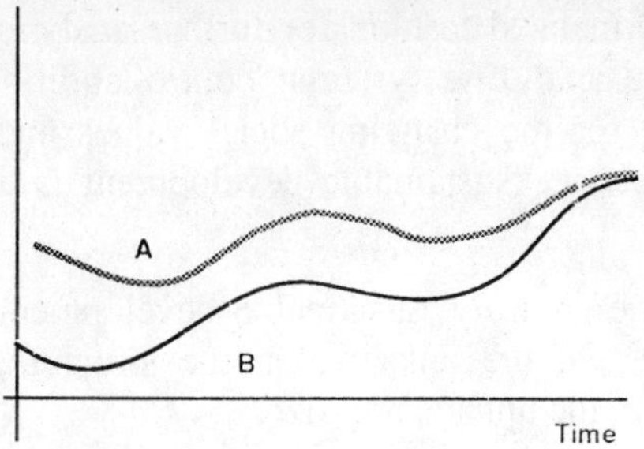

Fig.1a. Stability concept.

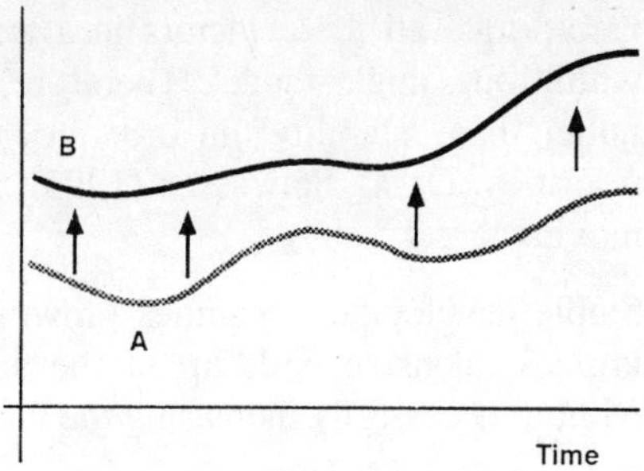

Fig.1b. Sustainability concept.

Robinson (1991) stresses that sustainability calls for maintenance of the dynamic capacity to respond adaptively. Instead of analyzing the system structure in order to understand its behavior and the state or dynamics that it will demonstrate, we are now investigating the flexibility of the system to adapt according to the changing goals and controls.

With the system displaying the behavior presented by trajectory A, suppose that due to some changes in the values and social perceptions we now want the system to follow trajectory B. If we can find the appropriate control mechanisms and change the system parameters or structure in such a way that the system will follow trajectory B, then we may call our system sustainable. It should be noted that in this case human decision makers take the responsibility of constantly managing the system in such a way that this equilibrium is preserved. Sustainability appears as dynamically maintained stability.

The stable state is being achieved due to some intrinsic system mechanisms—it is a function of the system. System sustainability assumes constant management. The system will be sustainable if there exists such a management scenario that will bring it to the desired state or dynamics. It will be unsustainable if the management scenario does not exist, or if it cannot be applied due to some external conditions (say, financial limitations) or the goal function fails to be modified in an appropriate way. In turning to the sustainable-development paradigm, humans in a way take a greater responsibility for the future of the system.

If we could create a socioecological model that would take into account not only the biological and physical processes, but would also describe the social and economic relations that are responsible for modifying and setting the priorities and value sets as functions of the other system components, then we could probably go back to analyzing system stability. In the sustainability analysis we admit that there is yet no adequate understanding of how and why decisions are made or what actually drives human priorities and values. We are not yet ready to incorporate all these factors in a totally formalized fashion for further analysis within one single model. Therefore, we are analyzing systems "controllability" rather than "stability" in order to provide for the changing social values and priorities. Or as Salwasser (1993: 587) puts it: "sustainable development is a moving target."

Stable development assumes growth. On the contrary, sustainable development implies intensive build-up of the structure and the links within the structure, without necessarily increasing the throughput, the uptake, and size.

Conventional vs. Ecological Economics

Scale, Population, Growth, and Appropriation of Natural Capital

The dominant paradigm in conventional economics at the moment is based on some assumptions about the world which, while very useful in dealing with the problems of efficient allocation of resources in the short-term, are less accurate and useful in dealing with the longer term, broader, and more difficult problems of sustainable development. In economic terms, sustainable development involves three hierarchically interrelated problems. These are maintaining: (1) a sustainable scale of the economy relative to its ecological lifesupport system; (2) a fair distribution of resources and opportunities, not only between the current generation of humans, but also between present and future generations and between humans and other species; and (3) an efficient allocation of resources over time that adequately accounts for natural capital (Daly 1990). The problem of distribution is thought by most conventional economists to be best left to the political process, and the scale problem is thought not even to exist as a problem due to infinite substitution possibilities and technological change.

In ecological economics we attempt to extend the domain of economics to cover the broader issues of scale and distribution, while integrating these problems with the allocation problem and extending the allocation problem to adequately internalize "natural capital."

Intervention is required because the market by itself is unable to accurately reflect the biophysical and ecological constraints. As stated by Daly (1984), "there is no more reason to expect the market to find the optimum scale than there is to expect it to find the optimum income distribution. Just as we impose ethical constraints on income distribution and let the market adjust, so must we be willing to impose ecological constraints on the scale of throughput, and let the market adjust." Daly's analogy with the "plimsoll line" on a boat clearly illustrates this. Suppose we are economists and want to maximize the load that a boat carries. If we place all the weight in one corner of the boat, it will quickly capsize. We need to distribute the weight evenly, and to do this, we invent a price system. The higher the waterline in any corner of the boat, the higher the price for putting another kilogram in that corner, and the lower the waterline, the lower the price. This is the internal optimizing rule for allocating space (resources) among weights (alternative uses). This pricing rule is an allocative mechanism only. In the real world, with lack of information and true uncertainty, it sees no reason not to keep on adding weight and distributing it equally until the optimally loaded boat sinks to the bottom of the sea. What is lacking is a limit (albeit dynamic) on scale, a rule that says "stop when total weight is one ton, or when the waterline reaches the red mark" (Daly 1984).

Fig. 2 illustrates this concept in relationship to two other prevalent visions of the future. Fig. 2a shows the "conventional" economic optimistic view of ever-continuing growth (in terms of the above definitions) of the human-made components of capital at the expense of natural capital. Environmentally minded individuals within the conventional camp argue that this growth can be used to fund preservation of some of the remaining natural capital, but only as a luxury, since natural capital is not necessary to the operation of the economy and could be driven to zero without causing collapse of the economy. Fig. 2b illustrates a more realistic (but pessimistic) view that shows over-expansion of the human economy causing collapse of the ecological life-support system and ultimately collapse of the economy which depends on it. The collapse may be more or less severe and allow for recovery afterward, but this is still not a very desirable vision of the future. The third vision (Fig. 2c) indicates the distinction between growth and development, the ecological "plimsoll line" (including uncertainty) and the possibility for continued development (in terms of the above definitions), if the physical dimensions of the economy are maintained below the planet's carrying capacity. This vision of the future encapsulates the essential characteristics of ecological economics.

During the early stages of industrial development, it was not so harmful to neglect the impacts on the ecosystem, because the size of the human population and the production and consumption patterns of resource use were small relative to the natural environment and did not significantly interfere with ecosystem functions, including their assimilative capacity for human wastes. Today, the situation is different. The world economy and human population have reached magnitudes at which the effects of its various activities can no longer be absorbed by ecosystems without significant changes and better adjustments of its material flows to the biogeochemical cycles of the biosphere (Ehrlich 1994).

Attempts to estimate where we are in relation to carrying capacity limits have been made on a global scale by Vitousek et al. (1986) and Rees and Wackernagel (1994). They have estimated that as much as 20%–40% of the global net primary production of natural terrestrial ecosystems are diverted to human activities and that human material demand now exceeds the long-term carrying capacity of Earth.

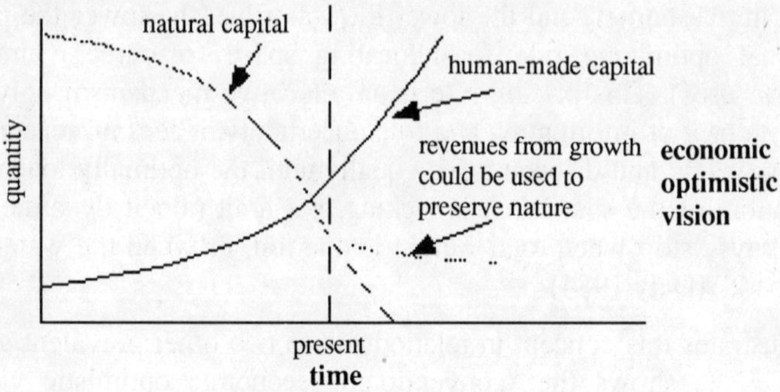

A. The economic optimist vision of the future with unlimited growth and the environment as a luxury good

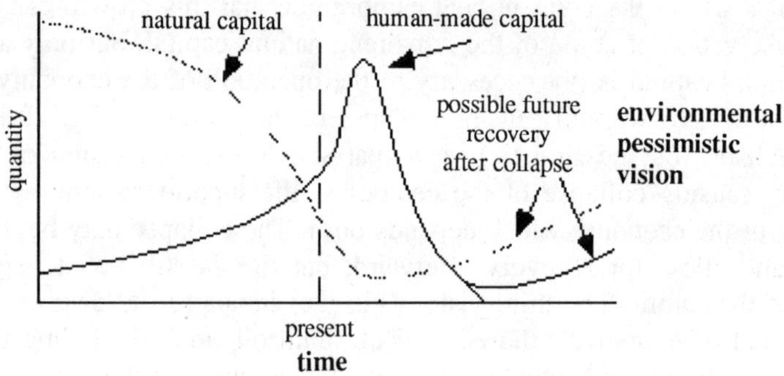

B. The environmental pessimist vision of the future with ultimate collapse of the ecosystem causing collapse of the economy.

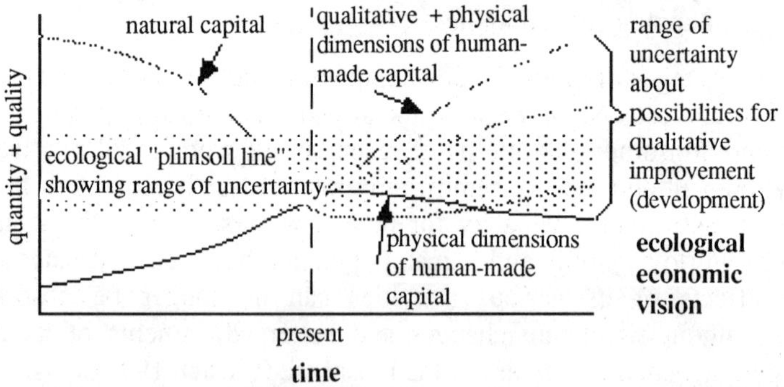

C. The ecological economic vision of the future emphasizing limiting the physical dimensions of the economy below the ecological carrying capacity while encouraging development (qualitative improvement) and acknowledging uncertainty.

Fig.2. Visions of possible future development.

Fixed Tastes and Preferences and Consumer Sovereignty

The conventional paradigm assumes that tastes and preferences are fixed and given and that the economic problem consists of optimally satisfying those preferences. In the short run (i.e., one to four years), this assumption makes some sense. Tastes and preferences don't change that fast. But they do change, and in fact there is an entire industry (advertising) devoted to changing them. Sustainability is an inherently long-term problem and in the long run it doesn't make sense to assume tastes and preferences are fixed. This is a very disturbing prospect for economists because it takes away the easy definition of what is "optimal." If tastes and preferences are fixed and given, then we can adopt a stance of "consumer sovereignty" and just give the people what they want. We don't have to know or care why they want what they want—we just have to satisfy their preferences as efficiently as possible. But if preferences are allowed to change over time and under the influence of education, advertising, etc., we need different criteria for what is "optimal," and we have to figure out how preferences change, how they relate to these new criteria, and how they can or should be changed to satisfy the new criteria.

One alternative for these new criteria is sustainability itself, or more completely sustainable scale, fair distribution, and efficient allocation. These criteria imply coming to a social consensus on a sustainable scale and fair distribution and then using both the market and other institutions like education and advertising in order to implement these social decisions. This might be called "community sovereignty" as opposed to "consumer sovereignty." If tastes and preferences can change, then who is going to decide how to change them? Two points need to be kept in mind: (1) preferences are already being manipulated every day; and (2) we can just as easily apply open democratic principles to the problem as hidden or totalitarian principles in deciding how to manipulate preferences. So the question becomes: do we want preferences to be manipulated unconsciously, either by a dictatorial government or by big business acting through advertising? Or do we want to manipulate them consciously based on social dialogue and consensus with a higher goal in mind? Either way, this is an issue that can no longer be avoided and one which we believe can best be handled using open democratic principles and innovative thinking.

Ecological economists stress that markets are not "free" in the sense of being independent. They always act within an institutional or societal framework. It is this framework that determines and directs the behavior of producers and consumers. Markets will contribute to sustainable development only if the societal signals are founded on such a perspective. Hence the importance of the underlying world view in society.

Unlimited Substitutability and Technological Change

Ecological economists speak of natural capital, human capital (and/or cultural capital), and manufactured capital. The latter two are sometimes referred to together as human-made capital (Costanza and Daly 1992).

A major assumption of the conventional paradigm is that natural and human-made capital are near perfect substitutes and that technological change will eliminate any resource constraints to unlimited growth.

This assumption of near perfect substitutability has little support in logic or in fact. Consider the following list of objections:

(a) Historically we developed human-made capital as a complement to natural capital, not as a substitute. It should be obvious that the human-made capital of fishing nets, refineries, saw mills and the human capital skill to run them does not substitute for, and would in fact be worthless without, the natural capital of fish populations, petroleum deposits, and forests.

(b) Manufactured capital is itself made out of natural resources with the help of human capital (which also consumes natural resources). Creation of the "substitute" requires the very thing that it is supposed to substitute for, and sometimes even more!

(c) A physical analysis of "production" reveals that it is really a transformation process—a flow of natural resource inputs is transformed into a flow of product outputs by two agents of transformation—the stock of laborers (human capital) and the stock of manufactured capital at their disposal. Natural resources are those which are being transformed into a product (the material cause of production); manufactured and human capital are those which are affecting the transformation (the efficient cause of production). The overwhelming reason for increasing the stock of human-made capital is to process a larger flow of natural capital, not to make possible a reduced flow. It is possible to reduce the waste of materials in process by investing capital in the recycling of prompt scrap, but this is marginal and limited.

The point made here is that the substitution of human-made physical capital for natural capital in the production of a given good is very limited and that on the whole, natural capital and human-made capital are complements in the production of any given good. There may remain considerable substitutability between human and manufactured capital (the two agents), or among various particular forms of natural capital (aluminum for copper, glass for aluminum) or even between nonrenewable natural capital and renewable natural capital. That is not in dispute. Nor are we disputing the possibility of substituting a technically superior product that requires less energy and materials to render the same human service (e.g., cars that get more miles per gallon and light bulbs that give more lumens per watt). The latter is efficiency-increasing technical progress (development) as opposed to throughput-increasing technical progress (growth). In this case there is a substitution of one human-made capital by another, but not a substitution for natural capital.

A frequently observed pattern, particularly in the modern world, is that ever increasing quantities of human-made capital substituting for natural resources means that ever increasing quantities of natural resources are being used elsewhere in the economy to produce that human capital (Regier and Baskerville 1986; Folke and Kautsky 1992). At some point the quantity of energy/matter used to produce the human-made capital will exceed the energy/matter saved by the substitution, after which further substitution of human-made capital will actually increase overall energy/matter inputs (Cleveland 1991).

Moreover, natural capital is increasingly becoming the limiting factor for further development (Costanza and Daly 1992; Daly 1994). Therefore, in order to sustain a stream of income, the natural capital stock must be maintained. This does not mean an unchanged physical stock, but rather an undiminished potential to support present and future human generations. A minimum safe condition for sustainability (given the huge uncertainty) is to maintain the total natural capital stock at or above the current level (Turner et al. 1994). An operational definition of this condition for sustainability means that the physical human scale must be limited within the carrying capacity of the remaining natural capital; technological progress should be efficiency-increasing rather than throughput-increasing; harvesting rates of renewable natural resources should not exceed regeneration rates; waste emissions should not exceed the assimilative capacity of the environment; and nonrenewable resources should be exploited, but at a rate equal to the creation of renewable substitutes (Barbier 1987; Costanza and Daly 1992).

Historical Continuity and No Threshold Effects

A third assumption of conventional thinking at the macro level is that the future will basically be a continuation of past trends and that there will be no major thresholds or discontinuities to disturb this pattern. This view stems from the relative stability of growth trends in western economies in recent history, but it is totally inconsistent with most of the research on the behavior of complex systems in general (Costanza et al. 1993; Holling 1994) and with the behavior of civilizations over a more extended period of history (Tainter 1988; Yoffee and Cowgill 1988; Ponting 1991). Because complex systems (including human societies) frequently exhibit threshold effects, discontinuities, and nonlinear and sometimes chaotic dynamics, the assumption of historical continuity is a very dangerous one. The past is not necessarily a good guide to the future in these cases. The consequence is to accept a much higher level of uncertainty in planning for the future and designing institutions that are flexible, adaptable, and resilient.

To understand the scope of the problem, it is necessary to distinguish between risk (which is an event with a known probability, sometimes referred to as "statistical uncertainty") and true uncertainty (which is an event with an unknown probability, sometimes referred to as "indeterminacy").

There are risks and uncertainties in reducing natural capital because of our imperfect understanding of the life-support functions, the inherently limited predictability of ecosystems and social systems (Holling 1994), and our limited capability to invent technical substitutes for natural functions. Complex systems like ecological and economic systems are fundamentally evolutionary and nonlinear in causation and limited in predictability.

Ecological Economics Perspective

Ecological economics is a transdisciplinary field of study that addresses the relationships between ecosystems and economic systems in the broadest sense, in order to develop a deep understanding of the entire system of humans and nature as a basis for effective policies for sustainability (Costanza et al. 1991). It takes a holistic "systems" approach that goes beyond the normal narrow boundaries of academic disciplines, developing new theory, tools, and techniques out of the dialogue to effectively deal with sustainability. It focusses more directly on the problems facing humanity and the life-supporting ecosystems on which we depend.

Humans have a special role to play in the system because we are responsible for understanding our own role in the larger system and managing it for sustainability (Costanza et al. 1991). This responsibility is not merely an ethical and a moral issue. It has to do with the fact that saving the environment actually means saving ourselves, including future generations, since we, as a biological species, are dependent on healthy ecosystems for survival. Thus, ecological economics is an anthropocentric field of study in the sense that it cares about the survival and welfare of human beings on this planet.

Ecological economics views the socioeconomic system as a part of the overall ecosphere, emphasizing carrying capacity and scale issues in relation to the growth of the human population and its activities, along with the development of fair systems of property rights and wealth distribution. The belief of many that we can continue on the same path of expansion, that technological progress will eliminate all energy, resources, and environmental limits, and that there is infinite substitutability between human-made and natural capital is considered to be a dangerous one, given the huge uncertainty. This blind faith in technology may be similar to the situation of the man who fell from a ten-story building, and when passing the second story on his way down, concluded, "So far so good, so why not continue?"

Research Agenda and Policy Implications

With this in mind and while recognizing that we need to take action now on the basis of imperfect information, and also recognizing that information about the structure and behavior of ecological and economic systems will always be imperfect, there are a range of critical issues that would greatly benefit from additional relevant academic research.

Scale and Sustainability: Maintaining Our Life-support System

A principal task of ecological economics is to seek ways to keep the scale of human society within sustainable bounds. We need to identify the present status of how large the economic subsystem is compared to the larger sustaining ecosystem and approach the issue of how large the economy can become without irreversibly damaging our future existence. The uncertainty of where society is in relation to the boundaries set by the functioning of the natural environment and the fact that most environmental problems suffer from uncertainty calls for a new approach to environmental research dealing more explicitly with uncertainty.

Modeling can serve as a useful tool to increase our knowledge about functioning of ecological economic systems. At this stage there is still hardly any model that would incorporate within the same framework both the ecological and economic subsystems at an adequate level of resolution.

Considerable work over the past decades has attempted to address aspects of the first of these issues, but for the most part this work has concentrated on either ecological or economic modeling approaches to the problem. The tradition of independent research has resulted in a body of environmental work completed by both sciences that overlooks the fundamental problems associated with their interactions. Namely, the studies are completed using different terms and units of measurement, and the studies neglect to account for the interaction between the processes and populations each seeks to model. For the most part, the ecological models have tended to ignore humans, except to the extent that they introduce apparently exogenous shocks into the ecological system, and economic models have taken stylized versions of "ecological facts" or relationships to connect human actions with the state of the ecosystem.

Both sciences recognize the problematic nature of these tendencies, and there have been some efforts to revise tradition, but they largely stay in the conceptual level. Still, models give the opportunity of investigating what is possible by running numerous scenarios. In this case the uncertain or unknown behavior of the pairing subsystem may be formulated as a set of rules, functions or measures, interpreted in modeling terms and fed into it as forcing functions. A set of scenarios may show the response of the system; however, the feedback to the projected components that are described by the scenario can be estimated only outside of the model.

A generalized structure of an ecological economic model is shown in Fig. 3. The recent efforts at MIIEE have resulted in a rather generic approach to the spatial modeling of landscapes and ecosystems. The problem now is to incorporate a means of economic analysis into the model.

Natural Resource Valuation

To achieve sustainability, we need to incorporate ecosystem goods and services into economic accounting. The economic values we seek for ecosystem management are much broader than financial values or merely the cash flow generated by a resource. For any good or resource to have an economic value it must meet two conditions: (1) provide at least some agents (but not necessarily all) with improved well-being, and (2) be considered scarce in that agents desire more than is currently available. Fish, wildlife, recreation, wetlands, old growth forests, native grassland communities, etc., all meet this definition of having an economic value to society even if none of these resources or the services they provide are marketed.

The case of the spotted owl highlights the "passive use" or existence/bequest values that ecosystems provide to members of the general public who may never set foot in the forest. Existence value is the satisfaction gained from knowing that a particular species or entire ecosystem continues to exist and function. Bequest value is the satisfaction gained from knowing that protection today will provide future generations with a particular species or ecosystem. Randall and Stoll (1983) describe the recreation use, existence and bequest values as "Total Economic Value," as it captures many of the motivations people have for caring about resources. Obviously, these motivations are quite broad and can arise from a variety of concerns and may partially include such important but overlooked sources of benefits as spiritual and cultural values. Using the contingent valuation methodology in surveys of the general public, Bishop and Welsh (1993) have shown that these existence and bequest values can be two to ten times larger than the direct on-site recreation use values. In some respects this is not surprising. While per-person visitors have much higher benefits than non-visiting members of the general public, the number of visitors is often limited to a few thousand and almost never larger than a few million. Yet, nearly every member of the general public could receive satisfaction from simply knowing an ecosystem exists. While the value per household may be small ($5–$40), there are literally millions of households depending on the geographic extent over which people care. Thus an important research topic in ecosystem valuation is just how wide is a ecosystem's drawing power? This of course depends on the uniqueness of that ecosystem. The Grand Canyon has a North American and possibly worldwide geographic extent, whereas a common type of wetland might have a value just locally, as similar wetlands exist in other counties or states.

Transforming society towards sustainability requires transdisciplinary, problem-oriented knowledge. Broadly trained environmental professionals are needed, in

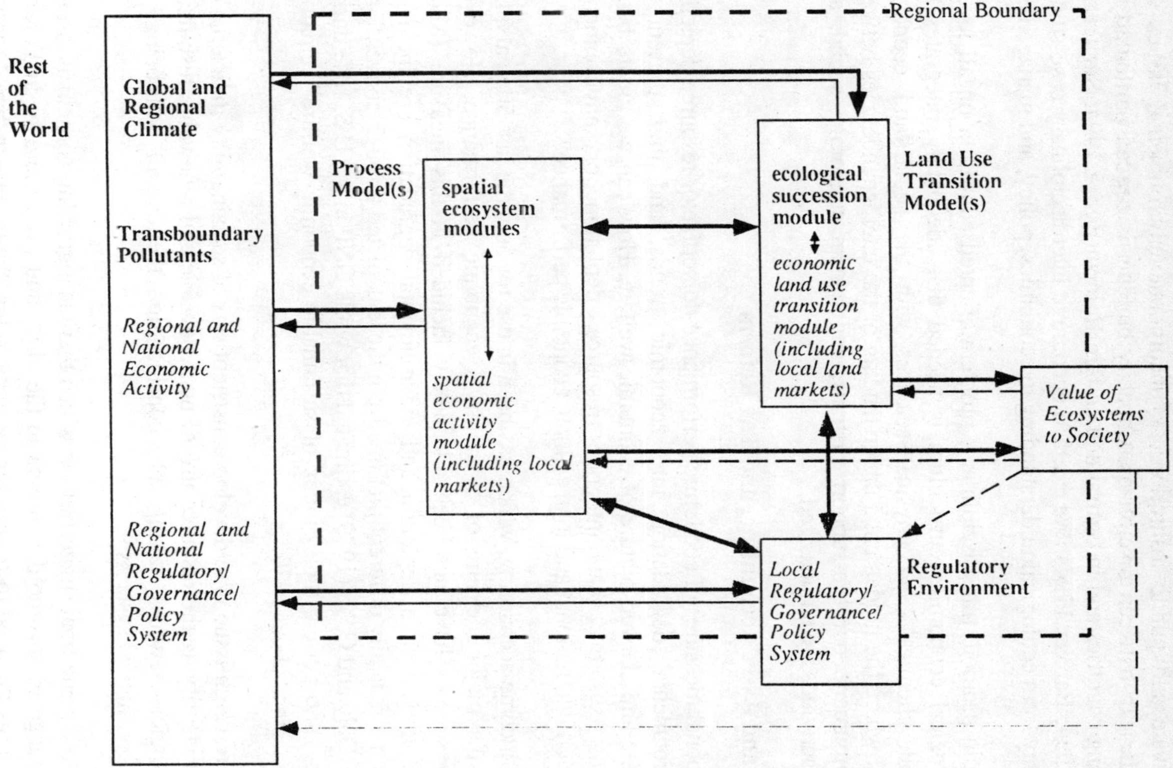

Fig.3 Integrated ecological economic modeling and valuation framework.

both private and public institutions, to incorporate environmental issues and bridge the gap between disciplines. Capacity building is needed particularly in developing countries and in central and eastern Europe (Zylicz 1994; Segura and Boyce 1994; Barbier 1994). We need to encourage interdisciplinary case studies all the way from ecological understanding to economics, policy, and management.

We should promote transdisciplinary, problem-oriented education on all levels. An ecological economics curriculum is needed that adequately integrates the biological, chemical, and physical sciences with the social and economic sciences, where the fundamental understanding of the interdependencies between the environment, our life-support systems, and human economic, social, and cultural activities are emphasized.

Income and Asset Accounting, and Tax Reform

Conventional measures of economic performance do not factor resource depletion or environmental degradation into economic trends, and thus present an incomplete and skewed picture of economic welfare. How can we create better accounting systems that integrate resource depletion or environmental degradation into national and international economic performance?

National income accounting, which has been in use for only the last 50 years, has become the major measure of a country's economic performance. However, natural resources have not been adequately included in this measure (Mäler 1991). When resource depletion and environmental degradation are included, results show radically different pictures from those depicted through conventional methods. Daly and Cobb (1989) estimated that while GNP in the U.S. rose during the 1956–1986 period, their "index of sustainable economic welfare" remained unchanged since about 1970.

A major research issue is to develop consistent sets of sustainability indicators on all levels identifying the resource use of nonrenewable and renewable resources, and ecological services, taking into consideration the overall well-being, or health, of ecological systems.

We may offer one final suggestion on which there is a growing consensus among a broad range of stakeholder groups in the U.S., and even more so in Europe. This concerns the need to reform tax systems to tax "bads" rather than "goods." Taxes have significant incentive effects which need to be considered and utilized more effectively.

The basic idea is to limit the throughput flow of resources to an ecologically sustainable level and composition, thus serving the goal of a sustainable scale of the economy relative to the ecosystem. We intend to gradually shift much of the tax burden away from "goods" like income and labor, and toward "bads" like ecological damages and consumption of nonrenewable resources. Such a shift would have far-reaching implications, and should simultaneously encourage both employment and ecologically sustainability. The more traditional goal of efficient allocation of resources is also served by this instrument because it raises the tax on "bads" and lowers the tax on "goods." However, like all consumption taxes, it is regressive. This could be counteracted by retaining the income tax at the extremes—a positive income tax for high incomes, a zero or even negative income tax for very low incomes, and a negligible income tax between the extremes. Of the three major goals of sustainable economic policy (sustainable scale, efficient allocation, and fair distribution), the ecological tax reform serves the first two quite well, and the third partially, requiring some supplement from an attenuated income-tax structure.

Institutional Changes and Property Rights

The scale and distribution problems cannot be solved inside the market allocation mechanism, even if markets are "perfect" in the sense of all external costs internalized. Rather, the scale and distribution problems must be solved outside the market, and the market can then be utilized as an efficient instrument to implement the solutions. Institutions that recognize the need for an ecologically sustainable economic development are needed

The practical and institutional barriers to sustainability need to be identified and institutions with the flexibility necessary to deal with ecologically sustainable development need to be designed and implemented to provide the proper framework within which the market can act. The uncertainties faced by ecological economists on the ecological side are compounded by the uncertainties on the economic and social sides and vice versa, and over the next few critical decades, those uncertainties are unlikely to be significantly reduced. Therefore, two related major challenges for ecological economists are (1) to help society identify and evaluate the "premiums" worth paying for environmental insurance, and (2) to develop strategies for moving toward sustainability that are relatively insensitive to uncertainties—"robust" strategies that are helpful regardless of how uncertainties are ultimately resolved (Ehrlich 1994; Costanza 1994). This will most likely involve systems of governance and decision-making (1) that can involve stakeholders (including nature and future generations) in a more effective way, (2) that distribute property rights more effectively, and (3) that are adaptive and flexible (Golley 1994; d'Arge 1994; Viederman 1994).

Hence, ecological economics is seeking an appropriate "political economy" for sustainability. Ecological economics argues that unfettered markets and "economic growth-dominated" policy strategies are definitely unsustainable and carry with them significant environmental degradation and damage-cost burdens. Thus, market interventions in the sense of democratically based systems to keep the scale of the economy within sustainable limits and to insure a fair inter- and intragenerational and interspecies distribution of resources are fundamental prerequisites for an efficient, a practicable sustainable development strategy. The market can be an amazingly efficient servant in the cause of sustainability if the scale and distribution problems are solved first, but it is a poor and unsustainable master if left to its own devices.

Presently, institutions with the flexibility necessary to deal with ecologically sustainable development are lacking. At the same time as global connections are becoming more evident, local and regional adaptive capabilities are eroding, and much environmental knowledge held by indigenous people is being lost. Establishing rules at one level without rules at other levels will create incomplete institutional systems that may not endure in the long run (Ostrom 1991; Berkes and Folke 1994). Public participation and the involvement of local citizens is needed for creating consistent institutional systems capable of handling sustainability (Golley 1994).

Centralized control and state property in many cases seem to be appropriate for fair distribution, appropriate scale and hence sustainability. Under this arrangement a certain area is owned and managed by the state, which defines the management goals and methods and carries the responsibility for fulfilling them. Private capital is allowed in this territory only if it complies with the rules and regulations set up at the state level. From the environmental viewpoint centralized control may look like one of the most effective strategies for managing the commons. The decision-making may be based on recommendations of well informed and educated, analytic and scientific services. Tax money can be centrally planned and used to support the most crucial or critical elements of the system.

The problem is, however, in implementing the decisions. The central controlled systems lack the flexibility to adapt to changing conditions and to generate proper incentives that would organize the system as necessary. Centralized control is based on flat and complicated bureaucratic hierarchies, which certainly lack the optimality of smaller systems with intrinsic feedbacks and self-regulatory mechanisms. Many links have to be doubled with additional controlling and checking agencies to fight corruption and negligence inherent to officials that do not benefit directly from the results of their performance.

Private properties are not burdened by these complications. Employing some of the most natural human instincts, they possess the necessary feedbacks, that optimize their performance for the benefit of the proprietor. However, the same instincts of greed and avarice that turn out to be so effective for improved economic operation, become dramatically dangerous in the ecological perspective. By mastering the short feedbacks that promptly reflect the decisions made into the performance achieved, private ownership lacks the knowledge and understanding of the long feedbacks that influence them via the operation of the higher level systems, regions, and biosphere that they are nested in. These long feedbacks are distanced in time and space and therefore, they are rather difficult to perceive without special training and knowledge.

Communal properties are usually based on certain traditions, cultural postulates, and taboos. In some cases these not only stipulate the environmental load and the scenarios of resource uptake, but also control the parameters of the human population. The communal arrangements work fairly well in a sustainable fashion as long as the communities are not exposed to the two other property rights systems. They also turn out to be fairly vulnerable as soon as private or state ownership, or both, come in direct contact with them. This may be attributed to both the aggressiveness of the former two and the brittleness of the latter.

Ecological economics attempts to merge the benefits of all these different property rights organizations for the sake of systems sustainability.

The use of a broad range of policy instruments should be promoted based on criteria of equity, scientific validity, consensus, and environmental effectiveness. We should institute regulatory reforms to promote appropriate use of financial, legal, and social incentives, and implement pollution taxes, taxes on natural capital depletion, regulatory targets, and make use of tradable permits (Clark and Munro 1994; King 1994). Within the baseline regulatory framework (ambient quality standards and targets, etc.) economic incentive instruments have a vital "cost-effectiveness" role to play. Regulations and incentives should be designed and implemented as entirely complementary, mutually supportive devices. The implementation of this set of sustainability-enabling policy measures can both serve to correct for current market and policy failures, as well as being impeded by such failures and the social and political rigidities that underpin them. Both ecological, economic, and other information will be needed to find the optimal trade-offs between more information for a better understanding and the costs for not acting, thereby making sure that information is not wasted and uncertainty reduced optimally. We need to promote the employment of positive sustainability incentives based on the precautionary principle, and the polluter-pays principle (Costanza and Cornwell 1992; Costanza 1994; Zylicz 1994). In decisions of scale and distribution, individual freedom of choice must yield to democratic collective decision-making by the relevant community.

Case Study

How can we apply some of the above ideas to the situation in Buryatia and the Baikal region? The situation in particular regions and localities in Russia is aggravated by the overall uncertainty about the trends of Russian economy. As Russia in general moves toward a market economy, attempting to substitute private ownership for state property, the situation across different regions and territories varies immensely, creating a patchy disharmonious economic and political background. The former system of rule in the U.S.S.R., based on state ownership, proved to be inefficient both economically and environmentally (Feshbach and Friendly 1992).

In many respects the current situation is reminiscent of the period of wild capital accumulation that was underway in Western Europe some two centuries ago. Legislation was poor and there were none of the ethical and moral principles that guide and restrain the capitalists of today. This unruly early stage might be an inevitable stage in building markets and establishing the capitalistic mode of production. If this is the case, then we might not need be overly concerned about the existing distortions and criminal expansions in Russia today.

However, there is a major difference that we cannot ignore, one which primarily comes from the state of the environment. The carrying capacity of ecosystems at the time of the early capitalist expansion in the West was much higher than at present. Now in many cases they are already on the verge of collapse. There is simply no reserve for irresponsible exploitation of any kind of resources. There were no global climate change or ozone depletion problems facing the early capitalists. There were no nuclear energy risks like those of Chernobyl. At this stage we can hardly allow a repetition of history, when markets were built through an exercise of greed and immoral competition on the brink of legality or even beyond it.

This means that in Russia there is hardly time to wait until the drive for sustainability grows organically among the population. If this is only starting to happen in the West now, we can only imagine how long it would take to wait for the rest of the world to go through all the phases of market economy to move toward sustainability, and in what state the world environment would be by that time.

Of even more concern are such unique regions as Lake Baikal and its watershed. Such regions badly need strong national and even international involvement and control. Such ecosystems should definitely be protected and preserved, and this can be done only if they become elements of sustainable systems. And the creation of such systems is to a great extent a responsibility of the international community united with the environmentally concerned forces within the national and regional systems.

Sustainable systems may be built on two different principles. On the one hand, sustainability can be based on natural self-adaptation and self-regulation mechanisms that we find in ecological systems. These systems have the ability to organize their structure, growth, and development of their components in such a way that they keep on existing over the lifetime that they are assigned for, and even have the reserve to withstand some of the external perturbations as long as they do not surpass a certain level. This matches the major requirements of the sustainability paradigm. We will call such systems Type 1 Sustainable.

On the other hand, sustainable systems, which ecological economics is mostly trying to focus on, are the systems that would include the developing component of Homo sapiens with its requirements and priorities. These are the systems that may ensure sustainability by investing appropriate effort and capital to environmental management, creating efficient incentives and institutions to realize the goals of ecological economics, and to manage the value sets and priorities that define the goals. Though to a different extent, all these systems are still based on human management, they are created by humans and can persist only while controlled and sustained by humans. These systems need certain effort and investment to be launched; however, they also produce revenue in generating additional man-made capital, at the same time supposedly preserving the natural capital. These we will term Type 2 Sustainable Systems. Obviously, the underlying landscapes in these systems will be significantly modified and the pristine ecosystems will give place to the artificial human-managed systems. The extent of this change may vary, and there is no strict border between the two system types. However, the driving mechanisms and therefore, the management strategies would be quite different, and the transfer from one type of sustainability to the other eventually is quite a costly, lengthy, and difficult process, in most cases even hardly reversible.

So, in all cases there should be a decision about what parts of the area should be preserved as pristine ecosystems for conservation not only of wildlife and vegetation, but also of cultural and traditional ways of life for indigenous peoples. It should be determined which regions will be designated for sustainable development within artificially designed and managed ecological and socioeconomic systems, within the conventional tourism industry, and within development that will invest sufficiently in the cleanup procedures, and thus ensure healthy ecological conditions in the region at the larger scale.

In the Baikal region there have been already several attempts at zoning and regional planning. According to the most widely accepted concept, central and buffer zones should be identified. They are close but do not completely coincide with the suggested sustainability criteria. Their demarcation in many cases is fairly mechanistic and does not really take into account the implementation phase and the requirements of sustainability.

The proportions of the two types of systems are prescribed by social agreement very much dependent on the resources available. Type 1 systems at present cannot be preserved on their own—they need additional investment to perform their protection. There is no way to rely upon natural sustainability if heavy poaching and unlawful forest cutting is to continue. There may be different mechanisms to provide protection, but hardly any that would be free. It should be also realized that the majority of strategies for this type of sustainability are best based on centralized control and public property rights, which is a loss in the accounting balance and further burdens the budget.

Type 2 Sustainable Development is desirable for the rest of the territories adjacent to Baikal. This type of sustainability is capable of generating revenues and becoming cost effective, and it may also partially support the protected regions. However, Type 2 sustainability may be eventually associated with some very high risks. As we have mentioned above, it is not clear whether humans will continue to provide timely technologies and management practices that will substitute for the natural control and restoration mechanisms and compensate for growing human intervention into natural processes and functions. As the scale of human activity grows and the rates of capital, material, and resource cycling increase, the uncertainties grow as well, with the consequences and prices of errors becoming exceedingly high. Therefore, one of the obvious restrictions for the development of the Baikal region is a total ban on all activities that have the potential of ruining the Lake's ecosystem as a whole. The price of compensatory procedures in this case will be too high, probably infinite.

For the rest of the newly developed enterprises, among the instruments suggested by ecological economics is the flexible environmental-assurance-bonding system designed to provide for uncertainties and risks involved in the development of the region. In addition to direct charges for known environmental damages, an assurance bond equal to the current best estimate of the largest potential future environmental damages would be levied and kept in an interest-bearing escrow account. The bond plus a portion of the interest would be returned if and only if the firm could prove that the suspected damages had not occurred or would not occur in the future. If damages did occur, the bond would be used to rehabilitate or repair the environment, and to compensate injured parties. In this way the burden of proof is shifted from the public to the resource user while at the same time providing a strong economic incentive to investigate the true costs and develop efficient pollution control technologies (Costanza, 1992).

Resource- and energy-intensive activities should be heavily charged, creating additional incentives for recycling in industry and organic farming in agriculture. All this definitely decreases the profits and lowers the chances of creating totally self-sufficient economies in the region. The only hope comes from the other constituent of sustainability, the one aimed at shifting priorities and human value sets.

The two sustainable systems should interact closely. The Type 2 systems would be serving as a buffer for the other type. The paradox is that even though "more" sustainable in essence, based on some natural control mechanisms and therefore seemingly more self-sufficient, the Type 1 systems are more vulnerable to external invasions, and they have less intrinsic mechanisms to protect themselves from unfavorable human-made perturbations. Therefore, they need the additional investments to protect themselves.

The less Type 1 systems exist proportionally among the others, the more external support they need to protect themselves. On the one hand, this might be explained by the simple mathematical relationship between the length of the border and the area of a convex system: the larger the area, the less the proportion of the

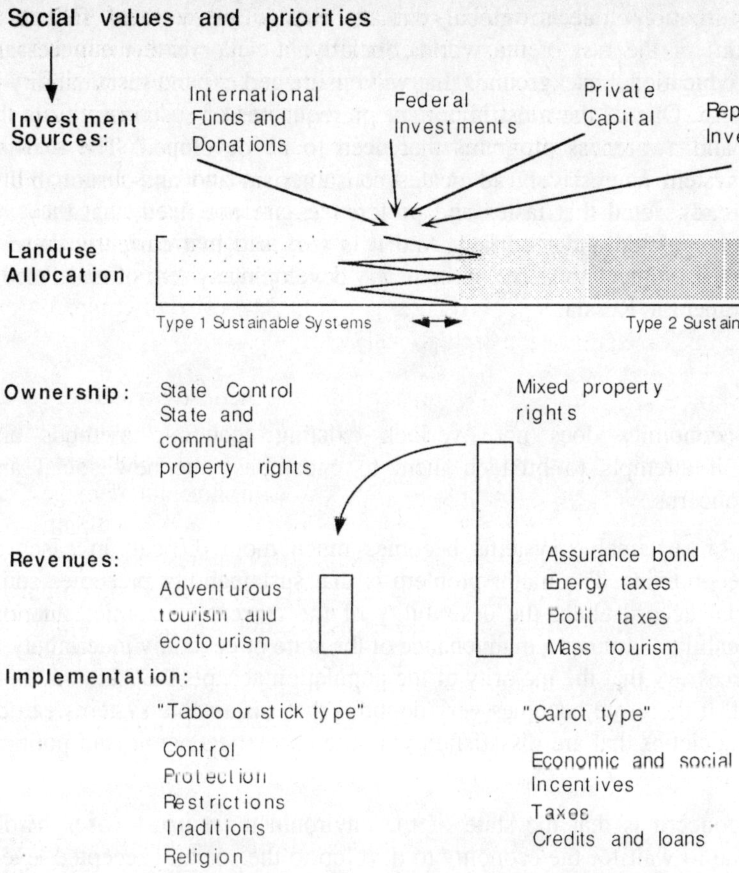

Fig. 4 Merging the two types of sustainability

bordering elements. On the other hand, designing sustainability is very much a positive feedback process. Type 1 sustainability both implies a certain level of ecological conscience and education and stimulates their further development. However, at first, building sustainable systems requires investments. Protecting sustainability of Type 1 systems requires additional investments that may be difficult or simply impossible to generate within them.

Some of the specifics of the two types of sustainable development in a regional level are presented in Fig. 4. Social values and priorities in local, national and international levels define the amount of investment that the societies are ready to make for sustainability. It also defines what fraction of the whole territory can be set aside and protected as a Type 1 sustainable system and what fraction will be developing in a Type 2 manner, serving as a buffer for the former both in geographic, social, and economic terms. Geographically and ecologically, Type 2 sustainability provides clean environment. Economically, it shares a portion of its revenues to protect Type 1 regions from direct effects (poachers) and indirect effects (perturbations due to global climate change or regional fallout of contamination) of the rest of the world. Socially, it will create the necessary cultural and educational backgrounds that will ensure and expand sustainability in the longer term. One of the most important prerequisites of sustainability is the educational and awareness programs that need to be developed. The existing advertising system aggressively advocates consumerism and anti-sustainability. We have already noted that tastes and preferences are not fixed, that they are constantly manipulated and modified. And it is very sad that currently there is little place for ecological concerns in the newly developing system of mass media and advertisement in Russia.

Conclusions

Ecological economics does not overlook existing economic methods and approaches; it attempts to broaden them to encompass the new social and ecological concerns.

The design of sustainable systems becomes much more difficult in cases of developing economies. The major problem is that sustainability presumes some level of social accord about the desirability of the current economic situation. Since sustainability assumes a maintenance of the state of economy indefinitely, it is at least necessary that the majority of the population accepts the state that is to be sustained. It therefore becomes very doubtful that sustainable systems can be designed in societies that are dissatisfied with the current economic and political state.

Our major concern is that the state of the environment in many cases hardly leaves us time to wait for the economy to develop to the socially accepted levels and for the market to produce the adequate prices based on ethical and moral concerns. This is especially crucial for such unique and ecologically important regions as Baikal. If such territories are to be preserved or developed as model territories for sustainability, there should be a certain external influx of capital and externally applied incentives to mold the desired mechanisms and institutions. There should be a merge of private and state property rights with state control based on wide public participation.

The existing international institutions can hardly meet the demands of sustainable development, providing little or no adequate incentives and driving mechanisms that would promote and support sustainable regional systems. While it might be appropriate for the international community to wait for the developing world to find its own approaches to sustainability and to build national strategies of sustainable development, it might be necessary to accelerate the process externally when dealing with unique and biospherically valuable sites. This becomes especially important when national approaches to sustainability tend to diverge from those generally accepted by the scientific community due to some political, economic, or even linguistic arguments.

The international bodies of such influence are hardly in place, and the mechanisms that would combine extremely delicate economic and political pressure with urgent economic and scientific aid are still to be created.

The only other chance for sustainability in the deprived, developing, or restructuring economies may be in focussing on the other end of the sustainability conditions. Keeping some of the existing traditions instead of breaking them up in a rush towards market economies; rebuilding some of the traditional infrastructures; focussing on the historical and cultural priorities, and relying heavily on religious, in this case Buddhist, visions and values may certainly help in designing sustainable systems at least on a smaller scale.

References

Barbier, E. B. 1994. Natural capital and the economics of environment and development. pp. 291–322 in: A. M. Jansson, M. Hammer, C. Folke, and R. Costanza (eds.). Investing in natural capital: the ecological economics approach to sustainability. Island Press, Washington DC. 504 pp.

Berkes, F. and C. Folke. 1994. Investing in cultural capital for sustainable use of natural capital. pp. 128–149 in: A. M. Jansson, M. Hammer, C. Folke, and R. Costanza (eds.). Investing in natural capital: the ecological economics approach to sustainability. Island Press, Washington DC. 504 pp.

Bishop, R. C. 1993. Economic efficiency, sustainability, and biodiversity. Ambio. 22: 69–73.

Botkin, D.B. 1990. Discordant harmonies, a new ecology for the twenty-first century. Oxford University Press, New York, NY, USA.

Clark, C. W. and G. R. Munro. 1994. Renewable resources as natural capital: the fishery. pp. 343–361 in: A. M. Jansson, M. Hammer, C. Folke, and R. Costanza (eds.). Investing in natural capital: the ecological economics approach to sustainability. Island Press, Washington DC. 504 pp.

Cleveland, C. J. 1991. Natural resource scarcity and economic growth revisited: economic and biophysical perspectives. In Ecological Economics: The Science and Management of Sustainability, ed. R. Costanza. New York: Columbia Univ. Press.

Costanza, R. 1994. Three general policies to achieve sustainability. pp. 392–407 in: A. M. Jansson, M. Hammer, C. Folke, and R. Costanza (eds.). Investing in natural capital: the ecological economics approach to sustainability. Island Press, Washington DC. 504 pp.

Costanza, R., and L. Cornwell. 1992. The 4P approach to dealing with scientific uncertainty. *Environment* 34: 12–20.

Costanza, R., H. E. Daly, and J. A. Bartholomew. 1991. Goals, agenda and policy recommendations for ecological economics. pp. 1–20 In: R. Costanza (ed.) Ecological Economics: The Science and Management of Sustainability, ed. New York: Columbia Univ. Press.

Costanza, R., and H. E. Daly. 1992. Natural capital and sustainable development. *Conservation Biology* 6: 37–46.

Costanza, R., L. Wainger, C. Folke, and K.-G. Mäler. 1993. Modeling complex ecological economic systems: toward an evolutionary, dynamic understanding of people and nature. *BioScience* 43: 545–555.

Daly, H. E. 1984. Alternative strategies for integrating economics and ecology. In Integration of Ecology and Economics: An Outlook for the Eighties, ed. A. M. Jansson. Department of Systems Ecology. Stockholm: Stockholm University. Reprinted in Daly, H. E. 1991. Steady-State Economics. Covelo, CA: Island Press.

Daly, H. E., and J. B. Cobb. 1989. For the Common Good: Redirecting the Economy Toward Community, the Environment and a Sustainable Future. Boston: Beacon Press.

Daly, H. E. 1990. Toward some operational principles of sustainable development. *Ecological Economics* 2: 1–6.

Daly, H. E. 1994. Operationalizing sustainable development by investing in natural capital. pp. 22–37 in: A. M. Jansson, M. Hammer, C. Folke, and R. Costanza (eds.). Investing in natural capital: the ecological economics approach to sustainability. Island Press, Washington DC. 504 pp.

d'Arge, R. C. 1994. Sustenance and sustainability: how can we preserve and consume without major conflict. pp. 113–127 in: A. M. Jansson, M. Hammer, C. Folke, and R. Costanza (eds.). Investing in natural capital: the ecological economics approach to sustainability. Island Press, Washington DC. 504 pp.

Ehrlich, P. R. 1994. Ecological economics and the carrying capacity of earth. pp. 151–168 in: A. M. Jansson, M. Hammer, C. Folke, and R. Costanza (eds.). Investing in natural capital: the ecological economics approach to sustainability. Island Press, Washington DC. 504 pp.

Feshbach, M., and A. Friendly, Jr., 1992. Ecocide in the USSR : health and nature under seige New York, NY : BasicBooks, 376 pp.

Folke, C., and N. Kautsky. 1992. Aquaculture with its environment: prospects for sustainability. *Ocean and Coastal Management* 17: 5–24.

Golley, F. B. 1994. Rebuilding a humane and ethical decision system for investing in natural capital. pp. 169–178 in: A. M. Jansson, M. Hammer, C. Folke, and R. Costanza (eds.). Investing in natural capital: the ecological economics approach to sustainability. Island Press, Washington DC. 504 pp.

Holling, C. S. 1994. New science and new investments for a sustainable biosphere. pp. 57–73 in: A. M. Jansson, M. Hammer, C. Folke, and R. Costanza (eds.). Investing in natural capital: the ecological economics approach to sustainability. Island Press, Washington DC. 504 pp.

King. D. M. 1994. Can we justify sustainability? New challenges facing ecological economics. pp. 323–342 in: A. M. Jansson, M. Hammer, C. Folke, and R. Costanza (eds.). Investing in natural capital: the ecological economics approach to sustainability. Island Press, Washington DC. 504 pp.

Mäler, K.-G. I.-M. Gren, and C. Folke. 1994. Multiple use of environmental resources: a household production function approach to valuing natural capital. pp. 233–249 in: A. M. Jansson, M. Hammer, C. Folke, and R. Costanza (eds.). Investing in natural capital: the ecological economics approach to sustainability. Island Press, Washington DC. 504 pp.

Ostrom, E. 1990. Governing the Commons: The Evolution of Institutions for Collective Actions. Cambridge, U.K.: Cambridge Univ. Press.

Ponting, C. 1991. A green history of the world: the environment and the collapse of great civilizations. Sinclair-Stevenson, London. 432 pp.

Rees W. E. and M. Wackernagel. 1994. Ecological footprints and appropriated carrying capacity: measuring the natural capital requirements of the human economy. pp. 362–389 in: A. M. Jansson, M. Hammer, C. Folke, and R. Costanza (eds.). Investing in natural capital: the ecological economics approach to sustainability. Island Press, Washington DC. 504 pp.

Reiger, H. A., and G. L. Baskerville. 1986. Sustainable Redevelopment of Regional Ecosystems Degraded by Exploitive Development. In Sustainable Development of the Biosphere, eds. W. C. Clark and R. E. Munn. London: Cambridge Univ. Press.

Robinson, J.B., 1991. Modelling the interactions between human and natural systems. International Social Science Journal 130: 629–647.

Salwasser, H., 1993. Sustainability needs—more than better science. Ecological Applications, 3(4), p.587–589.

Segura, O. and J. K. Boyce. 1994. Investing in natural and human capital in developing countries. pp. 479–490 in: A. M. Jansson, M. Hammer, C. Folke, and R. Costanza (eds.). Investing in natural capital: the ecological economics approach to sustainability. Island Press, Washington DC. 504 pp.

Tainter, J. A. 1988. The collapse of complex societies. Cambridge University Press, Cambridge. 250 pp.

Turner, R. K., P. Doktor, and N. Adger. 1994. Sea-level rise and coastal wetlands in the U.K.: mitigation strategies for sustainable management. pp. 266–290 in: A. M. Jansson, M. Hammer, C. Folke, and R. Costanza (eds.). Investing in natural capital: the ecological economics approach to sustainability. Island Press, Washington DC. 504 pp.

Viederman, S. 1994. Public policy: challenge to ecological economics. pp. 467–490 in: A. M. Jansson, M. Hammer, C. Folke, and R. Costanza (eds.). Investing in natural capital: the ecological economics approach to sustainability. Island Press, Washington DC. 504 pp.

Vitousek, P. M., P. R. Ehrlich, A. H. Ehrlich, and P. A. Matson. 1986. Human Appropriation of the Products of Photosynthesis. *BioScience* 36: 368–73.

Yoffee, N. and G. L. Cowgill (eds.). 1988. The collapse of ancient states and civilizations. University of Arizona Press, Tuscon. 333 pp.

Zylicz, T. 1994. Implementing environmental policies in central and eastern Europe. pp. 408–430 in: A. M. Jansson, M. Hammer, C. Folke, and R. Costanza (eds.). Investing in natural capital: the ecological economics approach to sustainability. Island Press, Washington DC. 504 pp.

Concepts in Nature Conservation for Sustainable Development

Johann Schreiner
North German Academy for Nature Conservation
Hof Moehr
D-29640 Schneverdingen
Germany

1. Nature conservation as a strategy for safeguarding the natural life basis

Nature conservation is a collection of measures produced by society to provide lasting security for its life basis. The aims of these measures are to protect, care for and develop both nature and landscape in both populated and unpopulated areas in order that:

– the capacity of the landscape to perform its essential functions,

– the usefulness of the natural resources,

– the wildlife and,

– the variety, individuality and beauty of both nature and landscape are lastingly secured both as a life basis for man and as the prerequisite for his recreation within both nature and the landscape.

Analysis of the above aims shows that:

– nature conservation constitutes a demand for land area and usage that will inevitably come into conflict with other claims for land and usage (Erz 1978);

– the demand for land for nature conservation is generally put at 100% and that this,

– in accordance with the establishment of priorities both qualitatively and quantitatively, must be altered according to area and time, and that

– the legal securing of certain areas, thus reserving them for nature-conservation objectives, does not alone guarantee the complete achievement of these objectives, as some natural resources are not extensively ascertainable (e.g., air) and

– certain effects do not always ensue (e.g., climatic change) and that

– areas from which damage to the natural resources air, water and climatic change emanates must also be subject to nature-conservation objectives.

Therefore, it is inevitable that for considerations regarding sustainable development we must qualify the land-use demands made by nature conservation. This should, as

NATO ASI Series, Partnership Sub-Series, 2. Environment – Vol. 6
Sustainable Development of the Lake Baikal Region
Edited by Valentin A. Koptyug and Martin Uppenbrink

already established, encompass 100% of the landscape and so it is necessary to qualify all areas.

2. Qualification of the land use demands for nature conservation

2.1 Basis of the qualification

The basis for the following considerations is supplied by Erz (1978) with the presentation of the demands for land and intensity made by nature conservation. The qualities and quantities of different intensities of demand as stated there are not measurable as a continuous entity. Erz (1978) shows that:

- the intensity of land use (forestry, fish farming, settlements, industry, transportation etc.) and nature conservation are in conflict with one another,
- the demand for land is decreasing along with an increase in the intensity of protection and conservation,
- as a percentage, it is difficult to differentiate between field areas, due to increasing intensity of protection,
- a qualification of demand for land must cover all land.

2.2 Land types and their conservation function

In detail, land types with the following functions can be distinguished:

2.2.1 Areas for a representative safeguarding of natural resources (e.g., reserves)

A representative safeguarding of various noteworthy characteristics of such natural resources as soil, water, and organisms in their natural and historically aged variety, which can be bound to a certain area, is necessary for both existential and ethical reasons.

With regard to the protection of organisms, the reserve should serve as a centre of regeneration and spreading, in the form of an "area of breeding surplus", and should momentarily take on the functions of a "retreat", in which breeding is ensured and a better adaptation to the cultural landscape can be developed (Erz 1981).

With regard to the resources soil and water, the reserve should achieve a "continued representative safeguarding" of noteworthy characteristics and should guarantee the undisturbed further development of such qualities over a minimum area. In this way reference areas were provided for research. Therefore, in order to establish areas with the sole function of conservation (e.g., reserves), in which it should simply and exclusively be allowed to remain with an operational continuation of forms of land use which are permanently preservable, as many areas as possible should be seized and protected with as many nature-regulation mechanisms as possible.

2.2.2 Areas which represent the safeguarding of unique elements of the landscape (e.g., natural monuments)

Outstanding elements of the landscape (nature's single-handed creations), should be safeguarded for aesthetic, cultural, scientific and educational reasons. They serve both to enrich the landscape and as "loupes" for the imparting of knowledge about history, local history and natural history to a wide-ranging public. The sole use of these objects is for the stated purpose.

2.2.3 Areas for the safeguarding of ecosystems which cannot be reestablished

– The recoverability of an ecosystem is determined by three factors:

– period of time of development,

– site factors,

– situation in relation to other areas of the same or similar type.

The last factor refers exclusively to the concrete object. It is dependent on the location of the relevant area in the structural system. This ensures the opportunities for immigration of the typical species of the biotope. It cannot be taken into general consideration. The scale for the measurement of the development period can be either the actual length of time taken to develop (object-based), the "age" of the ecosystem in question, or the length of time necessary for the redevelopment of an adequate ecosystem (type-based).

For the considerations over land-use demand, use of the second approach is prevalent, as it allows a certain degree of standardisation.

Such a time scale for the development period in Central Europe was put forward by Kaule (1988). He divided the habitats into development periods lasting from 1–15, 15–50/80, 50/80–150, 150–250, 250–1,000, and 1,000–10,000 years, thereby lending support to the opinion that ecosystems which are over 150 years old find themselves in such a state of development that today it is impossible in Central Europe for replacement (in the biological sense) to occur. The recoverability of an ecosystem with regard to its site factors is type-based. It competes with the possibilities of re-establishment through the reconstruction of a site in its characteristic pastureland.

2.2.4 Areas acting as buffer zones for reserves

"Buffer zones" should protect reserves from all kinds of damage from adjacent areas, and at the same time should capable of acting as a "temporary habitat". In naturally developed, structured landscapes, as a rule, the existence of buffer zones for the different biotopes is not required. There the necessary balancing function between different biotopes, in the form of transitional biotopes (ecotones), is in existence. Ecologically, buffer zones play a special role, as in landscape arrangement, especially in Central Europe, very precisely defined borders have been placed between biotopes needing protection and intensively used areas (Heydemann 1983).

2.2.5 Areas serving as a reserve network towards the safeguarding of genetic exchange

Species populations are genetically different from each other and can therefore settle in habitats with differing environmental conditions and/or react to changes in environmental conditions (within certain limits). If an exchange of information through the individual exchange between singular populations of a species is no longer possible, then a one-sided genetic alignment, which expresses itself in a reduced ecological malleability or a reduced ability for resistance, can be the solution. "Discoveries" which are necessary for survival in a population can then not be passed on to the remainder of the population.

On the other hand, genetically isolated populations have the chance of an original, evolutive further development in the direction of developing new ecotypes, subspecies and finally species. An increase in genetic variety is the result. With the qualification of nature- conservation's demand for land, the establishment of a reserve network for the protection of genetic exchanges will be given priority. On that matter, Heydemann (1981) points out that even such large areas of land (over 500 ha) with ecosystems in continued existence should be placed in ecological, genetic isolation if within a distance of at least eight to ten kilometers a similar ecosystem cannot be offered. According to the type of ecosystem, clustered elements (stepping stones) and linear elements (connecting structures) are suitable to the networking of reserves.

2.2.6 Areas with special duties for the safeguarding of genetic variety

Highly mobile animal species (such as migratory birds) rely on being able to find resting areas along their migratory path as well as breeding areas. The safeguarding of the breeding ground is only possible if the corresponding resting areas are also secured.

As not all of them lie either in buffer zones and networked areas or participate in a recoverable ecosystem, it can be necessary to secure further fields outside these areas.

2.2.7 Areas with specific purposes for the safeguarding of the natural resources soil, water and air

The virtually "ubiquitous" natural resources—earth, air and water—can be spoilt by measures in all areas. This implies the overall obligation to safeguard and protect these resources. Certain areas with special local conditions are particularly sensitive to generally practised forms of land use, and some have a particular conservational function in comparison with neighbouring fields. Particular attention is dedicated to this under the aspect of conservation of natural resources. In particular, in Central Europe it deals above all with the following areas:

– Buffer zones bordering surface waters:

Surface waters are, with regard to the entry of nutrients and biocides from adjacent fields, particularly sensitive ecosystems. In many cases, agricultural use extends right up to the river bank.

To remedy the matter, portions of natural shore not in agricultural use will have to be developed.

– Agriculturally exploited peaty soils:

Peaty soil is characterised by a high water saturation and a high percentage of organic substances. Ventilation of this soil type through drainage leads to the breakdown of organic substances and thus to a loss of the natural resource soil.

Kuntze (1973) establishes that through agricultural usage, approximately 1 cm (0.5 cm through usage of green land) of high moor is used up. For a fen, he doubles these figures under Central-European climatic influences. He shows, for example, that through aeration of approximately 30 cm of a fen cross section due to drainage, independant of dung, precipitation and bacterial supply, approximately 250–500 kg N/ha willbe mineralised per year. The end products of this oxidation breakdown are above all nitrate (NO_3^-) and dinitrogenoxide (N_2O) which on their part lead to a pollution of the natural resources water and air.

– Areas in danger of erosion in inclining locations:

Soil erosion due to water on agriculturally used fields is dependent on the following factors:

precipitation, soil, slope length, slope inclination, crop rotation and cultivation methods. Depending on the soil-classification number, in Central Europe levels of erosion of between 2–14 t/ha per year are seen as tolerable (Auerswald 1983). The tolerability is however not a measurable "scientific" value. It is dependent on the soil type and indicates how much soil can be eroded away in the long term, without too great a risk to the impairment of the soil fertility.

– The effect of this erosion is not only the loss of the natural resource soil but also, as 1/3 of the eroded soil reaches surface waters (Auerswald and Schmidt 1986), pollution of water due to fertilizers and biocides. In accordance with the aims of nature conservation, in order to secure the lasting usefulness of the natural resources, it is necessary to find soil-conserving ways of using the land.

– Periodically flooded, agriculturally used fields in inundation areas:

If agricultural plots of land in inundation areas are used as arable land, then the danger arises that due to the flooding, soil, together with the attached fertilizers, will be eroded away. The consequence is not only a loss of the natural resource soil but also pollution of the natural resource water. Here it is necessary to safeguard meadows on agriculturally used land in inundation areas, or to promote afforestation.

– Agriculturally used land in areas of water-supply priority:

Water-conservation areas have, above all with regard to the aim of the long-term safeguarding of the usability of natural resources, a special significance.

It must be the aim in these areas to measure both the intensity of any use, and in particular, agricultural use, and the fulfilment of nature-conservation aims for the lasting safeguarding of the usability of the natural resource water.

2.2.8 Areas for overall ecological balance within agricultural areas

It is the aim of integrated plant protection, with the employment of all justifiable scientific, technical, ecological and toxicological methods, to bring harmful organisms below the agricultural harm threshold. So the utilisation of the natural demarcation factors is firmly in the foreground (ANL & DAF 1991). An essential part is played by predatory and parasitic organisms. A large number of these cannot satisfy their need for land on the agriculturally used areas alone, and so a shared habitat is allocated outside these areas, in areas striving for ecological balance.

Knauer (1989) characterises such areas as hedges, field and path edges, non-extended agricultural routes, running water, embankments, trees and bushes, ditches, ponds, marshy fields, undergrowth, bogs and fallow fields. Many of these fields are not only effective as habitats for particular animal species, but also as influences on the micro-climate, the water budget, or the area's structure in terms of plant protection.

Fields named in the previous sections 2.2.1–2.2.7 fulfil these criteria. Thus, the proposed relatively coarse network will be supplemented with the previously named "smaller structures", in order to leave no insuperable barriers in the areas in between, and to achieve a total ecological balance in agricultural areas.

2.2.9 Areas for overall agricultural, forestry and fish-farming use

For the safeguarding of the production of food and raw materials, those areas where overall agricultural, forestry and fish-farming land use are of prime importance are necessary, alongside those where the prime aim is conservation. Just as areas with the predominant function of conservation can be awarded a certain share of usage, so can certain requirements with regard to the guarantee of nature-conservation aims be placed on areas with predominant usage functions.

The safeguarding of the lasting usability of the natural resources water and soil is of primary concern, as is the precaution that the natural regeneration possibilities are not over-burdened. Usage must be so successful that no toxic pollution or other such damage of either the scattered, valuable biotopes or the neighbouring sensitive ecosystem takes place and that through it no obstacles which prove irreconcilable obstacles for wildlife are established (BFANL 1989).

2.2.10 Areas for industry, traffic, settlements and recreation

In order to safeguard man's social and economic demands, areas for industry, traffic, settlement and recreation are necessary. As in section 2.2.9, there are, however, certain requirements with regard to the guaranteeing of nature-conservation aims here, too.

Basically, this concerns on the one hand the prevention of immediately harmful effects (e.g., emissions) to man (as defined by technical-environmental protection), and on the other hand, the effects which lastingly spoil either the capacity of the landscape to perform its essential functions, the usefulness of natural resources, the wildlife or the variety, individuality or beauty of the landscape.

3. Quantification of land-use demands for nature conservation

According to the qualification of the land claims for nature conservation in chapter 2, a quantification for Bavaria was carried out by Schreiner (1990) as an example.

Table 1: Results of the approximate quantification of land claims for nature conservation in Bavaria:

function of land type	actual area (ha)	corrected area (ha)	% of state area
representative safeguarding of natural ressources (Reserves)	406 000	374 000	5.3
safeguarding of unique landscape elements (nature monuments)	7 000	-	-
safeguarding of unrestitutionable ecosystems	350 000	280 000	4.0
buffer zones for Reserves	313 000	249 000	3.5
Reserve Network for safeguarding of genetic exchange	288 000	288 000	4.1
particular safeguarding of genetic variety	7 000	-	-
particular safeguarding of the natural resources soil, water and air	1 112 000	1 046 000	14.8
overall ecological balance in agricultural areas	381 000	57 000	0.8
overall agricultural, forestry and fishfarming usage	6 061 200	4 179 700	59.3
industry, traffic, settlements and relaxation	581 000	581 000	8.2
total	**9 506 200**	**7 055 200**	**100.0**

Table 1 shows the overall results. The corrected area is the actual area in which multiple uses of the same area have been eliminated. Areas in the scale of 1<194> have been listed, but have not been taken into account in the corrected areas, as the trust limits of the other values are larger. The addition of listed areas in table 1 is problematic also for the corrected areas as the protection intensity as well as the relationship of conservation function to use function, of area category to area category is different. The stated aims of nature conservation, as given in chapter 1, are each fulfilled differently by the land categories.

4. Practical execution of the land-use concept of nature conservation for a sustainable development

The previous considerations and data on the land claims of nature conservation should provide the basis for concepts for a sustainable development with which they may be established in certain regions.

– Based on the installation of mapping biotopes and species it is necessary to legally secure a network of reserves of an effective size and distribution as national parks and nature reserves. The requirement for the fulfilment of the demands (exclusively nature-conservation land use) is at the same time, as claimed by Erz (1978), the conviction of the right of suitable ownership. Management and development plans for all of these areas should allow both the representative safeguarding of the ecosystem types and a further natural dynamism.

– Supplementary natural monuments must be identified. At the same time, unique occurrences of plants and animals (mostly with a small radius of activity) must be preserved. As natural monuments have no economic use whatsoever, they must also be placed in suitable ownership.

– Non-restorable ecosystems are subjugated to general protection. Interventions in these habitats must be, in view of their avoidability, subject to particularly strict examination. Special financial programmes should ease the economic restrictions in properties.

– The reserves are to be supplied with buffer zones.

The care and development plan for each reserve area must also apply to the accompanying buffer-zone statements. The devised aims and measures then either comply with the right of ownership (e.g., protective plantings) or are secured by a special financial support programme (e.g., to promote extensive use).

– It is necessary to develop a reserve network for the safeguarding of genetic variety. The basis for this is a stocktaking of the available, existing areas.

On this basis, new aims must be developed for newly created areas. These can either be calculated on the available land of local government or other such comparable entities, or it can be planned within the framework of compensating or substitute measures for environmental impacts.

– Areas with particular duties for the maintaining of the genetic variety are to be secured in a suitable way. Where the fundamental concern is resting habitats for waterfowl and waders outside of reserves, the disturbing influences in the area are prevented through local decrees.

– Areas with special duties for the safeguarding of the natural resources soil, water and air are to be carefully used. Apart from the use of legal means, special financial support programmes for uses suited to the site and for the restoration of natural habitats are to be offered. The desired land area and the level of the compensation payments shall stand in a market-economy equilibrium. With certain types of land (e.g., bank areas), a safeguarding through right of ownership with public ownership can be useful.

– In the agricultural areas, a sufficient number of fields for overall ecological balancing must be secured. The aim is a closely networked biotope-linking system of protective strips, extensive fields, unused parcels of land and special natural landscape elements. In these areas, typical members of the agro-ecosystem can find a possibility for survival and can build a stable community (BFANL 1989). This aim is to be achieved by special financial support programmes and special land-use planning.

– Harmful effects (e.g., emissions) from areas used for agriculture, forestry, fish-farming, industry, traffic, settlements and recreation to man andenvironment must be reduced to a tolerable level based specifically on landscape planning by using clean technologies and technical environmental-protection measures.

5. Accompanying measures for a sustainable development

The fulfilment of the aims of nature conservation by the stated measures is the basis for a successful endeavour for the safeguarding of the natural basis of life. It is necessary to make a wide sector of the public aware and to acquire the necessary means. These endeavours are doomed to failure if accompanying measures are not met. The following measures should be particularly emphasised:

– legislative and administrative measures must develop nature conservation as a sociological issue of equal ranking and can thereby achieve a separate management of its estates;

– the new assessment of the “free products” air, water and soil in the market economy through an environmental tax law;

– a change in the economic framework conditions for agriculture and forestry, so that the preparation of a product “environmental quality” is worthwhile (Schreiner 1989);

– overall measures for the prevention of harmful effects from intensively used areas, i.e., agricultural, forestry and fish-farming areas as well as settlement, industry, traffic and recreation areas. Thus, it must be taken into account that borderline values (e.g., for emissions) are not only measured with regard to people but also to certain conditions of the ecosystem balance;

– accompanying research which, amongst other things, observes the sustainability of the development with regard to the fulfilment of nature-conservation aims (Schreiner 1994);

– accompanying education, training and public relations with the aim of establishing an overall agreement on the interests of nature conservation to the estates as opposed to the short-term calculated use interests.

6. Summary

Nature conservation is a collection of measures produced by society to provide lasting security for its life basis. As a basis for considerations about sustainable development the land-use demands for nature conservation must be qualified. In detail, land types with the following functions must be distinguished:

– areas for a representative safeguarding of natural resources (e.g., reserves);

– areas for the safeguarding of unique elements of the landscape (e.g., natural monuments);

– areas for the safeguarding of ecosystems which cannot be re-instated;

– areas acting as buffer zones for reserves;

– areas serving as a reserve network towards the safeguarding the genetic exchange;

– areas with special duties for the safeguarding of genetic variety;

– areas with specific purposes for the safeguarding of the natural resources soil, water and air;

– areas for overall ecological balance within agricultural areas;

– areas for overall agriculture, forestry and fish-farming use with the aim of safeguarding lasting usability and with the precaution that the natural regeneration possibilities are not overburdened;

– areas for industry, traffic, settlements and recreation with the aim of avoiding harmful effects to man and environment.

The practical carrying out of the land-use concept for a sustainable development requires

– a network of reserves of an effective size and distribution in a suitable ownership;

– mapping and legally securing of non-restorable ecosystems;

– mapping and guaranteeing the careful use of areas with special duties for the safeguarding of the natural resources soil, water and air by legal means and/or financial-support programmes;

– a sufficient percentage of natural and near-natural habitats within agricultural areas achieved by a special land use planning and financial-support measures;

– prevention of harmful effects (e.g., emissions) from areas used for agriculture, forestry, fish-farming, industry, traffic, settlements and recreation to man and environment, especially based on landscape planning, by using clean technologies and technical environmental protection measures.

– The practical carrying out of the land-use concept for a sustainable development requires accompanying legislative and administrative measures, research, education, training and public relations.

7. Acknowledgement

Many thanks to Miss Emma Hayselden, Sheffield University, who helped me to formulate the text in English.

References

ANL & DAF = Akademie für Naturschutz und Landschaftspflege & Dachverband Agrarforschung (1984): Begriffe aus Ökologie, Umweltschutz und Landnutzung. <197> Informationen 4. Akademie für Naturschutz und Landschaftspflege. Laufen, 125 S.

Auerswald, K. (1983): Vereinfachtes Schema zur Vorausschätzung des Bodenabtrags durch Wasser. <197> Institut für Bodenkunde, Weihenstephan. Manuskript.

Auerswald, K. & F. Schmidt (1986): Atlas der Erosionsgefährdung in Bayern. Karten zum flächenhaften Bodenabtrag durch Regen. <197> GLA-Fachberichte 1. Mucnchen. 72 S.

BFANL = Bundesforschungsanstalt für Naturschutz und Landschaftsökologie (1989): Leitlinien des Naturschutzes und der Landschaftspflege in der Bundesrepublik Deutschland. Beilage zu Bohn, U., K. Buerger & H.-J. Mader: Leitlinien des Naturschutzes. <197> Natur und Landschaft 64, Heft 9.

Erz, W. (1978): Probleme der Integration des Naturschutzgesetzes in Landnutzungsprogramme. In: Forderungen an die Naturschutzgesetze der Bundesländer aus der Sicht der angewandten Ökologie, der Planung und der praktischen Naturschutzarbeit. <197> TUB 2 10: 11–19. Berlin.

Erz, W. (1981): Flächensicherung für den Artenschutz—Grundbegriffe und Einführung. In: Flächensicherung für den Artenschutz. <197> Jb. Natursch. Landschaftspfl. 31: 7–20.

Heydemann, B. (1981): Zur Frage der Flächengrösse von Biotopbeständen für den Arten- und Ökosystemschutz. In: Flächensicherung für den Artenschutz. <197> Jb. Natursch. Landschaftspfl. 31: 21–51.

Heydemann, B. (1983): Vorschlag für ein Biotopschutzzonen-Konzept am Beispiel Schleswig-Holsteins—Ausweisung von schutzwürdigen Ökosystemen und Fragen ihrer Vernetzung. In: Integrierter Gebietsschutz. <197> Schr.-Reihe des Deutschen Rates für Landespflege. H. 41: 95–104.

Kaule, G. (1988): Faktische Möglichkeiten der Wiedergutmachung ökologischer Schäden. In: Die Rechtspflicht zur Wiedergutmachung ökologischer Schäden. <197>Laufener Seminarbeitraege 1/87: 15–19.

Knauer, N. (1989): Praktische Umsetzung naturschutzpolitischer Vorgaben in der Landwirtschaft. <197> Laufener Seminarbeiträge 3/87: 34–44.

Kuntze, H. (1973): Moore im Stoffhaushalt der Natur—Konsequenzen ihrer Nutzung. <197> Landschaft + Stadt 2: 88–96.

Mac Arthur R. H. & E. O. Wilson (1967): The Theory of Island Biogeography. <197> Princeton University Press, Princeton, N. J..

Schreiner, J. (1989): Naturschutzfachlicher Flächenanspruch und Landwirtschaft. In: Naturschutzpolitik und Landwirtschaft. <197> Laufener Seminarbeiträge 3/87: 45–51.

Schreiner, J. (1991): Flächenansprüche des Naturschutzes: Qualifizierung und Quantifizierung.<197> Jb. Natursch. Landschaftspfl. 44: 9–39.

Schreiner, J. (1994): Naturschutz als angewandte Wissenschaft. <197> Z. f. Kulturtechnik und Landentwicklung 35: 281–291

A Proposal for a Set of Aggregated Indicators of the Environmental Dimension of Sustainability

Bedrich Moldan
Charles University, Centre for Environmental Scholarship
Petrska 3, CZ 11000, Praha 1
Czech Republic

I. Background

The concept of sustainable development was accepted unanimously by the global community of nations (cf. Rio Declaration on Environment and Development, UNCED 1992).

Its substance, however, is not much more than an indication of a general direction. It is a "lowest common denominator", the broadest outline of the developmental policy of nations. By no means is it a definite comprehensive strategy or a blueprint for a new sustainable world. Sustainability has meanings different for different regions and nations.

While it is beyond doubt that the idea of sustainable development is integrated and holistic, it has at least three fundamental dimensions. Probably the original and most basic of the three is the ecological dimension. In this context, sustainability means the balanced and stabilized functioning of global geophysical, geochemical, and biological systems. Of prime importance are the critical subsystems on which the overall stability of the planetary biosphere depends. The key aspects of this ecological dimension are the protection and maintenance of geophysical structures such as climate or the stratospheric ozone layer, of species biodiversity and ecosystems, of natural resources for sustained human use by present and future generations and of a habitable human environment.

Second is an ethical dimension. Ecological ethicists emphasize the value of a modest and frugal life, demonstrating respect for nature and for life in general. They speak about human values, intergenerational equity and sustainable living (IUCN 1991). Politicians also point to this aspect, stressing justice and equity among nations and among people and calling for enhanced human development in general, and the eradication of poverty in particular. Human rights and other social activities focus on freedom, cultural diversity, rights of minorities, and the role of major societal groups (such as children, young people, women, and indigenous peoples).

The third important aspect of sustainable development is the economic aspect. An enormous amount of work is now devoted to the exploration of this field, which many consider to be the most important (cf. Pearce and Warford, 1993, Proceedings 1994). According to Munasinghe (1993), the economic approach is based on the Hicks-Lindahl concept of the maximum flow of income that can be generated while

NATO ASI Series, Partnership Sub-Series, 2. Environment – Vol. 6
Sustainable Development of the Lake Baikal Region
Edited by Valentin A. Koptyug and Martin Uppenbrink

at least maintaining the stock of assets or capital which yield this benefit. An underlying concept of optimality and economic efficiency is applied to the use of scarce resources. The main problems concern identifying the kinds of capital to be maintained and the ability to find appropriate substitutes for them. An important role is also played by uncertainty, irreversibility, and the possible catastrophic collapse of the ecological processes.

One thing is absolutely certain: in no single country, and even less in the whole world, does the contemporary evolution of human civilization have a sustainable character. Thus, the idea of sustainable development is a challenge, an urgent call for exploration into ways in which it may be accomplished. Moreover, it is absolutely clear that one of the first prerequisites of progress in this direction is the ability to be able to see whether any progress is being made at all. This implies that a type of measuring stick must be available.

There have been many efforts to develop indicators of sustainable development (see, e.g., Moldan 1993, UNEP/UNSTAT 1993 and Tunsstall et al. 1994). A consensus seems to be emerging that the three fundamental aspects of sustainability (ecological, ethical and economic) should be measured separately. Considerable progress has been made in the development of indicators of the ethical aspect of sustainability. The furthest advanced work is in the Human Development Index, introduced by UNDP in 1990 (UNDP 1990), which has since been successfully applied. There is a significant body of work dealing with specific aspects of this dimension (e.g., O'Connor 1992).

As part of the economic approach towards sustainability, various indicators have been produced focusing on the correction of conventional economic indicators such as GDP for various aspects of sustainability. One remarkable result is the recent manual on integrated environmental and economic accounting (UN 1993).

The same is true for the ecological dimension of sustainability. In the field of indicators, much excellent work can be cited, e.g., the Environmental Data Reports by UNEP (UNEP 1993), the series of World Resources Reports (WRI 1994), a great deal of work by the OSCD (OSCD 1993), and the recent initiative of Eurostat in launching a project on "Environmental Pressure Indexes" (Eurostat 1994).

Several ecological concepts underly individual proposals for indicators, such as:

– carrying capacity (e.g., IUCN 1991)

– critical loads (e.g., Downing et al. 1993)

– environmental space (e.g., NEPP 2 1994)

– global and national environmental impact (WRI 1994)

– biological-physical sustainability (WRI 1994).

This proposal is intended to complement these efforts and to bring in a new element. It should be viewed as a part of the SCOPE project "Indicators of Sustainable Development" (SCOPE 1994).

II. The Concept of the Total Earth Load

1. The General Concept

The proposed concept is based on notions of the whole earth and of the overall load which it bears. The proposed set of indicators must cover all of the critically significant environmental phenomena which are influenced by the interaction between humans and the totality of the earth. The proposed indicators are intended to be applied on the national level.

This proposal—elaborating on results of the planning meeting for the above-mentioned SCOPE project—distinguishes among four categories which cover interaction with the:

(i) geosphere,

(ii) biosphere,

(iii) natural resources,

(iv) human environment.

The notions of geosphere and biosphere correspond to the current International Geosphere-Biosphere Programme of thc International Committee for Scientific Unions (ICSU)— A Study of Global Change (IGSP 1988). Interaction with natural resources regards the direct or indirect use by humans of all types of natural assets, including raw materials, space, soil, water and natural dispersion services. The human environment category points to interactions between individuals and the environment in which they live.

The idea of the Earth's "load" corresponds to the OECD's "pressure" (OECD 1993) or Hammond/Dahl's "impact" (WRI 1994). The present proposal prefers the term "load" because it implies both pressure or impact as well as a concept of what is basically bearable, or what is mostly presumed to be sustained. Indeed, the load could be too great to be sustainable.

2. Important Criteria

The indicators of Earth's total load must fulfill many criteria. Some of these are well worked out and will not be repeated here (cf. Adriaanse 1993, OECD 1993, WRI 1994). Some features, however, must be stressed:

(i) Global significance. Only phenomena of global significance should be considered. However, this notion must be understood in two different senses. First, phenomena which occur on a global level should be covered, such as the climatic system. Second, phenomena that are so widespread that they can be recognized as globally universal must also be regarded.

(ii) Exhaustivity. Indicators must not leave out any truly significant phenomenon. This does not mean that every single aspect of such a phenomenon must be covered, but that the picture which results from the indicators must not be substantially distorted.

(iii) Selectivity. A very careful selection of topic must be made, because the number of actual indicators must be kept small. Therefore, only absolutely essential elements should be covered.

(iv) Policy or action orientation. It has been stressed repeatedly that attributes of sustainable development indicators must not be only quantitative, measurable, descriptive, and representative, but it is critical that they be policy-oriented. The indicators must point to certain actions. The real meaning of presented numbers should be obvious enough to elicit a feeling of urgency to do something (or complacency if conditions are satisfactory).

In my view, there are two basic ways to accomplish this. First, the indicator values may be compared to a certain benchmark (critical value, standard, or limit). The meaning and consequences of the value which is equal to the benchmark value will be clear, and therefore the implications will also be clear of the nearness of a value to the benchmark and when it is above or below the benchmark value. Second, an ordering of evaluated subjects could be made (in our case countries) which immediately indicates where countries fall in relation to one another.

(v) Political acceptability. If the proposed indicators are not used by national governments, then they are totally useless. (Note, for example, the negative reaction of Brazil and China to the Draft Chairman's Text on indicators for sustainable development, which prevented discussion of this issue and an eventual decision of the Commission on Sustainable Development of the UN during its second substantive session in New York in May, 1994).

III. Four Categories of the Earth's Total Load

1. Geosphere Load

Important phenomena:

- climate system,
- ozone depletion,
- disturbance of biogeochemical cycles of major nutrients,
- acidification.

Indicators:

- emission of greenhouse gases (in CO_2 equivalents according to GWP, per capita/yr)
- emission of ozone-depleting substances (composite index as formulated by environment, per capita/yr)
- emissions of NO_2, NH_3, SO_2 (kg + kg + kg, per capita/yr)
- flow of N-substances into major lakes, rivers, estuaries and coastal areas (kg N leaving national land-based borders, per capita/yr)
- flow of total P (dissolved and undissolved) into major lakes, rivers, estuaries and coastal areas (kg P leaving national land-based borders, per capita/yr).

2. Biosphere Load

Important phenomena:

- Loss of natural ecosystems (terrestrial, marine, wetlands)
- Loss of biological species
- Loss of ecological stability of the landscape (habitat loss, land-use conversions)
- Loss of biological and genetic diversity among wild and/or cultural and semi-cultural organisms and ecosystems (e.g., semi-natural forests).

Indicators: to be devised according to the methodology being developed under the Biodiversity Convention and the SCOPE Sustainable Biosphere project.

3. Natural-resources Load

Important phenomena:

- Agricultural practices
- Forestry practices
- Water management
- Fisheries
- Use of energy
- Use of raw materials
- Production of waters
- Recycling of key materials.

Indicators:

- Percent of appropriated products according to the methodology by Vitousek (1986) (percent)
- Topsoil loss (kg per capita/yr)
- Desertification (ha/yr: methodology according to a convention on desertification still to be devised)
- Deforestation (loss of natural or semi-natural forest areas; does not apply to forest plantations) (ha/capita/yr)
- Sustainability of forest yield (% of unsustainable production of the total production)
- Water withdrawal (% water resources used of the total resources)
- Sustainability of fisheries yield (% unsustainable catch of the total catch)
- Use of energy from nonrenewable sources (fossil, nuclear, (BTU/capita/yr)
- Use of mineral raw materials (total minerals, including oil, coal, gas, metals, building and construction materials, kg/capita/yr)
- Production of hazardous wastes (kg/capita/yr)
- Recycling of paper, glass, plastics, metals (average percentage of nonrecycled material in four categories).

4. Human Environmental Load

Important phenomena:

- Hygiene
- Potable water
- Medical services
- Air pollution
- Noise
- Industrial pollution
- Environmentally caused diseases
- Unsafe waste deposits
- Disasters

Indicators:

– Percent of population not served by waste-water treatment (at least biologically treated, %)

– Percent of population without access to safe drinking water (%)

– Percent of population without access to medical services (%)

– Percent of population living in areas exceeding WHO ambient-air standards for any of the major pollutants (TSP, NO_2, SO_2, Pb, VOC, CO_2, O_3) more than 100 days per year (%)

– Percent of population living in areas with levels of noise above international standards (%)

– Percent of industrial plant units of any kind with more than 50 employees without biological or advanced chemical treatment of waste water (%)

– Percent of deaths caused by environmentally related diseases (cholera, dracunculiasis, diarrhea, malaria, neoplastic diseases) (%)

– Unsafe landfills with any kind of waste (cubic meters/capita) - Victims of natural and man-made disasters (no./yr of victims of industrial and natural disasters, property loss in US$/yr).

IV. Additional Indicators

While the indicators of the earth's total load approximately apply to the "pressure" term in the OECD "pressure-state-response" scheme, the additional indicators (corresponding to "state" and "response") serve to complete the picture for individual countries and for the state of the global environment.

1. State of the Global Environment

The proposed indicators should characterize the state of the global environment by means of global values of key variables:

– average global temperature

– atmospheric concentration of key greenhouse gases

– atmospheric concentration of key ozone-depleting substances

– state of the ozone layer

– intensity of UVB radiation

– pollution of key oceanic areas

– volume of the global cryosphere

– frequency of catastrophic events of atmospheric or hydrologic origin (number of hurricanes, floods, and other naturally occurring disasters, according to WHO methodology)

2. State of the National Environment

– Population density

– Arable land per capita

– Use of industrial fertilizers (kg/ha)

– Total agricultural land per capita, livestock density (no./ha)

– Natural ecosystems (excluding deserts, % of total land)

– Forested areas—open and closed forests (% of total land)

– Cold and warm deserts (% of total land)

– Affected land (by urbanization, industrial activity, transport, derelict land) (% of total land)

– Biodiversity (no. of plant species, no. of animal species)

– Water resources (total) (cubic meters/capita/yr)

3. National Response

Indication whether the following activities are:

– in full operation or broadly applied

– in limited operation

– planned for immediate implementation

– nonexistent

– national sustainable-development or environmental-protection institution on a cabinet level or similar

– national sustainable-development or environmental-protection plan or strategy approved by the cabinet and/or the parliament, progress monitored

– regular research and monitoring institutions covering all major environmental sectors

– comprehensive set of legal environmental regulation with appropriate enforcement

– punitive measures to enforce regulation (imprisonment, fines)

– economic instruments for environmental protection (taxes, charges, fees)

– full public access to environmental education and information

– public and overall expenditures for environmental improvement (% of general budget, % of GDP)

– access to international treaties (treaties ratified, ratification of key treaties, namely on climate, biodiversity, the Basel Convention)

– government sponsorship of environmental NGO's (US$/yr)

– priority given by the public to environmental issues according to public opinion polls

Note

Simple calculation of the country rating for each category of indicators:

1. Countries are ranked by the values of indicators, from the "best" country (= 1), e.g., the lowest emissions of CO_2 equivalents, to the "worst" (= total of countries).

2. Each country obtains for each indicator a certain number of points (P), according the following formula: $P = R \cdot 100/N$ where R = country ranking, N = number of countrics counted

3. The country rating corresponds to the average value of points in the given category (e.g., geosphere load, the "best" country would have the lowest value).

References

Adrisanse, A. (1993): Environmental Policy Performance Indicators: A Study on the Development of Indicators for Environmental Policy in the Netherlands. Dutch Ministry of Housing, Physical Planning and Environment. Sdu Vitgeverrij Koninginnegracht, May, 1993, 175 pp.

Downing, R.J., Hettelingh, J.P., de Smet, P.A.M., eds. (1993): Calculation and Mapping of Critical Loads in Europe: Status Report 1993, RIVM Report no. 259101003, Bilthoven, 163 pp.

Eurostat (1994): Eurostat Environmental Pressure Index Project: Call for Tender. Official Journal of the European Communities, C 1k86, 8.7.1994, Brussels.

IGSP (1988): The International Geosphere-Biosphere Programme: A Study of Global Change. IGSP Report no. 4, Stockholm. 200 pp.

IUCN, UNEP, WWF (1991): Caring for the Earth. A Strategy for Sustainable Living. Gland, Switzerland, 227 pp.

Moldan, B. (1993): Discussion Paper: Indicators of Sustainable Development. The Bureau of the Commission on Sustainable Development. United Nations, New York (mimeographed), 20 pp.

Munasingha, M. (1993): Environmental Economics and Sustainable Development. World Bank Environment Paper no.3, Washington, D.C., 112 pp.

NGPP 2 (1994): The Netherlands National Environmental Policy Plan 2. Ministry of Housing, Spatial Planning and the Environment. The Hague, 242 pp.

O'Connor, J. (1992): Profiles of Poverty and Revised Social Indicators of Development. World Bank, Washington, D.C. (mimeographed), 5 pp. + 10 attachments.

OECD (1993): Environmental Indicators: Overview of Work Programme and Publications. OECD State of the Environment Division. Paris. (Mimeographed.) 35 pp.

Pearce, D.W., Warford, J. (1993): World Without End: Environment, Economics and Sustainable Development. Oxford University Press, Oxford (523 pp.)

Proceedings of the international symposium (1994). "Models of Sustainable Development. Exlsusive or Complementary Approaches of Sustainability?" Paris, March 1994, vol. 1, vol. II (1212 pp.)

SCOPE (1994): SCOPE Planning Meeting of the SAC for the Project on Indicators of Sustainable Development, Prague, 22–23 April 1994 (mimeographed), 10 pp.

Tunstall, D., Hammond, A., Henninger, N. (1994): Developing Environmental Indicators. A Report on the World Resources Institute Workshop on Global Environmental Indicators. Washington, D.C., December 7–8, 1992. World Resources Institute, Washington, D.C. (33 pp.)

UN (1993): Integrated Environmental and Economic Accounting. DESIPA, Studies in Methods, Series F, no. 61, United nations, New York, 182 pp.

UNDP (1990): Human Development Report. Oxford University Press, Oxford (189 pp.)

UNEP (1993): Environmental Data Report 1993–1994. Blackwell, Oxford, 408 pp.

UNEP/UNSTAT (1993): UNEP/UNSTAT Consultative Expert Group Meeting on Environmental and Sustainable Development Indicators, Geneva 6–8 Dec. 1993, UNEP, Geneva (mimeographed), 15 pp.

Vitousek, P.M., Ehrlich, P.R., Ehrlich, A.H., Matson, P.A. (1986): Human Appropriation of the Products of Photosynthesis, BioScience, vol. 34, no. 6, pp. 369–373

WRI (1994): World Resources 1994–1995. Oxford University Press, Oxford, 402 pp.

Economic Conditions and Requirements of Sustainable Development in General and for the Lake Baikal Region in Particular

V.V. Kuleshov
Institute of Economics and Industrial Engineering of SB RAS,
Novosibirsk, Russia

The article focusses on three components of sustainable development: economic (scale of production and efficiency of resource use); environmental (environmental protection and attitude towards it); social (extent to which the population's interests are satisfied). This three-fold approach is regarded as an economic approximation of sustainable development. The Lake Baikal Region is taken as a model territory, and the intention is to work out mechanisms of integration of: interests and means; different forms of property; intraregional management functions; organizational and managerial forms of productive activity, e.g., major enterprises as well as small businesses.

1. The meaning of the term "sustainable development" is very vague. This is confirmed by a variety of indicators used to define this term.

Sustainable development is understood in a different way in different countries, according to whether they are advanced (developed), developing, etc.

In a general sense it is a positive development acceptable for the entire society: it does not threaten the environment, minimizes social tension and ensures a decent standard of living for everybody. On the other hand, the notion of "balanced development" has long been in use.

The concept of balanced development dates back to the ideas of self-supporting rural economies formulated by Mahatma Gandhi in the 1930's. They are characterized by decentralized decision-making and local resources-based production to meet local needs.

In the USSR, the ideology of efficient balancing formed the guiding principles of planning the national economy.

2. To keep things in perspective, we think it useful to refer to the development logic of market economies, towards which Russia is moving today. When limiting the analysis to the 20th century and developed countries, it will be possible to reveal a certain sequence of stages in their development. Nominally, the following stages (epochs) are distinguished: of mass production, of mass production and consumption, of post-industrial development, of sustainable development (the future). Each epoch is characterized by a particular set of factors relevant to all the epochs, reflecting their most typical features. The most important of them are

NATO ASI Series, Partnership Sub-Series, 2. Environment – Vol. 6
Sustainable Development of the Lake Baikal Region
Edited by Valentin A. Koptyug and Martin Uppenbrink

producers' general economic formulae, the achieved standard of living, state regulation priorities, ratio of extensive to intensive factors of development, the economic role of technological progress, attitude towards the environment, etc.

Generally speaking, one can argue that during the transition period large-scale production with high specialization, a swollen bureaucracy and resource-wasting methods seem less attractive than small environmentally friendly production switching from standardization to coöperative mechanisms. This does not at all imply the introduction of new elements into the old system, but rather changes in the concept of economic development.

3. In abstract terms, development is a system of three coordinates:

– economic (scale of production and efficiency of resource utilization);

– social (the extent to which society's needs are met);

– environmental (environmental protection and attitude towards it).

The system of administration generates and implements certain goals (with some restrictions). In the past, economic and social goals and targets were a top priority.

Sustainable development is a development balanced with regard to the three coordinates: economic, social and environmental.

In our opinion, the formula is rather simple: social stability depends on economic stability, whereas environmental stability is based on both.

This is a triad, since a feedback is observed. The environmental component impacts the content of the economic one, at least as regards two aspects (requirements):

– imposing increasingly stringent restrictions on technologies, the economy, social institutions, behavior of social groups, communities and even individuals;

– minimizing the use of non-renewable natural resources to meet basic needs of society and individuals.

The social component evolves through continuous changes in the way of thinking of individuals (their mentality) from the economic towards environmental imperative.

4. With the above factors, it can be stated that in the second half of the 1980's, the Russian economy, based on the "overtake and surpass" principle, was mainly at the second stage (featuring essential elements of the first) and, probably, very near to the end of that stage. Ecological requirements became much more severe. To meet the new requirements, the operation of hundreds of enterprises was halted. This strongly affected the economic and social components, although for the most part the requirements were completely justified.

The situation is that the whole triad has dramatically slid backwards (in time) to the level of the 1960 to 70's (figs. 1–4).

Figure 1 describes the three-year dynamics of industrial production in several facets, including the alleged stabilization of decline in the summer of 1964. Figure 2 features a number of forecasts and theoretical generalizations made for different industries in the first quarter of 1994. Figure 3 provides a forecast for the early summer of 1994 for a wider range of industries. The forecast is of a short-term nature. Figure 4 characterizes the efficiency of the reproduction processes in the Russian Federation (RF).

Figures 1–3 describes the current situation and short-term forecasts, while figure 4 outlines long-term prospects of the country's development, including the ecological situation. There is an opinion that a decline in industrial production might be of benefit to the environment. In terms of the above triad, the following conclusion is justified: environmental damage from production in crisis conditions will be on the increase.

DECLINE IN INDUSTRIAL PRODUCTION*

(1992-1994)

7 MONTHS

In % to the respective period of the previous year

1994	- 25.8%
1993	- 16.2%
1992	- 18.8%

MONTHLY AVERAGE

rates of decline

1994 (7 months)	- 2.8%
1993 (II half-year)	- 2.2%
1993 (I half-year)	- 0.3%
1992 (II half-year)	- 2.3%
1992 (I half-year)	- 1.1%

PERIODS OF 1994

In % to the respective previous period

I half-year	- 12.0%
I quarter	- 10.0%
II quarter	- 4.0%

SUMMER OF 1994

June (the first decade of May - holidays)	- 0.0%
July (1.0% - growth)***	- 6.5%**
August	-

** Data from the State Committee on Statistics. ** "Izvestiya", 24 Aug. 1994. *** "Rossiyskaya gazeta", 24 Aug. 1994.*

Figure 1

APRIL OF 1994

According to many experts, our country has entered a new depression stage of economic crisis at which the ongoing production decline does not depend much on the financial policy of the state.

Many experts share the opinion that it is infeasible to stop the decline till production falls to its crisis point, expectedly by the end of the current year.

Importantly, ***volume of production at its lowest point will account for approximately 40% of 1990 level,*** including:

consumer goods industry	25%
machine-building	30%
ferrous metallurgy, forest complex and construction materials industry	40%
nonferrous metallurgy and chemical industry	45-50%
food industry	55%
fuel and energy complex	65%

Conclusion: ***As world experience shows, first and foremost investment will help to overcome economic crisis.***

Figure 2

It will be caused by the following:

– inadmissibly excessive wear and tear of technological, transport and sewage equipment;

– use of off-grade resources and low-quality materials;

– less investment in nature use, since respective funds will be redirected towards economic and social needs;

– forced return to using low-grade coal to supply heat in Russian cities.

An important expected development could be that logging will dominate in the exploitation of forest resources in the regions with developed transportation networks, and there is a danger of reviving drift floating. Economically most developed regions are expected to intensify ploughing up and cutting. The agrochemical industry will collapse, and natural regulation of ecosystems in such

JUNE OF 1994

FORECAST*

Production will stabilize when, compared to 1991 level, it amounts to (in%):

Coal industry...............45	Forest industry....................50
Oil industry..................70	Wood processing industry.....25
Metallurgy...................35	Agriculture (excluding subsidiary farms)..60
Machine-building.........25	Food industry (excluding alcohol)..............50
Chemical and petrochemical industry.20	Railway traffic.......................55
Consumer goods industry.......................15	Air traffic..............................30

* *"Ekonomika i zhizn'", 1994, No. 24*

Figure 3

EFFICIENCY

1. Labour productivity in the entire industry accounts for 24%, in agriculture 12% of the USA level and keeps on declining.

C o m p a r e: In the USSR in the late 80s, the respective figures were about 60% and 20%.

2. Capital assets of enterprises are depreciated and almost worn out physically.

PATTERN OF CAPITAL ASSETS IN THE NATIONAL ECONOMY

Lifetime (years)	Proportion (in %)
up to 5	2
6 - 10	7
11 - 20	28
21 - 40	34
over 40	29

As a result, the technical level of only 10 - 12% of all enterprises meets qualitative requirements.*

3. The structure of industrial production is increasingly degrading to the advantage of primary industries.

* *"Ekonomika i zhizn", 1994, № 24*

Figure 4

regions will be unbalanced. The collapse of the agrochemical industry is very likely to result in decreased agrochemical pollution. Migration of the population from northern regions will lead to production decline and, consequently, to a decrease in basic pollution under the threat of massive discharges.

The matter is not that, in terms of the triad, we have been thrown back to the 1960's level, but we also have changed the principles of economic activity and the mechanisms of management. State funding of any investment projects (programs) is being dramatically reduced. It is imperative to search for new forms and sources of financing.

All this takes time. Thus, for a long time to come the ideas of sustainable development in Russia can only be the subject of discussion and academic research (both as regards the whole country and its particular regions).

We have to experience the transition period. In this respect, some elements of sustainable development imply the following conditions to be fulfilled during this period: figure 5.

In this sense, the discussion about Lake Baikal is not an exception. If the task is to protect the environment around the Lake and the Lake itself, then this requires the following:

– to define what exactly should be protected and, as precisely as possible, permissible changes in the environment during a specified period of time;

– to consider the population's needs in relation to the destiny of the Lake and its surroundings, at least in the form of restrictions; therefore, it concerns not only people living near the Lakc, but also those consuming what is produced in the area. It is important to take potentially unemployed people into account;

– to formulate restrictions arising from the fact that regions affecting the lake and its surroundings will not introduce any changes or bring about specific and/or limited changes; these should be multilevel restrictions ranging from the global level to the level of particular regions, directly influencing the Lake and its surroundings;

– to formulate tasks aimed at defining the roles of participants involved in whichever way in accomplishing the goals (provided their interests are balanced), and their responsibility.

When considering the triad as an economic approximation of sustainable development, the central element in the development of the Lake Baikal Region as a model territory of sustainable development will be the working out of the integration mechanism. By this we mean the integration of:

– interests and means:

– of the present and future generations;

– of governmental responsibilities (the world community, the country, the region).

MAIN CONDITIONS (REQUIREMENTS) OF THE TRANSITION PERIOD TO THE MARKET

ECONOMIC
Retaining jobs already created.*

SOCIAL
Preserving the current number of inhabitants in the region.

ENVIRONMENTAL
Maintaining the present condition of the environment.

* *However, it is wrong to believe that the future structure of production will be determined by the current production capacities. It will be determined by the market.*

Figure 5

INTERESTS OF SUBJECTS OF DIFFERENT LEVELS IN SUSTAINABLE DEVELOPMENT OF THE LAKE BAIKAL REGION

(expert assessment)

ADDRESS OF INTERESTS	NATURE OF INTERESTS Economic	Social	Environmental
WORLD COMMUNITY*	-	-	+
RUSSIA	+	+	+
BAIKAL REGION**	++	+++	+

* *First of all developed countries.* ** *Interests of the population get specified with a focus on national and cultural development of ethnic groups.*

Figure 6

– different forms of property.

– functions of intraregional management (the region is regarded as a single management entity).

– organizational and managerial forms of productive (not exclusively production) activity—for instance, major enterprises and small businesses—etc.

5. From the academic point of view, it is sensible to set very concrete tasks, i.e., principles of a program-oriented approach should be implemented.

In the case under review, the program should be in keeping with the interests of the world community, the country and the region (the local population). Figure 6 shows how they overlap.

The state should introduce legislatively mandated rules and boundaries of activity in this region, including the Law on Lake Baikal, systems of licensing etc.

The following proposal has been put forward. (The information below was provided by V. Gukov, Candidate of Economic Sciences, Section Chief, the Department of Regional Economics and Location of Productive Forces of Eastern Siberia.) The state "purchases" the Lake Baikal basin and the Lake itself as a water resource through a state act. An approximate price may be estimated in the following way. If the Lake Baikal water is valued at a price of sea water desalinated by nuclear reactors at 12 cents per cubic metre, the overall price will total 2.76 trillion dollars. When valuing the Lake Baikal water at a cost of water produced from the ice of icebergs shipped from Antarctica, the price will be 0.5 trillion dollars.

The natives of Alaska, in exchange for their permission to use their land for setting up oil production, got the right to keep part of the land inviolable as well as a monetary compensation of up to $ 965.2 million, which was paid them over ten years beginning in 1972.

Through the above Act, the Fund of Future Generations, or the Lake Baikal Fund, would be established. The newly established fund (or the system of funds) would have its capital in the form of a portfolio of shares (possibly including foreign shares) and would be getting dividends (capital assets are not to be spent) which might be directed to social programs, as well.

To illustrate, what are the concrete mechanisms to realize the ideas of sustainable development in Alaska? Today two main lines of transforming the present-day industrial economy are being pursued: setting up small businesses and permanent trust funds.

The institution of corporate ownership (trust) has been already introduced in civil legislation of the RF through the Decree of the President of the RF No. 2296 dated 24 Dec. 1993 "On Corporate Ownership (Trust)".

The following types of trust funds have been defined:

1. A regional fund formed out of capitalized profits from the exploitation of natural resources.

2. A trust fund formed out of government financial resources (the nominal "purchase" of Lake Baikal as a water resource) with subsequent investments.

3. A trust fund created through financial contributions from the world community (under the same conditions).

As any other model, the proposal for trust funds is of a hypothetical nature, so the latter may fail to function properly. Therefore, whatever the situation may be, the program for the development of the Lake Baikal Region should be based on state support.

Influence of Environmental Factors on the Health of the Population of the Lake Baikal Region

M.F. Savchenkov
Regional Center of Medical Ecology, Irkutsk, Russia

A.A. Maiboroda
State Medical Institute, Irkutsk, Russia

S.I. Kolesnikov, B.P. Lemeshevskaya, V.S. Rukavishnikov
East Siberian Research Center of the Siberian Branch of the Russian Medical Academy of Sciences, Irkutsk, Russia

A.K. Tulokhonov
Baikal Institute of Rational Nature Management of SB RAS, Ulan-Ude, Russia

In the report, data on the environmental contamination of the Lake Baikal Region, including the state of water resources, and on the aggravation of the population's health in this region are presented.

It may be asserted that under modern conditions of technology any kind of economic activities, and especially industry, has a negative influence on the biosphere and public health. In the Irkutsk region the leading role in this respect belongs to the chemical industry, petroleum processing, non-ferrous metallurgy, deep wood processing, and power industry. At present, there are considerable areas of contamination by emissions of industrial enterprises and thermal power plants (Table 1).

Table 1 Environment pollution in the Baikal and Transbaikal regions

Territory	Area, in thousands square kilometers	
	total area	polluted area (over 2 MPC*)
Irkutsk region	767.9	41.1
Chita region	431.5	10.0
Republic of Buryatia	351.3	4.5

* MPC - maximal permissible concentration

NATO ASI Series, Partnership Sub-Series, 2. Environment – Vol. 6
Sustainable Development of the Lake Baikal Region
Edited by Valentin A. Koptyug and Martin Uppenbrink

The most contaminated are industrial centers. In the Russian Federation, 41cities are considered as having a high level of atmospheric pollution, and six of them are situated in the Lake Baikal Region. Especially conspicuous is the city of Bratsk with its extreme ecological situation.

The state of the population's health is noticeably influenced by the quality of potable water. For medico-ecological assessment, it is important to know the quality of water at the sites of water intake from which water is supplied to the population. High water quality with respect to all parameters is consistently observed in Lake Baikal only at the origin of the Angara and Belaya rivers (with the exception of single parameters that exceed the norms in the spring by 1.5 times). Water at the water-intake site in the Oka River and in the Irkutsk reservoir in spring and summer is polluted with petroleum products (whose maximum values exceed the maximum permissible concentrations nearly by a factor of three). In the water of other sources (the Olkha, Kitoi, Biryusa, Oka and Lena rivers), in the warm period of the year, an elevated biochemical oxygen consumption (BOC) of up to 1.5 to 2.4 times the maximum permissible concentrations (MPC) and chemical oxygen consumption (COC) up to 1.2 to 3.6 MPC are found. In the Lena River, the maximum COC reaches 195 mg/l, which exceeds the standard for category I water bodies by 13 times. It should be noted that mean annual values of organoleptic and chemical indices in all the above mentioned bodies of water, except for the Lena River, are below MPC.

The water of the above water bodies (except for the Lena River), according to combined organoleptic and chemical indices, has a moderate degree of contamination, which, according to the current norms of SanPin 4630-88, does not preclude its use as potable water, on condition of appropriate treatment in purification works. The water of the Lena River has a high degree of contamination; it is not fit for drinking or any other domestic use.

All the above mentioned bodies of water are characterized by high saturation with soluble oxygen throughout the year. This promotes oxidation-reduction processes and reduces the negative influence of low temperatures on the rate of self-purification of bodies of water.

Noteworthy is the low water mineralization in 8 out of 10 water bodies of the region. It is only in the Lena and Olkha rivers that the mean annual concentration of solid residual varied in a range of optimum values (200–400 mg/l). Especially low is mineralization in the Biryusa and Angara rivers and in Lake Baikal, where mean annual concentrations of solid residual amount to 66.7, 87 and 93 mg/l, respectively. This is considerably lower than the recommended minimum of 100 mg/l. Low water mineralization is considered by science as a negative factor.

In all bodies of water, except for the Lena and Olkha Rivers, there is a deficit of calcium, magnesium and sodium salts and of such vitally important microelements as fluorine and iodine (Table 2).

Table 2 Concentrations of salts and microelements (mg/l) in water bodies

Element	Average annual	Minimum permissible	Optium
Ca	11-15.5	20	30-50
Mg	2.8-5.1	10	
F	0.1-0.37	0.5-0.8	
Na+K	3.4-5.8	30	30-50

This may be the cause of a high incidence of cardiovascular diseases, endocrine pathologies (endemic struma), tooth decay, bone diseases, and other pathological conditions.

Specialists are beginning to gain a better understanding of the hygienic significance of water chlorination. Apart from the useful function of bacteriological disinfection of water, chlorination bears a great danger for the population due to formation of chloro-organic compounds. Wherever chlorine is used for water disinfection, phenol concentrations exceeding the MPC by 2 to 4 times has been detected (the Belaya, Biryusa, and Angara rivers, the Irkutsk reservoir, and Lake Baikal). This bears the risk of formation in the drinking water of especially dangerous polychlorinated organic compounds of very high toxicity and having carcinogenic and mutagenic effects.

Thus, water at the main sites of water consumption of the largest water bodies of the Irkutsk region (with the exception of the Lena River) is fit only for centralized water supply on condition of appropriate purification and preparation. Due to the low water mineralization of the majority of the bodies of water, and in order to improve its quality, it seems necessary to construct mixed water pipelines in which both subterranean— more mineralized—and surface waters are used. The transition from chlorination to other disinfection methods is also important.

A worsening of the population's health is manifested, first of all, in disturbances of the immune system, when teratogenesis, mutagenesis and carcinogenesis are activated. A correlation between water quality and public health is observed. For the last 20 years, an annual increase of incidence of lung diseases has been recorded. The mean mortality from these diseases in Russia is 62.7 per 1000; in Eastern Siberia this figure is 65.6 and in the Irkutsk region 66.0. The frequency of abnormal pregnancies and the rate of child mortality, especially in the first year of life, have greatly increased. In Russia, this index is 19.3 per 1000 on the average, and in the Irkutsk region 22.2. The incidence of development abnormalities and congenital malformations has greatly increased: in the Irkutsk region it amounted to 2.8 per 1000, in the Chita region 1.6 per 1000.

Table 3 Multiplicity of higher morbidity

Diseases with respect to mean values in Russia	Bratsk	Irkutsk	Usolye-Sibirskoye	Ust-Ilimsk
Hypertension	1.0	1.6	1.8	0.4
Ishemic heart disease	1.8	1.2	1.7	0.2
Diseases of upper respiratory pathways	1.6	0.9	1.2	0.7
Diseases of respiratory organs	1.4	1.1	1.2	0.6
Diseases of digestive organs	2.2	0.2	2.0	0.8

Population morbidity in the urban areas of the Irkutsk region is considerably higher than on the average in Russia (Table 3).

A common phenomenon for industrial cities of the Irkutsk region is increased allergization of the population. It has been found that the incidence of allergodermatoses has increased as a result of aggravation of the ecological situation.

In 1982, the specific weight of allergodermatoses in the total skin pathology amounted to 19.6%; lately, it has risen as high as 40%. The incidence of allergic diseases in adolescents during the last 20 years has increased 2.5-fold.

Emissions into the atmosphere, apart from the direct toxic effect on the population and on forest areas (death of forests due to the impact of the Bratsk Aluminum Plant), lead to industrial fogs and screening of ultraviolet radiation. In the city of Angarsk, e.g., such losses in winter amount to 45%; repetition of fogs is recorded 4 to 5 times more frequently than in previous years. This results in an ultraviolet deficit with a weakening of the organism's resistance to infections and development of ecological immunodeficiency, especially in children.

A direct correlation has been established between atmospheric pollution and the incidence of infectious diseases (influenza, whooping cough, dysentery, viral hepatitis, measles). Air pollution has a pronounced negative influence on female reproductive functions.

Unfortunately, the current classification of diseases is not sufficiently oriented to the tasks of ecological expertise and is therefore not fit for the prediction of consequences of ecological pressure and ecological transformations. On the basis of the current bibliographical data on ecology and of the experience of clinico-hygienic and anthropo-ecological studies, the following groups of indices may be sufficient as basic markers of ecological risk:

1. Indicator pathology that shows a high degree of dependence on environmental pollution (feminine and masculine infertility, congenital malformations, genetic defects, perinatal mortality, psychoneurological disorders in children, occupational diseases, oncological diseases, allergies, intoxications, etc.);

2. Ecologically dependent pathologies that reflect the mean degree of environmental-pollution impact (spontaneous abortions, neonatal mortality, total child mortality, secondary immunodeficiency, infant mortality, chronic bronchitis and chronic pneumoniae in children, chronic liver diseases, increased vulnerability of organismal systems on meteorologically unfavorable days, etc.);

3. Ecologically aggravated pathologies with a moderate dependence on environmental conditions (pregnancy disorders, increase of risk of major diseases, especially of respiratory and cardiovascular pathologies).

The above ranking shows that the most informative are indices that have to do with fertility and early stages of ontogenesis (perinatal period and age below 7 years).

According to the data of the East Siberian Science Center of the Siberian Branch of the Russian Academy of Medical Sciences, the Committee for Health Protection of the Irkutsk region, and the Regional Statistical Service, the most unfavorable shifts occur in perinatal pathologies: congenital malformations have begun to occupy an important place in the structure of infant mortality.

In ecologically unsafe cities (Usolye Sibirskoye, Bratsk, Shelekhov), there is a high incidence of childless marriages, spontaneous abortions, pregnancy pathologies (mainly gestoses) and, as a consequence, deviation from the normal neonatal mass distribution curve—birth of an increased number of underweight newborns (delay of fetal development), or fetuses with excessive weight. It has been demonstrated (L.I. Kolesnikova, 1993) that ecological stress brings about considerable changes in the female organism, which are directly reflected in the menstrual cycle (hormonal control), course of pregnancy and the whole reproductive cycle. Especially unfavorable are these processes in the extreme North and in the indigenous population (Buryats). Influence of environmental factors on pregnancy complications is proved by the data of the Institute of Pediatrics of the Siberian Branch of the Russian Academy of Medical Sciences (E.V. Osipova et al., 1993) and of the Institute of Geochemistry of the Siberian Branch of the Russian Academy of Sciences on the altered microelement metabolism in such women. This is characteristic also of female inhabitants of the Ust-Ordynsk Buryat Ethnic District.

Studies carried out by the Institute of Pediatrics in Bratsk, Shelekhov and in the Chita region have discovered a correlation between the condition of the psychoneurological sphere and the “pressure” of environmental conditions: as many as 8 to 90% of children have some abnormalities, with the incidence of neuropsychopathology in children increasing with age.

In ecologically unsafe zones, a high percentage of children suffer from secondary immunodeficiencies from an early age. Thus, in most of the cities studied, immuno-allergic pathology was found in 64 to 72% of children below school age (V.V. Gorodisky et al., 1993). The early outbreak of child infections, especially diphtheria, is associated with defects in this immune system.

In this way, the influence of ecological factors is reflected first of all in reproduction processes. It is known that this period of optogenesis is the most controllable, and that there are possibilities of preventing serious disturbances that can inflict a palpable social and economic harm.

Serious attention must be given to thorough examination of women of reproductive age before onset of pregnancy and to ensuring their health, including correction of microelement metabolism by means of administration of sorbents, vitamins with microelements, etc. Extremely necessary is the setting up of a perinatal service with a special attention to prenatal diagnosis of developmental defects (genetic counselling, echolocation, biochemical investigations) for prevention of birth of children with malformations, and also the setting up of centers of neuro-psychosomatic pathology, family rehabilitation and prophylaxis, centers of immunodiagnostics, prevention and treatment. Among necessary measures, one has to include carrying out mass vaccinations of all children in the region with the "Kodivac" vaccine produced by the Institute of Epidemiology and Microbiology of the Siberian Branch of the Russian Academy of Medical Sciences.

It is evident that these measures are only partially preventive because they do not remove the primary cause. Nevertheless, they are badly needed for the prevention of aggravation of the population's health in the Irkutsk region.

It would not be right to attribute the present public-health situation only to the unfavorable ecological conditions. No doubt it is a result of a large set of factors, including socio-economic ones; however, for some cities with intensive industry and environmental pollution, the ecological factor is the most important.

II. Main Problems of the Lake Baikal Region Development and the Role of Science and Culture for Their Solution

The Complex Federal Program for the Protection of Lake Baikal and Rational Use of Natural Resources in the Region: Mechanisms of the Program's Implementation

V.I. Danilov-Danilyan
Ministry of Environmental Protection and Natural Resources of the RF, Moscow, Russia

The paper considers main priorities of the program until the year 2000, its economic mechanism is suggested, and expenses planned. An important condition for completing this program may be all possible development of international cooperation.

The Complex Federal Program of Lake Baikal protection and rational utilization of natural resources of its basin was drawn up by the Ministry of Environment Protection and Natural Resources of the RF together with the Council of Ministers of the Republic of Buryatia, with the administrations of the Irkutsk and Chita Regions, and with concerned ministers and services, as an execution of the Decree of the President of the Russian Federation No. 295 "On Urgent Measures for Governmental Support of the Economy of the Buryat SSR" (25 March 1992).

Baikal is a unique creation of nature. The sensitivity of its basin and ecosystem to external influences is extremely high, and the existing adverse tendencies to change its basic elements require most urgent actions for their prevention. At the same time, the development of new economic conditions, decentralization and liberalization of the economy greatly exacerbate the contradictions between ecological demands and interests of economic development in the region.

It must be noted that in order to protect the ecosystem of Lake Baikal and to manage natural resources of its basin, a number of governmental decrees were enacted from 1969 to 1992. As a result, a complex of water-protection measures was adopted which effected a diminishing of the pollution of the Lake Baikal water-catchment area to a certain degree, a regulation of water utilization for industrial needs, an improvement in forest management and control of noxious emissions into the atmosphere. Unfortunately, the governmental decisions were not fulfilled in full measure. Since the time of their enaction, economic and environmental problems have been exacerbated.

The program under consideration belongs to the type of comprehensive regional programs of federal importance. It is based on current nature-protection legislation, decisions taken by the President and Government of the Russian Federation on the problems of Lake Baikal, international agreements, current specifications of norms for environmental protection, and other documents.

NATO ASI Series, Partnership Sub-Series, 2. Environment – Vol. 6
Sustainable Development of the Lake Baikal Region
Edited by Valentin A. Koptyug and Martin Uppenbrink

The program should take into account the maximum compliance of environmental measures with the level of development of productive forces in the region and the structure of control of its economy, and, also, the deficit of local budgets replenished by subsidies and grants from federal sources.

The program is envisaged to be realized by the year 2000.

Its priorities are based, first of all, on the restoration and conservation of natural and cultural complexes of the Lake Baikal Region in accordance with an ecologically safe strategy of its economic development. Its main stress should be laid on the prevention of negative consequences of industrial and agricultural activities, by means of proper equipment of pollution sources with waste treatment plants, and on conservation of unique coastal landscapes and the ecosystem of Lake Baikal.

Environmental safety of the economy must be mainly provided by means of applying low-waste or wasteless technological processes, highly efficient water-supply systems, methods of sewage and effluents purification, and development of a system for monitoring the environment.

In industry, it is planned to close or convert enterprises which are especially hazardous to the environment into ecologically safe or permissible production, to carry out technological re-equipping and reconstruction on the basis of national and international standards, and, also, to move them outside the region to ecologically more stable zones.

An important condition for preventing damage to the environment is the preferential use of environmentally safe fuels: low-sulfur and low-ash coals, natural gas, transition to electricity and heat supply in the coastal-protection zone, centralization of heat-power production.

The most important nature-protection measures in the transport system include the employment of diesel fuel and nonleaded gasoline and natural gas (this especially concerns trucks and buses), with simultaneous optimization of transportation and freight traffic, especially in cities, and, also, increasing of roads with solid pavement.

Talking into account the volume of atmospheric pollution of the Lake Baikal basin with industrial wastes of the Irkutsk-Cheremkhovo industrial center, it is planned to carry out a complete gasification of this center.

In agriculture, the following steps are envisaged: nature-protecting antierosion organization of the territory, introduction of protective soil treatment based on increasing non-moldboard, subsurface and minimal ploughing; creation of special equipment for soil protection; security of ecologically safe storage and use of fertilizers and means of plant protection; gradual decrease of chemical use and development of biological methods of plant protection.

Rational use of land resources requires that the current ecological norms and environmental situation take into account an objective economic estimation of lands, especially in the case of their privatization, and required limitations of land use in the water-protection zone of Lake Baikal.

In the field of mineral-resources protection, recultivation of disturbed lands near main mining enterprises is envisaged.

Continued nature-protection measures in the social sphere of the region will be required. Setting up new settlements is possible only outside the boundaries of the coastal-protection zone and under strict observance of general town planning norms and ecological requirements.

In the field of biological resources, protection plans exist to introduce scientifically based management of hunting, fur farming and secondary forest resources.

A special trend in rational nature management is restoration of the biological resources, including fish, of Lake Baikal, which, together with direct reproduction functions, must be ensured by strict measures to prevent environmental pollution.

The purposeful formation and development of a united network of protected territories will be continued comprising all valuable multifunctional sites of the ecological system of Lake Baikal.

Especially important nature-protection problems also include the improvement of forest management in the region.

In addition to the complex of nature-protection measures, the program envisages stimulation of scientific and technological progress in the sphere of environmental protection, with the purpose of creating the required technological potential to ensure protection and rational use of natural resources.

Structurally, the program is divided into a unit of basic activities and a unit of support.

The basic activities unit includes measures for decreasing harmful influences on the environment and measures for the restoration of renewable natural resources of the region.

The total sum of capital investments alone (in prices of 1994) for the envisaged measures will total 1615.4 billion rubles, including:

the Federal budget	518.5 billion rubles
budgets of the states of the Russian Federation	46.4 billion rubles
industries	809.5 billion rubles
credits	241.0 billion rubles

The support unit of the Program includes research and development, organization of complex environmental monitoring and geo-informational support. Expenses on the support unit of the Program will total 25.6 billion rubles, including:

the Federal Budget	13.5 billion rubles
ecological funds	12.1 billion rubles.

Though highly important and long-range, the outlined measures for the environmental safety of Lake Baikal seem to be unprofitable, and therefore the main sources of their funding remain enterprises and Federal and local budgets. However, the share of budget financing will be decreasing as the new economic mechanism of nature management and development of ecologically safe, economically effective activities is realized and a part of the profit is directed to the fund for Program implementation.

All possible development of international collaboration in solving ecological problems is an important condition of implementation of the Program. Of considerable interest in the development of such collaboration, in the opinion of the Ministry, is the agreement with the World Bank on elaboration of a project of technical assistance in the field of management of natural resources, and a project of development of ecological tourism in the Lake Baikal Region. It is envisaged to implement this and other international projects within the framework of this Program.

The mechanism of implementation of the program includes elaboration of its legislative, institutional and financial support.

The “Law of Lake Baikal”, which is being debated by the Federal Assembly of the Russian Federation, will be the legislative basis of nature-protection activity.

The basis of the institutional support system is the Governmental Commission for Lake Baikal set up by the Decree of the Council of Ministers – Government of the Russian Federation No. 653 of 12 July 1993.

The Program also proposes an economic mechanism for its implementation which envisages concentration and efficient use of funds from various sources to form a special fund. The establishment of such a fund is also planned by the "Law of Lake Baikal" of the RF. The Governmental Commission for Lake Baikal is to be responsible for the realization of the fund’s activity. It also seems necessary to consider the setting up of a Baikal Ecological Bank.

Taking into account the existing difficulties in obtaining the required funds from the Federal budget, it is reasonable to put forward joint proposals for the introduction of the following elements for financing the Program as:

- reduction of the share of value-added tax and profit-tax

deductions;

- retaining part of the funds from shareholders in the local budget;

- exemption from taxes of part of profits designated for investment for nature-protection objectives, and the introduction of ecologically safe technologies;

- interest-free loans on easy terms from budget funds;

- exemption from taxes and customs duties on transportation/import operations for the acquisition of technologies, equipment and commodities intended to ensure the protection of Lake Baikal and rational use of the natural resources of its basin.

A great importance in the implementation of the Program will be also attached to international cooperation. So, with a grant of the World Bank, elaboration of a project of control of the region's natural resources begins. The total sum of the allocated funds amounts to about 10 million dollars, 600,000 dollars of which are to be used during the period from October, 1994 to March, 1995.

As a result of the implementation of the Program, problems in ensuring the ecological safety of Lake Baikal and improvement of the environmental conditions in the Lake Baikal Region must be solved, and, also, the foundations of efficient nature management laid, including creation of efficient legislative, economic, and administrative mechanisms, and ensuring the sustainable economic development oriented to the observance of all possible ecological limitations specific to the Lake Baikal basin.

The Role of the Irkutsk Region in Realizing Sustainable Development in the Lake Baikal Region

V.K. Yakovenko
Administration of the Irkutsk Region, Irkutsk, Russia

The paper describes specific features of the Lake Baikal Region, in view of its sustainable development, its strategy, first steps made by the Irkutsk region, and large projects of national significance.

With the development of political and economic relations between Russia and the European Community, the U.S., Japan and other countries of the Pacific region, the role of the Lake Baikal Region, and particularly of the Irkutsk region as a favorable geographical communication center with a developed economy and transportation network and a considerable potential of mineral and wood resources, will increase objectively. In view of this circumstance, the administration of the region is striving for open international cooperation and expects that this attitude will help ensure the dynamic development of the region. However, desires and declarations alone are not sufficient for realizing such coöperation. Real steps in this direction are necessary. With this in mind, the region has actively begun to prepare conditions for economic interaction to form the appropriate infrastructure.

Considering the Lake Baikal Region as a prospective territory for the realization of the idea of sustainable development, it is necessary to take its specific nature into account. The main features are:

1. The focal character of economic development and concentration of industrial enterprises along the Transsiberian railway.

   For example, in the Irkutsk region the main industrial potential, occupying an insignificant part of the territory, gravitates towards the transportation arteries, above all the railway line, and is located in some twenty industrial centers (Irkutsk, Angarsk, Bratsk, Ust-Ilimsk, Cheremkhovo, Sayansk, Tulun, Taishet, Zheleznogorsk, etc.).

2. Abundant natural resources such as fuel and energy (bituminous and brown coal, oil and gas, condensate), hydropower and wood, gold, mica, iron ores, semiprecious stones, Baikal water, etc.

3. Unlike other regions of the world, it has vacant lands fit for habitation and for all kinds of economical activity. For example, in the Irkutsk region, the population density is 3.4 cap per 1 square km, which is 2.5 times lower than that of Russia as a whole.

NATO ASI Series, Partnership Sub-Series, 2. Environment – Vol. 6
Sustainable Development of the Lake Baikal Region
Edited by Valentin A. Koptyug and Martin Uppenbrink

4. Great remoteness from the central regions of Russia, from international seaports and foreign countries (about 5,000 km) means higher shipping expenses and the necessity of deep processing of raw materials, the improvement of production quality and its competitiveness.

5. Necessity of renewal of production capital. In the basic fields of the economy of the Irkutsk region, the main financial base was created in the 50's and 60's. At present, these industries are morally and physically worn out, causing elevated operating expenses (for overhaul and routine repairs) and a greater ecological burden. An accelerated renewal of capital is required by means of introduction of new, highly efficient and ecologically safe technologies and equipment.

At present, the region is going through the most difficult period of its development, the period of entering new economic conditions. The most important of these is that each of us, all the people, must clearly realize that not only the government, the regional and other authorities, but also all of us are responsible for the quality of life, for the welfare of our families and our neighbors.

It is quite obvious that no administrative entity of the Federation can effectively develop its economy alone: it is possible only within a united economic sphere. Therefore, attempts at introducing efficient market conditions without due governmental control will be fruitless, and at the same time a necessary condition of such an endeavor is sufficient independence of all economic units and their responsibility for the activities in the region. With the Irkutsk region, as a part of Russia, now entering the international market economy on the verge of the 21st century, with the present geopolitical situation in the Asian-Pacific region, with the necessity of a radical renewal and consolidation of the material and technical basis of production and the requirement of ensuring a social orientation of the economy, the position of the Lake Baikal Region and the Irkutsk region in the system of the territorial division of labor will be radically changed, and they will take on great national and international economic and political importance.

The strategy of sustainable development for our region must envisage the realization of the following functions and tasks:

1. The region should assume the function of Russia's outpost at its south-eastern borders. The concept of an outpost is very broad. It must possess a necessary volume of industrial and human resources for ruling out a possibility of sharp aggravation of geopolitical conditions in the region which may endanger the interests of Russia. For this, the region must increase its productive energies at a high rate, which constitute its economic welfare, ensure a higher increment of productivity (qualitatively and quantitatively) as compared to the other potential pretenders to these lands, accumulate resources and manage them.

2. Transition to the market methods of economic development increases the role of the region as an important (virtually basic) basis for the development of Northern Siberia and the Far East. For example, the Republic of Saha (Yakutia) receives 80% of its food supply and 100% of its fuel and lubricants from the Irkutsk region, and this determines one of the trends of its economy.

3. It is necessary to increase the degree of development of the territory by means of developing the branches of industry and agriculture that contribute to the realization of social reforms and the increase of living standards of the people. The existing heterogeneity and the extremely low general level of development of our region and of our nearest neighbors, the Saha and Buryatia Republics, the Chita region, the zone of the Baikal–Amur railway, are consequences of severe climatic conditions and consequent elevated costs of economic and social facilities, and, also, a result of peculiar features of the previous history of Russia. The way this problem is solved must reflect the strategy of development of productive forces of the Lake Baikal Region, which is impossible to do without specifying its role in the social division of labor, without consolidating its increasing significance and formal recognition of its new status.

4. For a long time, the Lake Baikal Region (and, first of all, the Irkutsk region) has been an arena of implementation of large technological and economic programs of national significance in the forest, hydropower and chemical industries as well as in the gold, nuclear and aluminum sectors. In the near future the functioning of these industries and realization of new projects will determine the potential of the Lake Baikal Region and contribute to the implementation of important social reforms.

5. The region plays a global role in the conservation of stability and productivity of the ecological systems of Siberia and Central Asia. Here are the sources of great world rivers (Angara and Lena), the unique Lake Baikal, vast taiga areas crossed by the traditional routes of the seasonal migration of animals. The genetic pool of flora and fauna corresponds to the biological potential of the whole zone of landscapes of moderate northern latitudes.

6. Economic conditions which dictate the priorities in the development of these territories, including the Irkutsk region, should be taken into account.

In the area adjacent to Lake Baikal, a large economic complex has been formed which exerts its influence on all neighboring territories. The region possesses vast natural resources. There are deposits of gold, mica, salt, coal, rare metals, magnesites; also, there are large hydropower and forest resources and unique reserves of water of supreme quality, all of nationwide importance. Noteworthy are also oil and gas deposits, deposits of iron ores, gypsum, talcum and chemically pure cement limestones. These resources provided high rates of economic development and a large economic potential in the Irkutsk region for several decades (see Table 1).

Table 1 Relative significance of territories in the Lake Baikal Region's economy (%)

Characteristics	Irkutsk Region	Buryatia	Chita Region
Area	49.5	22.7	27.8
Population	54.2	19.9	25.9
Industrial output	82.1	9.5	8.4
Energy production	99.1	6.6	5.3
Timber	85.1	7.2	8.1
Lumber	80.3	9.4	10.3
Consumer goods	64.2	20.1	15.7
Corn	53.2	13.9	32.9
Potato	56.3	23.5	20.2
Vegetables	44.7	26.3	29.0
Meat	64.1	16.1	19.8
Milk products	66.4	14.1	19.5
Eggs	65.9	20.4	13.7
Retail turnover	58.5	20.7	20.8
Services	62.4	14.7	22.9

With the population comprising only 1.9% of the whole population of Russia, the region yields 6.2% of the national production of electric power, 3.6% of nitrogenous fertilizers, 12% of synthetic resins and plastics, 15% of caustic soda, 12% of wood, 10% of timber, 12.7% of veneer, 20% of cellulose, 35% of aluminum production. The income of the Irkutsk region's population is the highest in Siberia and the Far East.

The production of the region enjoys demand on the foreign market. Its share is 2.3% of the export receipts of Russia.

What has been done, and what is being done in the Irkutsk region to ensure sustainable development?

In the spring of 1993, a scientific-practical conference on development of the economy in the region was held in Irkutsk, where the plans for developing productive forces up to the year 2005 were discussed and adopted. For implementation of the long-term strategy of economic development of the region within the framework of the adopted plan, the administration immediately began to draw up a scheme of development and distribution of productive forces up to 2000. The scheme was submitted to the Ministry of the Economy of the RF and, on its basis, the government enacted Decree no. 111 of 12 February 1994 "On Measures for Implementation of the Federal and Regional Program in the Irkutsk region in 1994–1997".

In 1994, the international scientific conference "Baikal as a Laboratory of Environmental and Climatic Changes" was held.

A list of investment programs and projects was prepared, many of which were highly efficient and attractive to foreigners. In 1994, a scientific conference was held to attract foreign investments into the region's economy.

The elaborated strategy of socio-economic development has been included in the foundations of the plan of action of the regional administration to realize the economic reform while specifying the most urgent measures for leading the national economy out of the crisis. The following steps have been taken to help solve strategic and tactical problems:

– Conceptional approaches for drawing up territorial budgets were worked out which should consolidate local budgets with the help of local taxpayers and at the same time will be responsible for providing part of expenditures provided by the Federal budget.

– In accordance with the accepted strategy, the activity of the regional administration is directed at reduction of budget subsidies for financing the communal economy, public transportation and other sectors at the expense of the population, while simultaneously addressing the problems of the poorest part of the population. In this endeavor, differentiation of donor territories and recipient territories is envisaged.

– An extremely important sphere of administrative activities has become the policy of taxation, finance, budget, credit and prices. It is based on the principle of creation of favorable conditions for business and, above all, for goods producers. In the region, unlike other administrative entities of the Federation, the profit tax is left unchanged at 22% (in the part that goes to the regional budget), although, according to the President's decree, this tax may be increased to 25%.

The considerable production of electric power by hydroelectric stations permits the region to have the world's lowest prices for energy suppliers and not have to finance industry from the budget.

The survival and development of business under new conditions is also promoted by the following steps taken by the regional administration:

– Formation of financial-industrial groups. The first group was formed in the forestry sector, another in industry (mechanical engineering and oil refinement complexes), still another group represents air transportation between Chita, Buryatia and Irkutsk, and still another a construction complex.

– Stable work of the fuel and energy complex as a basic industry of the region is ensured.

– The conversion process at defense plants in the region is under way, with simultaneous investigation of possibilities for national and regional contracts.

– A foundation has been set up to support industrial enterprises producing socially important products.

– Small and medium-size business development is being promoted.

For the purpose of implementation of federal and regional programs and projects, an investment company, Finprom, was set up. It was stimulated by the investments for development of the regional infrastructure. Such important projects as the construction of an international airport, "Gates to the East", "Communication", an information program, are being realized.

The initial activities of the investment company has made it possible to increase the efficiency of using the road fund, to set up a security fund, a regional development fund, and to increase the share of the region's holdings in prospective joint-stock companies, to set up financial-industrial groups in the forest industry and other industries. Carrying out leasing operations by the financial-industrial company has provided high-efficiency equipment for the processing industry. Contacts of the Finprom with federal financial corporations make it possible to attract supplementary financial resources to the region's economy. In view of the great importance and attractiveness of the regional resources for potential investors (including foreign capital), we should attentively and seriously evaluate the available proposals made to this end and lend preference to those which promise the highest profits for the region, with maximum observance of the region's interests.

Apart from this activity, a number of very large projects of national significance have been drawn up, and their realization will help make the sustainable development of the region possible.

1. Gas Project

Development of the Kovykta gas condensate deposit, situated 300 km from Irkutsk, with gas reserves sufficient for more than 100 years, will permit the ecological situation in the Baikal region to be improved by means of gasification of the thermal power system. For example, the conversion of the Irkutsk thermal power plants to natural gas alone will reduce the annual emissions into the atmosphere by 400,000 tons and the more than 10 million tons of coal extracted in the Irkutsk region can be delivered to the regions of the Far East which are in need of fuel resources, and, also, it can be exported abroad.

2. Oil Project

A specific feature of the oil project is reduction and then complete cessation of the West Siberian oil supply for processing by the Angarsknefteorgsintez companies and the orientation to modernization and reconstruction of the region's large oil refinery and processing enterprises.

3. Sukhoi Log Gold

Realization of this project will make it possible not only to increase gold extraction, but also to heighten the economic efficiency by means of introducing new technologies of exploitation of bedrock deposits and ensuring year-round work.

4. Magnesite from the Savino Deposit

The project envisages support of metallurgy in the Russian Federation and the prospect of its development in the Irkutsk region, especially on the basis of organizing small- and middle-tonnage production of high-quality steel. Realization of the project will allow the development of mechanical and instrumental engineering.

5. Diamonds

Placer and bedrock diamond deposits in the region were found as late as the 50's. This is one of the most expensive projects, but it is also our future. And today, we are taking active measures to realize it.

6. Agro-Industrial Complex

The goal of the project is to increase the processing of agricultural products and to attract people to rural areas in order to provide the region with food (meat, milk, potatoes), and to increase the yield of alimentary and fodder grain. The project is aimed at reviving the Russian village and developing the rural industrial complex. For example, in recent years the rural population in the region increased by 30,000; this was caused solely by the migration of townsfolk to the country.

This project envisages solutions to the problems of improvement of soil fertility, introduction of intensive technologies, modern agricultural machines and equipment and development of social infrastructure in the complex for farms, livestock complexes and poultry farms, associations of farms, private plots and industrial enterprises.

7. Social Development

Here, our efforts are directed at the stabilization of the population's standard of living, with special emphasis on the support of poor people, prevention of a rapid growth of unemployment, securing the social sphere (first of all, functioning and development of pre-school institutions, schools and hospitals). In order to solve these problems, even today extraordinary measures are necessary: revision of the expenditures of the region's budget and the search for non-traditional solutions that would make it possible to increase salaries and ensure social payments in the budget sphere in due time.

8. Ecology

Nature-protecting activities are very important today. The focal development of industry and high concentration of production have made Irkutsk, Bratsk, Angarsk, Shelekhov, Zima, Sayansk the most polluted cities of Russia. A general understanding of the fact must be reached that the solution of ecological problems requires, first of all, the replacement of production technologies. Inspections, fines and expertises have no meaning unless they stimulate the introduction of "clean", resource-sparing technologies. All nature-protection programs must be obligatory and reviewed economically. The comprehensive ecological program of the city of Bratsk has been reviewed this way and approved at all levels.

The first stage of development of the regional ecologo-economic comprehensive program "Baikal" has been completed. It envisages federal financial support of measures for reduction of technological impact on the Lake Baikal ecosystem and for rational utilization of natural resources of the Lake Baikal Region. The program includes subprograms of landscape planning, reprofiling of the Baikalsk Pulp-and-Paper Plant, protection of communications from dangerous natural processes, reconstruction of the Circum Baikal railway (Kultuk-port Baikal fragment), development of ecotourism, the Pre-Baikal National Park, fish rearing, distribution of Baikal water, alternative power supply, monitoring of environment.

This program represents specification of the federal program for Lake Baikal, whose implementation will require about 600 billion roubles from the federal budget. We are becoming more and more convinced that the solution of the Lake Baikal problems will be a burden on us. This is not quite fair— it was not we who constructed the pulp-and-paper plant, yet Moscow does not allot money for revamping it. That is why the just and fair RF "Law of Lake Baikal" is necessary today.

Work on a series of local laws has begun to determine the framework for a federal document that will regulate the matters of nature management and protection in the Pre-Baikal area. Development of normative acts for the regulation of tourism and landscape planning is awaiting its turn. The latter envisages the drawing up of natural-economic plans of land-management-law rank, with large-scale functional zoning of the territory and determination of the best variants of utilization and protection of resources.

The solution of ecological problems must become the cornerstone of economic development and the formation of a new socio-economic complex in the Irkutsk region. Taking the unique natural and difficult ecological situation in the industrial Pre-Baikal districts into account, a clear interterritorial coördination of efforts, first of all with the Republic of Buryatia, is necessary when preparing the measures for the stabilization and improvement of the ecological situation in the

Lake Baikal basin. That is why the development of the interaction mechanism between the Irkutsk region and the Republic of Buryatia must be envisaged such that they are directed to the basic solution of problems of ecological recovery in the Lake Baikal Region, especially under the changing economic conditions and new forms of property.

We realize that a considerable improvement of the ecological situation in the Lake Baikal basin is possible only on condition of interaction not only between these two regions adjacent to the Lake and Mongolia, but also with the whole world.

This international symposium will make it possible to attract the interest of the world community to our region and give a stimulus to further drawing the attention of scientists, businessmen and policy-makers to the problems of the region.

Rational Branch and Territorial Structure of the Economy of the Republic of Buryatia

A.N. Zhiltsov
Government of the Republic of Buryatia, Ulan-Ude, Russia

In this paper the compliance of the structure of the economy of the Republic of Buryatia with eco-ecological requirements (standards) of sustainable development is considered. The basic aspects of the economic and social concept of future development of the Republic of Buryatia are presented.

First of all, let me greet, on behalf of the Government and peoples of the Republic of Buryatia, the participants of this Symposium and wish them success in its work.

Choosing the Lake Baikal Region as one of world's model territories of sustainable development is considered very important because of the uniqueness of Lake Baikal, the desire to make the coastal complex a World Natural Heritage Site, serious experience gained in nature-protecting activity in the Republic, and awareness of the necessity to put an end to a nature-destroying economy. During the period of 1972–1985, the Government of the USSR adopted three decrees on Lake Baikal which contributed to the implementation of nature-protecting measures in the Republic of Buryatia. I will mention some of them.

Before 1993, about 8% of all capital investments was assigned to nature-protecting purposes (improvement of technologies, purification works, etc.). No other region of Russia had made such investments in nature protection.

At present, 1.5% to 4.5% of the cost of industrial production is taken up by expenditures in support of nature-protecting measures. This naturally makes production more expensive here as compared with production of the same type in other regions.

Thanks to the persistence of the administration and scientists of the Republic, the exploitation of the Oshurkovo apatite deposit, which endangered Ulan-Ude with atmospheric pollution, was stopped.

For the first time in this country, a wide ecological certification of industrial enterprises and some of the farms was carried out. It allowed a quantitative and qualitative inventory of sources of water and air contaminants to be made, from the source to the end of the pipe, and to outline measures ensuring the decrease of industrial wastes. Part of the previously planned measures had been introduced before the crisis of the economy started in this country.

NATO ASI Series, Partnership Sub-Series, 2. Environment – Vol. 6
Sustainable Development of the Lake Baikal Region
Edited by Valentin A. Koptyug and Martin Uppenbrink

Thus, as a result of joint efforts of scientists and the staff of the Selenginsk Pulp-and-Paper Mill (SPPM), for the first time in the history of the pulp-and-paper industry, a completely closed water-circulation system was introduced in the mill, and in August, 1990, the ejection of polluted waste water ceased. So the SPPM is an enterprise which is ecologically acceptable to the conditions of the Lake Baikal Region today.

For several years now, various ecological training courses of different levels have been functioning in the Republic. All projects are subjected to ecological expertise.

Considering Buryatia as a possible model territory of sustainable development, it seems necessary to describe its economic problems and the degree of adequacy of the present economic structure of the Republic to the eco-economic requirements of sustainable development.

In previous years, the territorial division of labour, the specific development of productive forces in the Republic, determined the direction and character of economic specialisation. The economic system of the Republic specialised in mechanical engineering, wood processing, glass manufacturing and light industry, extraction and primary processing of non-ferrous ores, coal, etc. Dominant in the economy is industry. Its share in the gross national product in 1991 amounted to 51%, in 1993 to 44.2%, in transport and communication to about 22%, in building 13%, in agriculture 15%, in materials and equipment supply 13%. The shares of forestry, trade and other activities are small.

During the previous decade, the productive forces in the Republic were mainly developed extensively, and production increased by more than 1.4 times by building new and extension of already existing enterprises. Investment policy was accordingly oriented to an expenditure method entailing an increase of production potential on the basis of traditional technologies.

Extremely inefficient proportions between product and expenditures have been formed in some branches of the economy. Over 50% of industrial products was exported in previous years beyond Buryatia, and, at the same time, more than 2/3 of its needs in industrial production were satisfied by means of supply from other regions of the country. Mechanical engineering and metal-processing industries exported over 80% of their production, and, at the same time, the needs of the economy for machines, equipment and other technology were 3/4 covered by interterritorial exchange.

The most unfavourable situation occurred in consumer-goods production: 69% to 90% of the population's demand for various commercial products (other than food) was imported from the other economic regions of the country.

Geographically, Buryatia has economic relations with more than 70% of the regions of Russia and the other countries of the former Soviet Union. The asymmetry in manufacture and consumption in the economic complex of the Republic contributes to the export of raw-material resources and intermediate products in some sectors and to import in others. It retards improvement of efficiency in the use of production potential. The disproportions that have developed in the production potential of the Republic hinder the reproduction structure of the combined product. Thus, the share of intermediate products in the manufactured total product of the Republic amounts to 64.5% and is on the increase, while the specific share of final products has, on the contrary, been falling of late.

So in the structure of the economic complex of Buryatia, raw material and intermediate stages of production are dominant, which is not efficient. Therefore, when the economic reform started, the structure of material production turned out to be very unfavourable as compared to the other regions.

Under the conditions of continuing economic crisis, the share of savings corresponding to increment of production stock has considerably decreased. So in 1993, the decrease totalled 2.4% of the previous year's figure. The share of savings in the sphere of circulating assets is associated with increasing prices for raw-material increases.

Today, productive forces in the Republic are not equally distributed. The territory in the vicinity of the Trans-Siberian railway is most developed. The main economic and technical potential, the Ulan-Ude industrial complex, is situated here. It includes large industrial plants: aircraft, helicopters, locomotives and car repairs, ship-building, instrument-manufacturing plants, "Buryatfermmash", "Elektromashina", "Teplopribor", a metallic-bridge construction plant. There are also canned-meat plants, flour mills, confectioneries, fine-cloth combines, and pasta and candy factories. The ferroconcrete-construction plant, glass works, prefabricated-house plant, and brick works represent the building-materials industry.

At present, these enterprises are not incompetitive due to the high cost of production caused by high electric power and railway transportation tariffs, the special nature-management regime in the Lake Baikal Region, and difficulties in the consumer market. All this complicates solving economic problems.

Pollution monitoring in Ulan-Ude was done by 70 indices. Only 20% of the city area may be considered as slightly polluted. In the rest of the city, technogenic pollution exceeds the extreme permissible concentrations of many indices. With respect to other industrial centres of Buryatia, technogenic pollution is thus far not a cause of special concern.

In agriculture, the most intensively used are land resources in steppes and forest steppes in mountain hollows and valleys. They are situated in the southern and southeastern parts of Buryatia and are characterised by intensive development of agricultural land use.

Intensive use of land resources is also typical for the Barguzin and Tunkin valleys and also for the lands along the delta of the Selenga River. By natural and economic factors in the Republic of Buryatia, there are four agricultural zones—dry steppe, steppe, forest steppe and mountainous taiga. The results of the economic estimate of agricultural lands show a high level of potential net income for farms located in the delta of the Selenga River and belonging to the forest-steppe zone. Farms of the Mukhorshibir, Bichur, Jida, Zakamensk, Tunkin and Kyakhta districts have similar results with respect to potential net income.

The ploughing of virgin sandy soils, considerable increase in the sheep capita and excessive grazing pressure on pastures as well as intrusions of agrotechnology in land tenure have resulted in erosion processes. As a result, fine earth brings to Lake Baikal residues of fertilisers, pesticides and other wastes, and one can imagine the scale of damage inflicted by soil erosion not only to agriculture, but ultimately also to the Lake.

To ensure an agriculturally optimum land use in the framework of agrolandscapes in the territory of the special nature-management regime in the Lake Baikal basin, organisational and agrotechnical measures should be introduced aimed at changing the structure of land use and restoring soil fertility. The proposed measures are based on scientific developments of the Siberian Branch of the Russian Academy of Sciences, and of the Russian Academy of Agricultural Sciences.

An extremely important environment-protection role is assumed by the forests of the Lake Baikal basin and adjacent mountainous territories. They ensure the clear river run-off, its regulation, and protection of soil against erosion, supporting the existence of the unique basin. At the same time, forests can provide various products, above all large amounts of valuable wood.

The forests of the Lake Baikal Region are divided, according to the accepted classification, into two groups:

– forests fulfilling mainly protective and social functions;

– forests having protective importance and a limited use.

Calculated felling has been reduced as a result of strict requirements of the forest management system in the region, as well as by including larger areas of especially important protective sites, the introduction of felling regulations aimed at the conservation of protective properties of forest ecosystems, and the marking off in 1987 of the coastal protective belt around Lake Baikal.

On the whole, scientific research in the Lake Baikal zone is directed at solving the problems of maximum utilisation of forest resources with minimum intrusion into the forest environment.

The problems of crisis and transition to free-market conditions in Buryatia are the same as in the rest of Russia; I will not dwell on them. The analysis of the economic reform in 1993 shows that in Buryatia, just like in Russia as a whole, there are signs of a new economy with a transition to free-market conditions. The private sector is gaining in importance: at present, 41% of working people are involved in it. At the governmental level, price limitations have been practically removed, and prices are regulated by supply and demand. A new tax system has been introduced.

Economic and Social Plans for the Further Development of the Republic of Buryatia

The proposed strategy was worked out before the requirements of sustainable development of Buryatia were widely considered. It is not to be expected that after discussions in the course of this Symposium and summing up its results necessary amendments will be introduced. The previously outlined strategy envisages an active socio-economic policy under the conditions of limited financial and material resources, based on the selective approach to determine priority spheres in which a rapid positive effect can be obtained.

Sectors of the economy of the region are divided into two categories, according to the degree of controllability. The first includes enterprises whose production capacity depends less on external factors, i.e., coal mining, forestry and the wood-working industry, manufacture of building materials, the construction industry, mineral mining, agriculture, except for poultry and pig breeding, which depend on fodder supply from outside. The second category includes enterprises in which production depends on the supply of raw materials, accessories and parts from other regions. These are mainly mechanical engineering, light industry, food industries, transportation and communication.

The economic policy to put an end to the decline of production in Buryatia should be directed, in our opinion, first of all to the enterprises of the first category. An indicator of the revival of their economy can be foreign investments. In this connection, a possible range of industries has been determined whose production can be exported. These are industries involved in extraction and processing of gold, silver, alloys, quartzite, wood, as well as mechanical engineering, foreign tourism, etc.

An extremely important aspect in the economic stabilisation and economic balance is the fuel and power complex, and its development is intended to satisfy Buryatia's needs on the basis of improving the consumption structure and increasing power production, first of all, by means of reducing power expenditures and using non-traditional renewable power sources. It is necessary to carry out considerable work in reconstruction, technical re-equipment, extension of functioning and construction of new coal pits, increasing the level of mechanisation, introduction of advanced beneficiating methods and implementation of other measures for the improvement of the quality of the extracted coal. The development of the Tugnui deposit seems to have the most promise. With it, the region can produce up to 12–13 thousand tons of coal per year.

An important condition of economic progress in the Lake Baikal Region and development of its productive capacity is progress in the mechanical-engineering complex. The main thrust in the development of this complex is on the improvement of the territorial structure of this sector, a radical technical re-equipping, modernisation of construction and construction of new engineering enterprises, specialised, first of all, in the manufacture of machines and equipment with regard to their operation under specific conditions of the Lake Baikal basin. Priority development of the following sectors is envisaged in the specialisation structure of the complex:

– instruments engineering (including that for thermal control);

– aircraft industry;

– mechanical engineering, ship building, agricultural-machinery industry;

– repairs industry.

Plans for the further development of these industries are mainly being made on the basis of currently operating enterprises. In future, special attention should be paid to creating sophisticated instruments engineering (e.g., radio and radioelectronics), and other especially "clean" types of production that meet the requirements of the water-protection zone of Lake Baikal.

The development of the metallurgical complex must first of all refer to the economy and ensure the efficient use of the most critical and expensive mineral resources, and, also, include processing of scrap metal as a secondary and ecologically safe raw material. The Jida tungsten-molybdenum integrated plant, the Kyakhta beneficiation factory and the subsequent start-up of the Naran fluorite mine may be basic enterprises for the restructuring and re-equipping of this sector. Construction and start-up of the Ozerny integrated mineral-beneficiation plant using the Vitim deposit of lead and zinc ores is a possibility, too.

Reorganisation of the wood-chemical complex as one of the leading components of the economy in the region has been reduced due to difficulties caused by the limitations in the water-protection zone of Lake Baikal, and by the necessity of preventing considerable emissions of toxic wastes from corresponding industries into water and atmosphere. Conservation and simultaneous modernisation of the wood-chemical complex is vital because enterprises of this sector are important elements of the region's economic well-being which manufacture important products and ensure employment of many highly skilled specialists. Continuous and inexhaustible forest management will be supported by a strict observance of calculated, scientifically substantiated fellings and norms of lumbering assigned to each enterprise. Provisions are made for advancing rates of development of deep processing and utilisation of low-grade and leafy wood and all felling wastes, and, also, for extension of the varieties of wood production, including fibre-board and chip-board production by about double.

The main task in developing the transportation network is the formation of an efficient system of passenger and freight transportation in the Lake Baikal water-protection zone. Implementation of the programme of capital construction as a basis of successful economic and social development and realisation of a nature-protecting policy are planned to be carried out by means of a reorganisation of the construction complex. This includes developing the construction and construction-materials industries, along with a wide utilisation of local raw materials and industrial wastes in thermal power, mining, and other branches of industry.

To improve food supply to the population and satisfy its needs by means of local potato, vegetables, meat and milk production, to improve their quality and variety, provisions are made for the dynamic development of all sectors of the agricultural complex and increasing its production potential in the following directions:

– improvement of arable-land structure, introduction of soil-protecting cultivation systems, above all on hillsides and other erosion-susceptible lands;

– transition to landscape contour-strip farming, with introduction of necessary equipment and tools, increasing forest-protection areas;

– sowing of perennial leguminous plants and cereals on eroded and deflated lands;

– increase of nomadic animals in relation to the total amount (horse breeding, increase of cattle capita, yak breeding, reindeer breeding) at the expense of those directions of animal husbandry in which fodder supply is imported from other regions (pig and poultry breeding);

– revival of apiculture, etc.

The problem of processing of agricultural production has to be solved at increasing rates by means of building small food-processing plants, modernisation of creameries in the settlements of Petropavlovka, Sosnovoozersk, Turuntayevo, canned-meat plants in Ulan-Ude, Naushki and Severobaikalsk.

An increase of fish production is planned by modernising the Ust-Bargusin and Kabansk canned-fish plants, and by increasing the capacity of refrigerators at the fish stations by more than double.

Serious attention should be paid to the development of the recreational system. So far, it has not yet been developed on a proper scale, by modern standards. The organisation of the recreational system implies the formation of recreation complexes as well as single facilities for recreation and tourism: small hotels and car-camping sites on tourist roads, summer tent camping sites, mountaineering and skiing facilities as well as such for hunting, fishing and surfing, etc. So far, none of this is available on Lake Baikal.

The extremely rich balneological resources of the region (mineral waters and medicinal muds) are also being insufficiently exploited (about 5–8%). In spite of a large number of mineral springs in Buryatia, there are only two resorts, one sanatorium, and several "wild" (non-organised) resorts with hydrotherapeutic facilities. The uniqueness of mineral springs, their high therapeutic effectiveness, shortage of alternative treatment methods, high prices for transport and the inaccessibility of the balneological resorts of the Crimea, the Caucasus and the Carpat mountains are responsible for the wide use of the local balneological facilities, even at rather high prices.

Increasing the balneological potential and further developing the mineral-water resources in Buryatia will require the realisation of the following practical measures:

– modernisation of the existing resorts (Arshan, Sayan, Goryachinsk) and all sanatoria and hydrotherapeutic facilities;

– organisation of new sanatoria on the the basis of already known mineral-water resources (Zhemchug, Khongor-Ula, Garga, etc.);

– construction of dispensaries for extra-resort treatment in cities and district centres;

– wide use of hydromineral resources for bottling and sale of dietetic and medicinal mineral waters.

Numerous lakes, rivers, mountainous taiga landscapes surrounding Lake Baikal, unique flora and fauna, monuments of material culture and of nature contribute to developing all types of tourism. In order to develop the tourism potential of the Lake Baikal Region at the level of international standards, large capital investments and organisation of commercial and recreational activities are necessary.

In conclusion, the following remarks may be made. Creation of a socially oriented, efficient and competitive economy is closely connected with the structural policy implemented by the Russian Government. A real prerequisite for the development of Buryatia is the available economic potential, with its restructuring in the directions mentioned above. In the field of regional policy on the whole, we have to solve the problems of the dividing-up of state-owned property and of reforming the taxation and budget policies, i.e., stabilising the norms of tax distribution between budget spheres and development of the legislative basis for local taxation.

To solve the strategic problems faced by our society, Buryatia possesses a large potential represented by academic, university and applied sciences. The leading role of scientists of Buryatia and of the Siberian Branch of the Russian Academy of Sciences are to be noted in the design and implementation of measures for protecting Lake Baikal.

Despite the scant attention paid to science and its role by the Russian Government in recent years, we must maintain its potential. Special attention must be paid to the organisation of interdisciplinary studies, and to relations of science with the Government and with society on all problems of the environment and sustainable development. Studies on resource- and energy-saving wasteless technologies, human safety, etc. are to be developed more widely.

An important problem faced by the Republic of Buryatia is that of setting up a system of efficient management which could help co-ordinate an efficient, fair and controllable system of governmental rule, with individual rights and possibilities of economic activities. It should be underlined that a certain turn of the market system to sustainable development can be made only under governmental economic control. The main objective of such control is to stop treating natural resources as a "free-of-charge commodity" and establish the rule: "the user of a natural resource must pay for it".

Making the Lake Baikal Region a world model territory of sustainable development, apart from giving it a certain prestige, will actively support the development of the up-to-date technical and technological basis of production, the improvement of education and an increase on the cultural and scientific planes.

That is why we all support this idea, while at the same time fully aware that, under the present critical conditions, this is an extremely difficult task. We have too little time and means at out disposal to change the pattern of the present unsustainable development of the Republic of Buryatia. But we have no choice. We cannot solve first the problems of today and then those of tomorrow. Responsibility for the successful realisation of plans rests with the governments of the Republic of Buryatia and the Russian Federation. It is to be noted that the present cabinet of the Republic's government has been in office for somewhat more than two months. Noteworthy is also the fact that the beginning of its work coincided with the holding of this Symposium, which, as we all hope, will contribute to our efforts and supplement it with proposals of additional measures necessary for the realisation of sustainable development.

Perspectives of Development of the Selenga River Basin in the Territory of Mongolia

D. Baatar
Mongolian Academy of Sciences, Ulan Bator, Mongolia

The Selenga river basin in the territory of Mongolia is an important region for the economic development of the country. In this basin, which occupies over 20% of the whole Mongolian territory, about 50% of the population and 70% of the agriculture is located. Here are also the main industrial sites. I think that in the very near future exploitation of the Hubsugul phosphorite deposit will begin. Plans for construction of an electric power station on the Eghiingol River have been announced. It is believed that the Darkhan Integrated Iron-and-Steel Works, which recently began operations, will be enlarged in the very near future. The expected intensification of land utilization, of exploitation of water, mineral and other resources in the Selenga River basin in the territory of Mongolia will undoubtedly be connected with the problem of sustainable development of the Lake Baikal Region and will require the coördinated activity of both countries. On the initiative of the Mongolian and Russian Governments, agreements on the regulation of questions concerning water resources in the frontier area between Mongolia and Russia have been concluded.

Thoughtless exploitation of forests in this region, in particular their uncontrolled felling, have resulted in severe damage of the ecological balance in the last 25 to 30 years. It is only thanks to the social movement that these questions have attracted attention during the last three to four years. In my opinion, a very important step in this direction is the drawing up of the joint Mongolian-American-Russian program of sustainable development of Lake Hubsugul and the Selenga River basin. By now this program has been successfully worked out thanks to the efforts and energy of Dr.George Davis, director of the program. We hope that it will not remain only on paper.

Lakes Baikal and Hubsugul are unique reservoirs of fresh water that belong to all mankind, and in order to be sure that they exist and prosper for further hundreds of generations, we must ensure the conservation of these water bodies.

That is why I propose to discuss the question of setting up a continuously functioning international foundation for this purpose under the aegis of an international organization or a bank (like the World Bank).

NATO ASI Series, Partnership Sub-Series, 2. Environment – Vol. 6
Sustainable Development of the Lake Baikal Region
Edited by Valentin A. Koptyug and Martin Uppenbrink

Characteristics of the Sustainable-development Model for the Outlying and Contact Zones of the Lake Baikal Basin

B.P. Ivanov, I.P. Glazyrina, A.M. Vozmilov
Administration of the Chita Region, Chita, Russia

T.A. Strizhova, V.F. Zadorozhny
Chita Institute of Natural Resources of SB RAS,
Chita, Russia

We consider several features of the sustainable-development model for the outlying part of the Lake Baikal basin. The comprehensive system of environmental and economic management is discussed. This system should be oriented to the "individual" approach to priorities of areas and creation of "fine" tools of management of natural resources.

The economic development of Eastern Siberia is closely connected with its natural and resource complex.

The Chita region is a typical representative of a region with a raw-material-oriented economy. With its rich and diverse natural and resource potential, the region has been developing the extractive industry for a long time. The preference is given to mineral resources. The "raw-metarial stage" was a cause of many negative economic, social and ecological consequences. Primitiveness and conservatism of technologies of exploitation, for example, of natural-resources deposits results in a low coefficient of use of waste products, thus polluting the environment. Such localities are characterised by a low quality of life.

The concept of sustainable development is becoming more and more popular in the Trans-Baikal, and not only among scientists, but also in administrative circles. The Russian-American Comprehensive Programme of Land-use Policy for the Russian Portion of the Lake Baikal Region, which was implemented (1) in co-operation with George Davis in 1991–1993, greatly promotes it. This programme also attracted the attention of those who manage the economy at the professional level, and to the role and possibilities of the regional institutional and economic mechanisms of management. However, to turn the programme from a "programme of policy" into a "programme of action", it is necessary to develop and refine it.

To implement any programme of sustainable development, it is important to create a complex of economic mechanisms which would promote environmentally sound directions of development and prevent environmentally dangerous ones. The problems of nature management in the market economy could not be solved only at the republic or only at the regional levels. The Federal level should ensure

NATO ASI Series, Partnership Sub-Series, 2. Environment – Vol. 6
Sustainable Development of the Lake Baikal Region
Edited by Valentin A. Koptyug and Martin Uppenbrink

a regulative and legal basis for ecological and economic regulations, and, also, for property relations in the sphere of nature management (2). However, the structure of the environmentally oriented economy could be solved in the most effective way at the regional level. On this very level it is possible to create "fine instruments" of management, and to put into practise an "invividual approach" for certain types of territories. However, the existing system of regional management is not good for establishing such mechanisms. To solve this task, it is necessary to attract the scientific potential of the region.

From the first step of transition to the model of sustainable development, attention should be paid to the renewable natural resources. Their exploitation does not considerably change the quality of the environment. The importance of bioresources, and especially of recreational ones, is great and comparable with the value of mineral resources, but they are not developed enough because in society the ideas of natural-resources value prevail (3).

On the one hand, the transition to market relations promotes sustainable development, as it stimulates preservation of resources. But at the same time they are dangerous for the environment because nature-protective measures in the raw-material industry are not profitable. The change of priorities in favour of renewable resources decreases negative influences of "market forces" on the natural environment. With such strategy the high degree of cleanness and preservation of nature is the basis for the economic development of the territory, recreation, tourism and related spheres. This is the way which can bring together ecological and economic objectives (4).

One of the characteristics of the outlying areas is that we have much fewer capabilities for such profitable sectors as tourism than in the central area. So we have to provide comparative advantages for other sectors of sustainable economy. In addition, the environmental demands and regulations will be more rigid in the central area than in the outlying and contact zones. As a result, enterprises will be moved in future from the centre to the outlying areas. It is impossible to prevent this process, but we have to keep it under control. One of the main tasks is to avert destruction of the best forests in the outlying areas of the basin. It is also important to form the Regional Protected Areas Programme (the area of protected territories should be expanded up to 10 percent, of strictly protected no less 5 percent), including a "national river" status for the Chikoy and Khilok Rivers.

The Chikoy and Khilok Rivers, tributaries of the Selenga, transport the increasing amount of organic polluting matter from the basin periphery. The specific hydrological rate of these water courses—the disastrous character of rain freshets, flooding agricultural lands and settlements, the sharp reduction of drain during summer and winter minimum flowings— limits the self-cleansing possibilities of rivers. The basins of these tributaries are good for gold extraction and production of building materials. The biggest metallurgical enterprise in

Eastern Siberia is also located there. The Transsiberian railway (about 350 km along the Hilok basin) has heavily contributed to the pollution of the area.

Technogenic discharge of pollutants to rivers (carbon monoxide, hydrocarbons, nitric and sulfur oxides, etc.) of railway and motor transport alone amount to almost 4,000 tons every year. Accordingly 20.1 and 26.3 thousand tons of pollutants are released into the atmosphere and the surface (5).

More often, extremely high concentrations (exceeding sanitary norms) of heavy metals, organic pollutants, and others are registered in river waters.

The condition of river ecosystems is an integral indicator of well-being or of ecological destabilisation of the basin. The suggestion of an "individual approach" to territories is ideal for the management of river basins.

The main economic activity of the Chita area of the Lake Baikal Region is performed in the Khilok basin ecosystem. This ecosystem may be suggested as a model water-economic system, in which the degree of technogenic influence is valued, dominating factors and causes distinguished, management decisions worked out and then realised, thus vastly improving the basin's environmental conditions.

Realisation of the total inventory of natural resources, their quantity, quality and condition, is very important for the Lake Baikal Region. To analyse the issues of growth we need to take into account the extremely beneficial geographic situation of Chita with respect to the Pacific Region. Though the city is situated outside the Lake Baikal basin, its further development would be very important for the economic growth of the outlying areas of the Lake Baikal catchment area; for the improvement of the environmental situation; for increasing the standard of living. Development of hotel services in Chita city must be considered as a necessary condition for business in the eastern outlying part of the Lake Baikal basin. So this programme must be one of "first-stage programmes". The opening of the international airport in Chita may provide its profitability.

On the one hand it is obvious that it is necessary to increase the list of exploited natural resources, and in many cases to change the priorities in the exploitation of the natural-resources potential, for transition to the model of sustainable development. On the other hand, simply "turning on" market mechanisms is very dangerous in conditions of the above mentioned "starting position" without proper institutional support. We see numerous negative consequences of this process in the sphere of nature management. The complex system of ecological economic regulation is necessary, whose main components we are going to consider in this paper.

The recent practise of a nature-use payment system is directed mainly to solving problems of federal, regional and local budgets. It reflects only a general "federal" approach and scarcely regulates nature-use processes. It is necessary to make regional improvements on this system based on an "individual" approach to the area's problems. The payment system should encourage all nature users to make the best decisions from the environmental point of view. In particular, it should direct them to use renewable resources if possible. Finally, this system can become an important factor of the sustainable development of the region.

It is crucial that the price of land must be comparatively high if land use is connected with damages of soil and landscapes, for example, in case of gold-mining enterprises. Privatisation and sale of land without an agreement about conditions and restrictions of land use must be prohibited in the Lake Baikal catchment area. Functional zoning would be very helpful for decision-making.

The payment system for the restoration of natural resources must be based on the concept of the assimilation potential (2, 6). The principles of this concept are implemented in the practise of the pollution-rights market in the USA. We are sure that the introduction of such a market in the outlying areas of the Lake Baikal basin when enterprises are moved from the central zone to the periphery would be reasonable. It would help to provide the restoration of lost nature potential and ecological balance in the territories with destroyed forest, soil, water and biological resources.

The Baikal Fund which is proposed in the draft of the Lake Baikal Law could be the first financial centre. It would be reasonable to use the fixed part of federal nature-use payments to establish shares of regional governments in this Fund. But this bank alone is not enough; several units must be created to provide competition in this sphere. One of the ways: establishment of hypothecary banks. In the period of formation of market economy it is necessary to establish state control for the implementation of land-use policy. Shares in the hypothecary bank should be state-owned (at least during the first five years). It will allow the implementation of the policy of land use in accordance with eco-economic and social priorities. It does not mean that authorities should give permission for loans only on condition of land mortgage each time. The main principles of implementation of a desirable policy of land use will be written in the Bank Charter. The board of administration of the Bank will work in accordance with this Charter, and each half year issue a report to the local authorities. The role of the hypothecary bank in the sphere of antimonopolistic regulation of nature use should be expressed as in the allocation of preferential loans on a competitive basis (see the previous item), and in the regulation of land use to provide its most desirable development (by means of the system of privileges for loans on land security).

Long-term international loans may be a way of establishing "shares of local governments". The advantage is that the funds help to promote sustainable economy at the same time they are involved in the financial market and may be used more efficiently. Monopolisation in nature use in the case of "large void areas" is a serious danger for sustainable development in the Lake Baikal Region.

It is necessary to work out special tools which would be directed to support "alternative economy" in the areas of activity of large mining companies. The complex of measures should be directed to support small and medium-size businesses oriented towards priority directions of economic development, which are, first of all, environmentally sound. In the first place it is the establishment of enterprises and services, based on safe nature use and used or renewable natural resources. It is reasonable to create the register of the "risk group" of areas in view of nature-use monopolisation and to work out the basic principles of antimonopolistic control for areas which are included in the register (8, 9).

In conclusion, it is to be noted that today under the severe conditions of the Russian economy, international assistance in protection of the environment would be appreciated. The main task of the Russian side is to provide the efficiency of this support.

References

1. A Complex Programme of Land-use Policies for the Russian Portion of Lake Baikal Region, March, 1993.

2. Golub, A.A., Strukova, E.B. Social-economic principles of ecological policy//Economics and Mathematical Methods. 1991, vol. 27, no. 3, pp. 510–521.

3. Glazyrina, I.P., Zadorozhny, V.F., Strizhova, T.A. Priority direction of natural resources use. Geography, Ecology and Health, Chita, 1992, pp. 76–80.

4. Glazyrina, I.P., Strizhova,T.A. Including nature management in the market economy: regional tools. Theoretic and Applied Problems of Ecology, Bukhara, 1992, p. 78.

5. Glazyrina, I.P., Strizhova, T.A., Tsybekmitova, G.T. The problems of estimation of the assimilation potential and environmental capacity. Mathematical Problems of Ecology, Chita, 1994, pp. 32f..

6. Golub, A.A., Strukova, E.B. Economic evaluation of the assimilation potential. Economics and Mathematical Methods. 1988, vol. 24, no. 3.

7. Golub, A.A., Gofman, K.G. Economic machinery of management of global nature processes. Economics and Mathematical Methods, 1992, vol. 28, no. 5–6, p.687–694.

8. Collection of documents for the introduction of nature-resources payment. Ministry of Environmental Protection and Natural Resources of the RF, Moscow, 1991.

9. The Environmental Protection Law of the Russian Federation, 1992.

[illegible]

In conclusion, it should be noted that [illegible] conditions of [illegible] could be [illegible] the [illegible] to [illegible] the [illegible] of [illegible]

References

1. [illegible]

2. [illegible] Math. [illegible]

3. [illegible]

4. [illegible]

5. [illegible]

6. [illegible] and Mathematical [illegible]

7. [illegible] Mathematics and [illegible]

8. [illegible] for the [illegible] of [illegible]

Energy: World Trends, the Situation in the Russian Federation and Siberia, Recommendations on Sustainable Development of the Lake Baikal Region

A.P. Merenkov, L.S. Belyaev, V.A. Gorelov, A.A. Koshelev, B.G. Saneev, A.D. Sokolov, G.G. Lachkov
Siberian Energy Institute of SB RAS, Irkutsk, Russia

G.O. Borisov
Buryatenergo Company, Ulan-Ude, Russia

V.M. Gorlov
Soviet of Ministers of the Republic of Buryatia, Ulan-Ude, Russia

S.V. Kuimov
Irkutskenergo Company, Irkutsk, Russia

The paper reveals a dual role of energy in the problems of life support and sustainable development. It presents a brief characteristic of the main problems in the development of the fuel and energy complex of Russia compared with world trends. The possibilities for energy development in the Irkutsk region and the Republic of Buryatia are evaluated. The problems of nature-protecting energy supply of the Lake Baikal zone are considered. Specific tasks of the regional policy for Siberia which will make it possible to overcome the crisis and achieve the transition to sustainable-energy development are formulated.

A Dual Role of Energy in the Problems of Life Support and Sustainable Development

Energetics plays a many-faceted role in the problems of economic and social development, including relations between society and the environment.

On the one hand it is one of the main factors of life support, economic development and improvement of human living standards. Thus, it serves as a qualitative and quantitative indicator of the industrial development of society.

On the other hand, energetics currently entails large economic expense and environmental damage in the form of depletion of non-renewable natural resources and extreme burdens to all components of the environment. Suffice it to say that energy provides about 50% of the pollution and damage to the atmosphere, water and land which are caused by industry. Therefore, the

NATO ASI Series, Partnership Sub-Series, 2. Environment – Vol. 6
Sustainable Development of the Lake Baikal Region
Edited by Valentin A. Koptyug and Martin Uppenbrink

transition of energetics to sustainable development [1, 2] has particular importance for prevention of the global environmental catastrophe.

The following minimum set of requirements to sustainable energy development (at the national, regional and global levels) can be formulated:

– gradual transition to the use of inexhaustible energy resources or those sufficient for the foreseeable future;

– provision of the environmental cleanness, conservation and safety of power plants and technologies;

– attainment of acceptable economic indices of energy technologies, plants and systems;

– provision of a suitable standard of living for the population.

At present none of the energy resources completely meets these requirements. Fossil fuel is limited and disturbs the environment when produced and pollutes it when utilized (including emissions of greenhouse gases). The use of nuclear fuel, including fusion reactions, remains insufficiently safe thus far, and, in some cases, economically inefficient. Traditional hydropower resources are obviously insufficient with respect to their economic potential. Non-traditional renewable energy resources are very expensive today.

Natural gas can play an important role in the transition to sustainable development: its reserves on the planet have recently been estimated as being very large [3], and it is much better than coal and oil from the environmental point of view. In principle, we should turn toward a substantial increase of gas utilization in the 21st century, though it cannot guarantee long-term sustainable energy development.

According to current understanding, only nuclear and solar energy, which are practically inexhaustible, can be referred to as primary energy resources. With fusion reactors and breeders, using uranium and thorium, nuclear energy will last for many centuries. The problem is to make these sources acceptable with respect to safety and economic indices. At present, it is impossible to indicate the time when this will come to pass. Therefore, the routes the transition to the development of sustainable energy will follow remain vague. It is believed that such a transition may embrace the whole 21st century.

It is supposed that all traditional and non-traditional methods of energy production will be used during the transition period. The change in the proportions of various primary-energy resources will depend to a large extent on technological progress in the devising of new energy technologies and systems.

Thus, the forthcoming period embracing many decades must also be a period of transition of energetics toward sustainable development. It will surely be characterized by a large variety of conditions and possibilities of energy development in different countries and regions of the world, depending on the level of their economic development and on the availability and composition of primary energy resources.

In the statement of the international conference devoted to the problems of sustainable energy development [4], emphasis is laid upon two general goals: energy conservation and utilization of renewable resources. These endeavors are timely, though at present extremely expensive for the majority of developing countries. Therefore, the forthcoming decades may be characterized (to a different extent and in different countries) by:

– increasing use of natural gas as the most uncontaminated organic fuel;

– introduction of non-polluting technologies in coal combustion;

– further development of nuclear energy by creation of reactors with higher safety standards and solution of problems of radioactive-waste disposal.

Russia is concerned with this to a considerable degree. However, here it is necessary to specify and consider the general trends and requirements at the present level, taking into account its vast territory and diversity of climatic conditions, great inertia and huge demands for investments in re-equipment and further development of power engineering, and the worldwide significance of fuel and energy resources (FER) of Russia and primarily of Siberia.

State-of-the-art and Main Problems in Development of the Fuel and Energy Complex of Russia and Siberia Against the Background of World Trends

Russia possesses one of the largest potentials of FER in the world: about 37% of the explored world reserves of natural gas, 19% of coal and up to 10% of oil are located in its territory (10% of land; 2.6% of the global population). The contribution of Siberia amounts to almost 80% of all extracted and produced primary energy resources in Russia (Table 1). A long phase of aggressive quantitative growth and service of the centralized and militarized economy was completed for the fuel and energy complex (FEC) of Russia with the maximum level of FER extraction and production of 1,914 million tons c.e. in 1988 (about 20% of the world level).

The rapid fall in oil production shows that the “light” oil period in Siberia is completed. The decrease in natural gas production indicates that the crisis in the gas industry has also set in. At the same time this industry (with unique reserves

Table 1 Role of the Siberian FEC in Russia (in 1994)

Area, million square kilometers (%)	6.5 (38)
Population, million people (%)	24.1 (16)
FER production, million t c.e.(%)	1200 (83)
FER consumption, million t c.e. (%)	235 (25)
Production of oil + gas condensate, million t c.e. (%)	220 (70)
Production of gas, billion cubic meters (%)	556 (92)
Extraction of coal, million t (%)	164 (61)
Production of electricity, billion kWh (%)	245 (28)
Centralized production of heat, million Gcal (%)	345 (19)

of potential and commercial gas resources in Russia) must be practically prepared to increase gas production to any levels expedient for the solution of economic, regional and environmental problems [3].

From 1980 to 1990, the global consumption of energy resources increased from 9,240 to 11,100 million t c.e., i.e., by 20% or by an annual average of 1.8%. Such sufficiently moderate growth was a consequence of the fact that as a result of the energy crisis in the 1970's, the governments of many countries developed national energy programs which envisaged:

– investigation of energy- and resource-saving strategies of economic development and increase in the share of science-intensive industries;

– change in the structure of energy balance to decrease the oil share and to increase the use of nuclear energy;

– extension of efforts in the field of non-traditional energy sources;

– intensification of the government's role in energy management.

As a result, great changes in the structure of the world energy balance took place. The share of oil and oil products was reduced from 46 to 39% and will apparently continue to decrease. As compared to 1980, coal consumption in developed countries increased by 8% in 1990 and that of natural gas by a factor of 1.35, with its share in the energy balance increasing from 25 to 29%. The gas consumption in these countries is expected to grow additionally by a factor of 1.3

Table 2 Main trends in world energy development

	1980	1990	2000	2010
Characteristics				
(% in relation to the preceding decade)				
	Prognosis			
Consumption of FER (billion t c.e.)	9.240	11.100	13.6	13.3–16.0
(%)	100	120	122.5	96–118
Consumption (%) of:				
oil and petrochemicals	100	102	117	90–104
natural gas	100	135	155	113–136
coal	100	123	105	85–115
atomic energy	100	200	125	110-123
Structure of FEB (%):				
oil and petrochemicals	46	39	38	31–35
natural gas	19	22	27	30–36
coal	29	29	25	22–25
atomic energy	3	7	7	9–10

by 2000, and by 1.5 by 2010, with its share in the balance increasing to 33 and 37%, respectively. At the same time, from 1981 to 1990 the nuclear-energy consumption increased 2.8 times in developed countries and 15 times in the developing ones. Its share in the balance of the developed countries is expected to increase further from 7% in 1990 to 9 to 10% in 2010 (Table 2).

As to the structure of FER consumption in Russia and Siberia, in 1990 it was characterized by the following figures (in %): 42 and 30 of gas, 29 and 12 of oil and oil products, 19 and 36 of coal, 7.4 and 10 of electricity, respectively.

The comparative analysis of the energy balances of developed countries, Russia and Siberia shows some common trends in their change:

– decrease in the forecasted levels of energy consumption in comparison with planned levels;

– decrease in the share of oil used for energy;

– prevailing role of natural gas in energy-consumption growth;

– preservation of the role of coal and nuclear energy;

– low contribution of non-traditional energy sources in the next 15 to 20 years.

According to estimates of a number of institutions, the rates of FER consumption growth will be approximately the same for 1991–2000 (2%) and 2001–2010 (1.7%) and the level of specific energy consumption will make up about 2.1 t c.e./person per year, which exceeds the 1990 level by only 3.5%.

Besides, it is necessary to acknowledge that energy in the developed countries meets requirements of post-industrial society, while energy and the economy of Russia face sharp technical and economic, structural and technological, ecological and regional problems. Here the period from 2000 to 2010, when it will be necessary to increase production of FER (for providing the economic development and living-standard improvement), may simultaneously appear to be the most critical for replacing obsolete equipment.

Table 3 Possible development of the main fuel sources in Siberia

	1994 (actual)	1995 (expected)	2000	2010
West-Siberian oil and gas complex				
oil + gas condensate, million t	220	207	200–300	200–400
gas, billion cubic m	556	550	630–685	680–745
Eastern Siberian oil and gas complex				
oil + gas condensate, million t	–	–	2.3	12–16
gas, billion cubic m	–	–	6	14–28
Kuznetsk basin				
coal, million t	102,2	95	95–110	100–120
Kansk-Achinsk FEC				
coal, million t	38,8	35	50–70	60–90
electricity, billion kWh	5,4	5–6	10–15	21–25

The regional energy programs [5] for West and East Siberia (Table 3) reflect the tendency of regions toward self-sufficiency and look logical enough at first glance. However, the integration of these programs for Siberia and Russia with the general schemes of development of the FEC industries creates an obviously oversaturated picture.

These data show that the FEC of Siberia will continue to supply 80% and more of fuel and energy to other regions of Russia and for export. However, two principal questions arise here:

1. Are the planned volumes of fuel and energy really required?

2. Will the national economy be able to realize these sectoral and regional plans, even if all possible sources of financing are utilized?

Therefore, the complex analysis and the rational choice of variants from available alternatives, with respect to the strategies of economic development of Russia and Siberia, take on high priority.

Main Problems and Strategies of Energy Development in the Irkutsk Region

The region possesses a large fuel and energy potential. In 1992, it produced 63.0 billion kWh of electricity, 70.0 million Gcal of centralized heat, 5.2 million t of fuel oil and mined more than 20 million t of coal. The region plays an important role in supplying other regions: 10.0 billion kWh of electricity, 4.0 million t of coal, 4 million t of fuel oil were exported in 1992. On a per-capita basis, energy consumption in the Irkutsk region accounts for 16.2 t c.e. (one of the highest levels in Russia).

At the same time, as concerns the high-grade forms of fuel (gas, fuel oil), the Irkutsk region ranks far below the European regions of Russia (less than 10% as against 64–65%). This fact has a negative effect on the operating costs in industry and is a cause of inadmissibly high levels of environmental pollution.

Immediate development of the hydrocarbon resources of the Siberian platform and involvement of natural gas in the economy are means for radical improvement of the situation in the region. The most preferential consumers of natural gas here are:

– plants of the chemical complex, where gas can effectively be used both as raw material and technological fuel (4 to 5 billion cubic meters will be required annually for the Angarsk-Usolie-Zima chemical complex);

– other large plants (Bratsk Forest Processing Complex, Angarsk Cement Plant, etc.), where shortages of fuel oil can be replaced by gas;

– communal general service of towns and industrial centers, agricultural consumers;

– thermal power plants and thousands of boiler plants, where natural gas can replace low-grade solid fuel.

The overall demand for natural gas in the region, depending on its economic and environmental efficiency, is estimated at not less than 10 billion cubic meters for the year 2000, with further growth to 20 to 25 billion cubic meters per year. The efficiency of hydrocarbon resources development can be additionally improved by using the accompanying components: helium, ethane fraction, condensate, volatile hydrocarbons, mineralized brines.

A sharp drop in the rates of electricity-consumption growth and involvement of natural gas in the future economic activity of the region will lead to a decrease in coal consumption and, as a result, to a substantial change in previously formed ideas on required production. Therefore, starting from the year 2000, the coal industry may be faced with the problem of increasing the competitiveness of Irkutsk coals on the interregional Russian market of FER.

The primary goals of the electric and heat power industries are [6]:

– updating and replacement of the obsolete basic equipment at power plants;

– efficient development of cogeneration plants (in optimum combination with large boiler plants) to meet the demands for electricity and heat;

– intensive construction of electric networks to improve reliability of energy supply, and to export electricity beyond the territory;

– improvement of the environmental indices of thermal power plants, for example, by converting them to gas.

Provision of reliable energy and fuel supply of the outlying and isolated consumers (including the Lake Baikal zone) is the most urgent problem, especially as a larger part of the region is situated in the permafrost zone by its economic, geographical and climatic conditions that necessitate non-standard engineering, economic and institutional decisions (see below).

Clearly, the Irkutsk region will be an area with the lowest prices of energy coals and electricity in Russia in the future. On the one hand, even neglecting inflation, the prices will be increasing due to: rise of the cost of their extraction and production, increasing demand for Siberian energy resources in other regions of the country, necessity of self-financing of FEC re-equipment and development. On the other hand, the excessive rise of fuel and energy prices can aggravate the financial situation of many industrial enterprises and cause their bankruptcy.

The following export-import ratios can be considered as the most probable:

Electricity export

Mongolia and China are potential markets for electricity. Electricity supply to China can reach 10 billion kWh. A transmission line to the Far East with a branch line to the border of China through the Chita region is to be constructed for this purpose. The Irkutskenergo company has already presented the draft of this project.

Natural-gas export

Geologists estimate gas reserves of the Kovykta field at 1.5–2.0 trillion cubic meters. The reserves would suffice to organize production of 25–30 billion cubic meters annually. With such production, 10–15 billion cubic meters of natural gas can be exported to China and Korea.

Coal export

Bringing large volumes of natural gas into the balance of the region after the year 2000 can lead to large surplusses of Irkutsk coals. For example, if natural-gas production accounts for 11–12 billion cubic meters annually and the Mugun mine in the Irkutsk region is put into operation at full capacity (20 million t), the coal surplus can reach 10–16 million t, even provided that the region will abandon supplies from the Kansk-Achinsk coal deposit. Hence, in order to have a reliable coal market in the Far East, steps should be taken for arranging long-term relations with important consumers as well as creation of joint-stock companies of coal mines and the Far-Eastern coal consumers and, also, the railway. Moreover, it is necessary to study possibilities for the Irkutsk coal on the market of the Asian-Pacific region (Korea, Japan, etc.).

Main Problems and Strategies of Energy Development in the Republic of Buryatia

FEC in the Republic of Buryatia, which is an energy-deficient area of Siberia, is objectively less developed than FEC of the Irkutsk region. In 1992, Buryatia produced 4.4. billion kWh of electricity, 10.0 million Gcal of centralized heat, about 5.0 million t of coal. The energy shortage was met by the import of 2.6 million t of the Kansk-Achinsk, Cheremkhovo, and Azeya coals and 1.4 billion kWh of electricity from the Irkutsk region. At the same time, 2.4 million t of the Tugnui and Gusinoozersk coals were exported from Buryatia to the eastern regions.

The expected development of the economy and social sphere in Buryatia (both in the "moderate" and "optimistic" scenarios) predicts the following levels of FER consumption in the period under consideration. By the year 2000, the demand for electricity in Buryatia will amount to 6.5–8.3 billion kWh (i.e., an increase by a factor of 1.48 to 1.9) and for heat to 11–14 million Gcal (1.1–1.4). By the year 2010, electricity consumption can additionally rise by 1.5–2.0 times and will total 12.5–14.0 billion kWh and heat consumption will amount to 14.1–19.6 million Gcal.

The demands of the economy and social sphere can be satisfied by a two-fold increase in coal extraction, bringing it up to 11.8–13.8 million t, as well as by the commissioning of 3085 MW of new capacities at power plants, including 1100 MW at the Ulan-Ude cogeneration plant 2, 1075 MW at the Gusinoozersk condensing power plant, 900 MW at the Moksk hydropower plant.

At the same time, the energy sector of Buryatia faces urgent problems and crisis phenomena requiring major decisions in the foreseeable future. The coal industry is characterized by decreasing production, a drop in labor productivity and rise of the cost of extracting coal, a considerable lag in new construction (Tugnui mine) and reconstruction (Kholboldzhinsk mine), complications of geological and mining conditions, increasing pollution of the surrounding areas, water and air basins, etc.

The electric-utility industry historically forms the base of the economy of Buryatia [6]. At present the crisis situation here is explained by the following main reasons:

– the chronic lag in commissioning of new generating capacities and insufficient transfer capability of transmission networks that lead to limitation and, in some cases, even interruption of power supply of consumers in Buryatia, with the excessive power balance in Eastern Siberia as a whole;

– intensity of the fuel balance of power plants due to disorders in coal supply in winter;

– growth of the share of obsolete facilities of power plants;

– increase in the complexity of the environmental situation due to combustion of low-grade fuel.

As a consequence, in 1992 emissions of harmful substances into the atmosphere by power plants of the Buryat power system totalled 49 thousand tons, and release of sewage water into reservoirs amounted to 6,665 thousand cubic meters. The power plants of Buryatenergo produce more than 500 thousand tons of ash and slag annually. Neverthelss, these problems can become even more urgent, since in the lapse of time until 2010, consumption of fuel (primarily coal) for electricity and heat production will grow.

The energy development strategy in Buryatia for the period until 2010 is based on the following fundamental principles:

– realization of a complex of measures aimed at the increase of electricity self-balance of the Buryat power system to enhance the economic safety of Buryatia; – creation of an optimum, reliable scheme of the backbone network;

– systematic realization of measures on dismounting and updating the main power facilities of power plants, substations and transmission lines;

– introduction of advanced highly efficient and environmentally clean power units at power plants;

– further expansion of the coal base in Buryatia by the maximum development of the Tugnuisk coal mine;

– establishment of the mutually beneficial market of electricity between Buryatia and the Irkutsk and Chita regions.

Problems of Nature-protecting Energy Supply of the Lake Baikal Region

If fuel and energy supply of large towns and the industry of the Irkutsk and Chita regions and Buryatia which belong to the Lake Baikal Region is furnished as a result of the development of the large-scale power engineering described above, the populated areas of the Lake Baikal coastal area should be considered separately.

The coastal area of Lake Baikal is an especially suitable area for the pioneering realization of decisions on non-traditional energy forms and sources [7]. Actually, quite a number of factors favoring it are combined here:

– The natural complex of Lake Baikal has a particular value on the global scale. It obviously precludes any decisions having an appreciable impact on the course of natural processes and the quality of all components of the ecosystem.

– Energy consumption of a large part of the territory is characterized by low density.

– Construction of recreational objects (for rest, tourism, therapy) of local, national and international importance which are of a seasonal nature is a promising branch for the intensive development of the economy.

– There seems to be no possibility for the supply of high-grade "smokeless" fuel in sufficient volumes in the near future. The pipeline transport of liquid and gaseous fuel is problematical due to topography; a large-scale supply of such fuel by rail or water is expensive and potentially dangerous to the environment.

– There is no system of centralized electricity supply in a large part of the territory, and establishing one seems economically inexpedient due to low density of load (except for the 220 kV transmission line along the Baikal-Amur railway and the 500 kV transmission line under construction through Slyudyanka, which will supply the northern and southern industrial centers with energy basically from external sources).

At the same time, the potential of non-traditional energy forms and sources is rather high here. For example, solar radiation is abnormally large above the water area and in the coastal zone of Lake Baikal and reaches 1500, 1900 and 1800 kWh/square meter per year in the northern, middle and southern parts of the lake, respectively, at solar-radiance durations of 1900–2000 h/year in the north, 2200–2400 in the south and central area (whereas the figure does not exceed 2000 h/year at the health resorts of the Caucasus). In the Tazheran steppe (the southern end of Olkhon Island) there are natural salt reservoirs which are unique sources and accumulators of hot water in the lower layers, the heat of which can be easily extracted all year round. Out of 20 coastal points with several years of observations in 11 points, the mean annual wind velocity exceeds 3 m/s. Note that the maximum wind activity is observed at Lake Baikal in winter, when the energy demand is at a maximum. There are many worthwhile sources of terrestrial heat found on the eastern coast and at the north of the western coast. The energy potential of mountain rivers flowing into Lake Baikal (their number exceeds 300) is extremely large.

It should be noted that in accordance with the agreement between the Irkutskenergo company, the Siberian Energy Institute of SB RAS, the Irkutsk territorial department of hydrometeorology and the Baikal-Lena National Reserve, the work on constructing a polygon for testing the pilot installations of various non-traditional energy forms and sources (solar energy, wind, water flow and biogas) were started in 1993 on the central-western coast of Baikal (Cape Pokoiny). The hydrometeorological station and the cordon of the reserve will consume the energy produced.

Thus, the energy supply of small remote populated areas (including tourist complexes) should be aimed to the maximum extent at renewable energy sources, to be doubled by advanced diesel installations. To achieve cleanness and conservation of nature, it is worthwhile to go to higher expense, including the use of the international foundations. Later on, with improvement of the economic state of the country and its technological progress, the energy-supply schemes for towns and other settlements of the Baikal coast will be developed and improved.

Conclusions

The above shows that the role and value of the regional aspect considerably increase in combination with requirements of sustainable energy development in new socio-economic conditions. It becomes an obligatory condition of scientific analysis, requiring really valid decisions to be made which combine different interests. Hence, the following specific problems currently seem to be of principal importance:

1) Forecast of the priority in reconstruction and development of the existing fuel and energy sources and creation of new ones. For each of them it is necessary to work out the corresponding project and business plan as a single document with separation of stages to achieve the final goal, with indication of possible sources for financing and the nature of interrelations with the local governmental bodies. The Eastern Siberian oil and gas complex based on the oil and gas fields of the Krasnoyarsk territory, the Irkutsk region and Republic of Sakha (Yakutia) can be referred to as new prospective sources.

2) Comparative and comprehensive (socio-economic , energy, and environmental) estimates of basic alternatives in development of the FEC industries in Siberia and Russia (gas, coal, nuclear industries) with forecast of their possible realization and analysis of the requirements to the related branches as well as economic and regional consequences of different priorities.

3) Development of the concept and specific schemes for the transportation of large amounts of the Kansk-Achinsk, Kuznetsk and Irkutsk coals to the west and east by new specialized types of transport (main coal pipelines, special railways for coal transportation, etc.).

4) Development of a long-term scheme of electricity output by the Siberian power plants, involving the formation of an East–West backbone network of the unified power system of Russia and the optimum participation of condensing power plants of the Kansk-Achinsk FEC and hydropower plants of the Angara-Yenisey cascade, to supply electricity to the areas of the Urals, Trans-Baikal, etc.

5) Forecast of the appearance and development of competition at the markets of different fuel and energy forms.

6) Solution to the problem of transformation of the heat economy of Siberia as the most heat-intensive industry of FEC, with the largest potential for energy conservation.

7) Elaboration of a special program of fuel and energy supply to the northern and isolated areas, including the Lake Baikal coast.

8) Elaboration of a program for a marked increase in the use of advanced "small energy" and non-traditional renewable energy sources.

9) Preparation of recommendations and a series of steps for the provision of reliable fuel and energy supply to the regions and territories in various extreme locations.

10) Devising and introduction of management methods and mechanisms for FEC and its industries at the federal and regional levels during a transition period to market conditions, etc.

In this connection, the basic principles which are intended to stimulate the transition of power engineering in Russia and its regions to sustainable development will be:

– planning of long-term socioeconomic priorities for Russia and its regions;

– transition from technical and economic criteria of economic and power-engineering development toward socio-economic and regional criteria;

– complex analysis of the alternatives rather than simply lumping together the development programs of specific industries and regions;

– setting up government mechanisms to analyse the consequences of decisions made and to correct periodically the energy policy on three main levels: national, regional and local;

– openness and public expertise of projects.

The purposive and complex solution to these problems will best facilitate: 1) the overcoming of crisis situations, re-equipping and further development of energy in the Lake Baikal Region and 2) radical improvement of the environmental situation in the region, which in turn will allow the problems of its sustainable development on the whole to be stated and solved more objectively and successfully.

References

1. L.S. Belyaev. Search for the ways of sustainable energy development//Trudy konfer. “Laws of social development: features and criteria of future development models”. Novosibirsk, SO RAN, 1994 (in Russian).

2. W. Haefele. Energy in the transition period: main problems//Izv. AN SSSR. Energetika i transport. 1990, no. 2 (in Russian).

3. Reports of the International business meeting “Natural gas and energy ’93”. Special issue of the journal “Gazovaya promyshlennost”, 1993, no. 10 (in Russian).

4. Global Collaboration on a Sustainable Energy Development//Proc. Internat. Conf. Snekkersten, Denmark, April 25–28, 1991.

5. A.P. Merenkov, B.G. Saneev. Energy of Siberia and the Far East: state-of-the-art and main problems of development//Izv. RAN. Energetika, 1993, no. 1, p. 3–11 (in Russian).

6. V.A. Gorelov, G.G. Lachkov, V.M. Sobolevsky. Trends in development of the electric-utility industry of Eastern Siberia//Izv.RAN. Energetika, 1993, no. 5, p. 3–10 (in Russian).

7. V.N. Parmon, A.P. Burdukov, L.S. Belyaev et al. Small-energy and non-traditional forms and sources of energy: their role and place in the coming years and for the future. Report at the All-Russia Conf. on the economic development of Siberia. Section of FEC. Novosibirsk: SO RAN, 1993, 51 p. (in Russian).

Mineral Resources of the Lake Baikal Region and Sustainable Development

E.G. Konnikov
Buryat Institute of Geology of SB RAS, Ulan-Ude, Russia

M.I. Kuzmin
Institute of Geochemistry of SB RAS, Irkutsk, Russia

V.A. Lbov
Committee of Geology at the Council of Ministers of the Republic of Buryatia, Ulan-Ude, Russia

G.L. Mitrofanov
Eastern Siberian Institute of Mineral Resources,
Committee for Geology of the RF, Irkutsk, Russia

K.Sh. Shagzhiev
Baikal Institute of Rational Nature Management of SB RAS, Ulan-Ude, Russia

The Lake Baikal Region possesses great reserves of mineral and energy resources. However, their development is unsatisfactory. In addition, departmental separateness hinders the complex utilisaton of resources. Provisions are made for priority exploitation of gold and platinum group deposits. To decrease the burden on the environment, methods of working some deposits are proposed. Changes are suggested in the taxation policy to achieve efficient management of resources.

The Lake Baikal Region is extremely rich in natural resources, the most important of which is Lake Baikal, the world's largest reservoir of pure sweet water, and the manifold complex of minerals, including a number of unique deposits (fig.1). The current problems of conservation and protection of Lake Baikal must not hold up the economic development of the Lake Baikal Region, which still remains at a rather low level as compared to the other regions of Russia. This region possesses an abundance of energy resources (especially the Irkutsk region). That is why the mineral resources here could and should be utilised with a high economic efficiency. Only on this condition can the sustainable development of this territory be ensured.

Condition of the Mineral Resources and Their Use in the Mining Industrial Complex

The mineral complex of the Lake Baikal Region is manifold. There are large deposits of gold (Sukhoi Log, Zun-Holba, Irokinda, Korolon), nonferrous metals

NATO ASI Series, Partnership Sub-Series, 2. Environment – Vol. 6
Sustainable Development of the Lake Baikal Region
Edited by Valentin A. Koptyug and Martin Uppenbrink

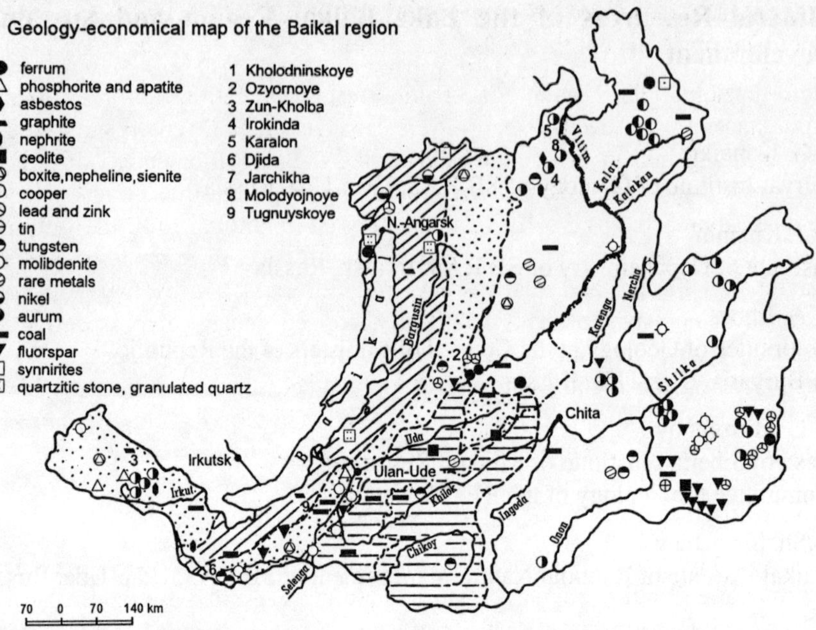

Fig. 1. Geology-economical map of the Baikal region

(Kholodninsk, Ozernoye), asbestos (Molodezhnoye), tungsten (Jida), molybdenum (Orekitkan, Maly Oinogor, Zharchikha), tin (Mokhovoye), beryllium (Yermakovka), niobium (Bolshetagninskoye), chemically pure quartz (Cheremshanka), phosphorous raw materials (Khara-Nur, Oshurkovo), coal (Tugnui, Cheremkhovo), fluorite, zeolites, and semiprecious stones.

To give an idea of the amount of mineral resources in the region: zinc reserves contained only in two deposits of Buryatia, at Kholodninsk and Ozernoye, make up 48% of all reserves of this metal in Russia. On the basis of the Kholodninsk deposit, it is possible to extract 6.5 million tons of ore annually for 27 years. The reserves of gold, magnesite, common and rock salt available in the Irkutsk region make up 10 to 15% of the total amounts found in the Russian Federation. The Molodezhnoye deposit of long-fiber asbestos will make it possible to cover the mineral deficit in the pipe and textile industries of Russia, and the phosphorous ore reserves of the Oshurkovo apatite deposit can satisfy the demand in these fertilizers not only for Siberia, but also for neighbouring countries. The Orekitkan molybdenum deposit can ensure the operation of the mine for 28 years, and the expenses on its construction will be repaid in five years. On the basis of nepheline sienites of the Mukhal deposit, an aluminium industry can be successfully developed. The aluminium reserves in the region will be practically boundless if the mining industry develops the "synnyrite" deposits of the Northern Pre-Baikal, which contain, apart from aluminium, excellent chloride-free potassium fertilisers, and are also a source of raw materials for glass, ceramics and concrete-manufacturing plants of the region.

Unfortunately, the state of natural-resources utilisation is extremely unsatisfactory. In 1990, there were twelve mining enterprises and one concentrating mill in the region, and during the period of economic crisis and Perestroika, the economic activities of the mining-industrial complex began to decline, approaching zero. In fact, the Jida concentration mill, the only one in Buryatia, ceased operations because of the absence of a tungsten-concentrate market. Most mines and quarries have reduced extraction by two- to three-fold because of nonpayment. This has nearly stopped geological prospecting in the region, because it is no longer financed from the state budget and is thus dependent on the mining industry. Approximately the same situation is in the Irkutsk region. Suffice it to say that in 1993, only 2,000 geological prospecting boring wells and 10 meters of prospecting excavations were made(!). This demonstrates that the increase of prospected mineral resources in the region has practically stopped.

A severe problem is the selected utilisation of mineral raw materials. Due to bureaucratic isolation, mining enterprises are not interested in practical utilisation of concomitant ore components. This results in irreversible loss of many valuable metals in dumps and creates a threat to the environment in many mining regions.

Choice of Priorities in Exploiting Mineral Resources of the Region

When elaborating the conception of development of the Lake Baikal Region, it should be taken into account that by now its natural complexes are undergoing a critical ecological challenge. That is why, when exploiting the mineral resources, ecological requirements must prevail over economic calculations.

In this case, the priority in exploitation of the mineral resources of the region must be given to gold:

- gold extraction does not require construction of large harmful processing works and a large-scale infrastructure and therefore causes a minimal burden on the natural environment;
- the major part of the gold-ore deposits of the region are situated beyond the borders of the Lake Baikal water-protection zone and do not constitute a contamination threat.

From the economic point of view, a favourable factor is the rapid recoupment of gold mining. Since ancient times, gold mining has been traditional in Eastern Siberia and the Lake Baikal Region. Recently, the importance of this region for Russia and Buryatia has increased, since placer gold deposits in Northern East Asia are exhausted and appear unprofitable. However, in connection with entering into the world gold market, exploitation of the Baikal placer deposits has turned out to be unprofitable, too, due to the low metal content.

That is why the main task now is the development of rich vein deposits. In the Irkutsk region, this is the Sukhoi Log deposit, and in Buryatia the Kholbinsk deposit. For a large-scale exploitation of these deposits, the funding is insufficient. Foreign investments are necessary for the mining industry of the region; however, the policy and mechanisms of this process are poorly developed in this country. Here we should profit from the world's experience. An example may be the experience of such countries as Indonesia, with its 25-year development of mineral mining. By adopting laws ensuring the profitability of foreign investments in its mining industry, Indonesia has become one of the leading countries with respect to the mining of gold, diamonds, copper and other minerals.

In addition to gold mining within the scope of local budgets, the development of metal deposits of the platinum-group elements (PGE) is now becoming very important for the local budget. The demand of industry for these metals has sharply increased (by 10 to 100 times) in connection with production of ecologically important catalysts. In the Northern Pre-Baikal area, in the region of the Kholodninsk polymetallic deposit, a very promising "Stillwater" type PGE deposit has been found. This is the third occasion such deposit has been discovered (the other two, one in the Kolsky Peninsula, and one in Karelia, are being explored now). This type of bedrock deposit is characterised by large-scale PGE reserves with isolated ore concentrations, like in the Norilsk deposits. The latest explorations of Buryat geologists in the Severobaikalsk plationoid deposit yielded evidence of useful concentrations at a profile level of such deposits (up to 10 g/t, PGE + gold). Unfortunately, the lack of funds in the Russian Geological Service precludes the exploration of this interesting site. Its economic evaluation requires cutting trenches and drilling boreholes, which can be performed only by special services.

The development of these ores is environmentally safe, as these ores contain practically no sulphide, but natural Pt, Pd, Rh, Ir alloys with Fe, Bi, Cu, Sn. The favourable geographic position of the deposit in the watershed of Lake Baikal and the Lena River makes it imperative to build a concentration plant in the Lena River basin to protect Lake Baikal from its harmful influence.

Diamonds have always been exported to forcign markets from Siberia; this takes second place after oil and gas. Today there are all possibilities of discovering kimberlite pipes in the Irkutsk part of the Siberian platform. In the sedimentary cover of the platform there are pipe-like intrusive bodies morphologically similar to diamond pipes. Over 100 finds of diamonds in alluvial sedimentations have been found here. There is a good chance of discovering a diamond province, the third in significance in Russia, but its prospecting requires large subsidies. Some foreign companies are interested in this proposed diamond project.

There are definite preconditions to consider Buryatia a diamond mining region. Buryat and Novosibirsk geologists presented a joint research programme on the exploration of carbonado deposits in the Archean metamorphic rocks, similar to a “kokchetav” type in Central Kazakhstan and the Dabeshan deposit in China at the beginning of 1994. This type of diamond bearing is very likely in the margin zones of the Archean block, where associated rocks of such mineralisation are found (garnet eclogites), and a high pressure in the metamorphic processes was obtained, which is a characteristic for diamond-bearing metamorphic complexes.

Environmental protection of the region has in part been brought to bear in developing the Ozernoye and Kholodninsk pyrite-polymetallic deposits. The exploitation of these deposits is highly profitable and favourable to its stable development. The Ozerninsk concentrating mill is being built outside the Lake Baikal catchment area, though within a special nature-management territory. To decrease harmful effects from ore processing, provisions are made to transport it to the Sherlovogorsk concentrating mill. For this purpose, construction of a railroad branch line from the Mogzon station to the Ozerninoye deposit has been begun. The proximity of Lake Baikal rules out the possibility of processing the Kholodninsk-deposit ore on site. Therefore, ore processing is possible only at the nearest concentrating mills in the Sherlova Mountain, connected to the Kholodninsk deposit by the Baikal–Amur railway. But the exploitation is complicated by its low profitability.

The evaluation of deposits is necessary to determine optimum conditions of management of mineral resources.

Considering the high demand for quality phosphoric fertilisers in Siberia, the exploitation of these deposits in the Lake Baikal Region should be investigated. Today bench tests provide technological processes of the apatite-ore processing at the Oshurkovo deposit in order to produce organomineral and long-life phosphate fertilisers without chemical conversion into complex compounds and with high utilisation efficiency. Organomineral fertilisers obtained from 32% of apatite concentrate mixed with coal and zeolites contain up to 92% phosphorous-mobile forms, humus, and many microelements. Such fertilisers may completely supplant traditional varieties, and what is important, they are ecologically clean. Their production would be possible by reconstructing the Transbaikalian Apatite Factory, which is currently not operating.

Highly efficient fertilisers equivalent to superphosphate can be produced at the phosphorite deposits in the Oka District in Buryatia. Technologies are available to produce phosphorite flour from phosphorite and oxidised brown coal at minimal capital expense by small-size enterprises. Humic acids dissolve 93% of the phosphorous in phosphorite flour. Field tests reveal that such flour provides an increase of crop yields from 30% to 90% and is environmentally safe.

As a result of the disintegration of the USSR, deliveries of manganese from the Ukraine and Georgia have ceased. In the Irkutsk region, in the Pre-Sayan trough, there are manganese deposits. High-grade manganese ores are mined in the Nikolaevsk deposit for the Vossibelement company. The prospecting of this deposit will provide its balanced development and supply manganese not only to this region, but also to the others.

Considerable recent attention has been focused on the problem of production of super-pure silicium and solar batteries, taking into account high economic demand for energy resources and unique raw materials, from the chemically pure quartzite of the Cheremshansk deposit in Buryatia. The solar batteries can provide cheap electric and thermal energy to the outlying and poorly populated areas of the Lake Baikal Region without expensive electrotransmission lines. A schematic process of non-traditional, environmentally safe production technology of super-pure metallic silicium was worked out by the Institute of Semiconductor Physics of SB RAS. At present it is necessary to examine it on laboratory and pilot scales.

The development of non-traditional energy sources in the Lake Baikal Region may also take place with the use of in-depth heat of the rift zone, whose resources are enormous in comparison with the rest of the country. Today the Nedra Research-Industrial Association, the largest boring association of the country, has worked out profitable projects of thermal power stations using boreholes. This programme is performed today in the European part of Russia, where heat flow is low. It is evident that it may be very profitable in the Lake Baikal Region. They can provide small villages with autonomous thermal power stations to meet the needs in energy, which will also allow the efficient development of the tourism industry on Lake Baikal.

The exploitation of rare and rare-earth deposits is of great importance for the economic development of the region. Much demand for these metals has increased in some highly developed industrial countries, though most of them have no deposits of these raw materials (Japan, the USA and some European countries). Today China allots 100 to 120 million dollars of its annual budget to this, exporting only rare-earth elements. It is necessary first of all to develop production of ferrum-niobium and lithium salts in the Lake Baikal Region. The latter in particular sharply decrease the exhaust of hazardous substances into the atmosphere during aluminium smelting at the Shelekhovo Plant.

High-ash coal is an environmental hazard in the Irkutsk region and in Buryatia. There are gas and oil deposits in the Irkutsk region which provide environmentally pure fuel to boiler plants. The specialists of the Siberian Energy Institute of the Siberian Branch of the Russian Academy of Sciences consider gas and oil as alternatives to coal which may prove favourable to the atmosphere of the region. However, this substitution will require large investment expenditures and is unreal in the near future.

Reorganisation of the Mining-Geological Complex Under the Conditions of Sustainable Development

The present state of the geological and mining services of this country (including the Republic of Buryatia) is unsettling. During the former—Soviet—period, these services created a powerful base of mineral resources, which made it possible to transform this enormous, mainly agricultural country into a first-class industrial power. It is well known that mineral resources are the basis of any modern economically developed society.

Meanwhile, it is becoming clear that the reorganisation of the geological-mining service in this country is inevitably leading to its destruction today. The essentially correct approach, comprising conducting exploration and prospecting, is dependent on mining enterprises (because they must, first of all, be interested in increasing the mineral resources). Under the present conditions of economic decay, including that of the mining industry, this has resulted in a sharp reduction of geological prospecting. The raw-material deficit, combined with its intensive export abroad, which is now going on on a large scale, will soon make this country face the necessity of importing them.

To avoid this prospect, it seems necessary to return to the system of financing geological prospecting projects from the national budget. However, this budget item should be formed by taxes taken from mining enterprises which are now paid by them as the price for the use of mineral resources, their restoration, etc. However, in order to avoid excesses in taxation, it is necessary first of all to thoroughly work out the legislation on the use of mineral resources. And one of the most important problems is, in our opinion, the development of principles of joint use of these resources by the citizens of the Federation and by the centre.

Rather promising seems the idea of setting up complex multi-profile corporations in mining centres that, within definite territories, would be responsible to the municipal or federal authorities for complete utilisation and processing of mineral resources, and for the observance of ecological regulations in their respective areas.

For the purpose of improving and developing the mining industry of the Lake Baikal Region, it seems expedient to give up creating mining giants at present and to be oriented towards setting up small- and medium-size enteprises (up to 100,000 tons of ore per year). As the experience of other countries (Finland, Sweden, Norway) shows, such enterprises turn out to be more profitable and make it possible to exploit small deposits with lower content of useful components in the ore. Besides, they have a much smaller environmental impact on the environment, which is important in an area such as the Lake Baikal Region.

Finally, a preventive measure to keep highly qualified personnel of the geological service, who are now leaving this industry in masses due to the reduction of prospecting projects and the drop in prestige of their profession and also because of the inadequacy of their small salaries to compensate hard labour conditions, may be a change of the profile of this service with a bias to the ecological sphere. In the Lake Baikal Region, there is an acute necessity of developing seismic forecast studies, as it is situated in a zone of high seismic activity. A considerable number of experienced personnel of the former Buryatgeologiya Geophysical Expedition can be successfully employed. In the Republic of Buryatia, a special research programme on seismic safety has existed for more than a year under the aegis of the Government of Buryatia. A successful implementation of this programme is greatly hindered by lack of funding. International co-operation with technologically developed countries of the European community could be of major assistance in this.

Ecological Modernization of Industrial Production

Horst Mierheim
Division of Environmental Instruments
Federal Environmental Agency—Umweltbundesamt
Mauerstrasse 52
D-10117 Berlin
Germany

The Principle of Precaution is a successful strategy to minimize or to eliminate harmful or hazardous substances as a result of industrial production. In order to prevent or to eliminate pollution from industrial production, the application of the use of Best Environmental Practice (BEP) for all sources and Available Technology (BAT) for point sources shall be promoted. To control whether the strategy is sufficient for the environment, Environmental Quality Objectives (EQO) should be defined. If the application of BEP and BAT does not lead to environmental acceptable results, e.g., if the EQO are exceeded, additional measures shall bc applied. To support this strategy, the introduction of an Eco-Management scheme is highly recommended.

1. Introduction

To realize an ecological modernization of industrial production a successful strategy needs to be transformed into environmental policy. The following presentation addresses the German strategy to minimize or to eliminate harmful or hazardous substances. The results will also be discussed in the light of supranational or international development. As one additional instrument to support the ecological modernization of industrial production, the European Union Eco-Audit Order will be taken into consideration.

2. The Precautionary Principle

Environmental policy in the Federal Republic of Germany has the following objectives:

- to protect human health and well-being from harmful effects on the environment and

- to protect flora and fauna as well as soil, air, water and materials from harmful effects resulting from human activities.

To realize these environmental objectives the German environmental strategy to minimize or eliminate harmful substances as a result of industrial production is the

NATO ASI Series, Partnership Sub-Series, 2. Environment – Vol. 6
Sustainable Development of the Lake Baikal Region
Edited by Valentin A. Koptyug and Martin Uppenbrink

application of the Principle of Precaution, which has already been agreed upon in most of the European countries and also outside the borders of Europe. To illustrate what is meant, the 1992 Helsinki Convention, the Convention on the Protection of the Marine Environment of the Baltic Sea Area[1] will be used. This Convention has been agreed upon by all Baltic Sea riparian states. Nearly exactly the same strategy is contained in the Paris Convention[2].

This Convention requires in general terms: The Contracting Parties shall apply the precautionary principle, i.e., to take preventive measures where there is reason to assume that substances or energy introduced, directly or indirectly, into the environment may create hazards to human health, harm to living resources and ecosystems, damage amenities or interfere with other legitimate uses of the environment even when there is no conclusive evidence of a causal relationship between inputs and their alleged effects.[3]

It is important to note that due to the precautionary principle measures are already to be taken when there is "only" reason to assume that emissions or discharges may create hazards to the environment even when there is no conclusive evidence of a causal relationship between inputs and their effects. The reason for such a stringent strategy is the fact that today in many cases it is still not possible to prove that the emission of a special substance relates to a clearly indicated impact on the environment. We will come back to this problem later.

---

[1]Convention on the Protection of the Marine Environment of the Baltic Sea Area, 1992, Helsinki, 9 April 1992 (Helsinki Committee), Bundesgesetzblatt, 1994, II, p. 1397 ff.

See also the similar definition in Art. 2 paragr. 2 (a), Convention for the Protection of the Marine Environment of the North-East Atlantic, Paris, 22 September 1992 (Paris Convention), Bundesgesetzblatt, 1994, II, p. 1360 ff.

[2]Convention of the Protection of the Marine Environment of the Baltic Sea Area, 1992. Helsinki, 9 April 1992 (Helsinki Committee), Bundesgesetzblatt, 1994, II, p. 1397 ff.

[3]This text has been changed by the author into a version related to the environment as a whole. The original text reads as follows: "The Contracting Parties shall apply the precautionary principle, i.e., to take preventive measures when there is reason to assume that substances or energy introduced, directly or indirectly, into the (marine) environment may create hazards to human health, harm to living resources and (marine) ecosystems, damage amenities or interfere with other legitimate uses (of the sea) even when there is no conclusive evidence of a causal relationship between inputs and their alleged effects."

Art. 3 paragr. 2, Convention on the Protection of the Marine Environment of the Baltic Sea Area, 1992. Helsinki, 9 April 1992 (Helsinki Committee), Bundesgesetzblatt, 1994, II, p. 1397 ff.

See also the similar definition in Art. 2 paragr. 2 (a), Convention for the Protection of the Marine Environment of the North-East Atlantic, Paris, 22 September 1992 (Paris Convention), Bundesgesetzblatt, 1994, II, p. 1360 ff.

In Germany, the Federal Immission Control Act of 1974 (BImSchG), amended in 1985 and 1990[4], provides an important legal basis for the implementation of the precautionary principle. Its primary goal is to decrease the total exposure of air pollutants and maintain the existing air quality in areas with relatively clean air. The Federal Immission Control Act is used here to demonstrate the German strategy of industrial modernization in a very suitable way.

3. The German Strategy

In order to achieve these objectives, the BImSchG contains the following two strategies, which are to be applied independently of each other:

- Emission Standards

According to the principle of precaution, emission standards are established to ensure that the Best Available (emission control) Technology (BAT) is applied in each individual case. That means that even in relatively low polluted areas (or waters), emission standards are applied requiring the Best Available Technology (technics, state-of-the-art - BAT).

- Environmental Quality Standards (EQS)

EQS are established to prevent tolerable concentration levels from being exceeded as a result of the construction of new installations or the extension of existing ones. In case of non-fulfillment, the competent authority instructs existing facilities on how to achieve compliance. New facilities are not permitted in that case; to achieve compliance their emissions must bc below the given emission standard according to the BAT.[5]

3.1 Environmental Quality Standards

EQS are established in particular by the Technical Instructions for Air Pollution Control (TA Luft)[6]. The standards are of particular importance for the licensing of industrial plants and the subsequent imposition of standards for existing plants.

There are eight ambient air quality standards for the protection of human health (dust, lead, cadmium, chlorine, hydrochloric acid, carbon monoxide, sulphur dioxide, nitrogen dioxide) and five standards expressly for the protection of flora and fauna against considerable encroachment and substantial impairments (dustfall, i.e.,

---

[4]Federal Immission Control Act (BImSchG), 15 March 1974, 14 March 1994 (Bundesgesetzblatt I, p. 880), 28 April 1993 (Bundesgesetzblatt I, p. 483).

[5]Art. 5 paragr. 1 BImSchG.

[6]Technical Instruction on Air Quality Control (TA Luft), 27 February 1986, Gemeindeblatt, p. 95.

non-hazardous dust; lead; cadmium; thallium; and hydrofluoric acid). Some eight additional air-quality standards are under discussion.

Environmental Quality Objectives for water and soil are also under consideration at the federal level.

However, a quality-control strategy based on a few EQS for some pollutants alone would not be sufficient. For the majority of pollutants, no quality standards exist because of the insufficient knowledge of the risks and effects of these pollutants. In particular, long-term effects and combination effects of several pollutants are very difficult to ascertain. Sciences that concentrate on the effects of pollutants can often provide no established answers to the question: how much pollution can the environment bear without damage?

3.2 Implementation of the Precautionary Principle by Using Best Available Technology—BAT

This dilemma is one of the important reasons why the principle of precaution has become the main basis for the German environmental strategy.

The principle of precaution serves to achieve the following purposes in particular:

- to develop planning tools that provide a safe margin from present immission limits so that emissions from future plants can be allowed;

- to preserve relatively clean and not yet polluted areas;

- to fend off risks as a precaution, i.e., to take action even where danger is only suspected;

- to reduce risks independent of danger and disturbances, e.g., as a prevention against lacking or uncertain knowledge.

On the basis of a general concept, prevention is required for all emitters at any time and at any place mainly because of the long-range dispersion of pollutants and of the difficulty or impossibility of attributing environmental pollution to one individual polluter. This concept is based on the BAT mentioned above. According to this principle, it is mandatory for all sources which are comparable in terms of size and type to apply the best available (emission reduction) technology (which is technically feasible and appropriate). The Federal Immissions Control Act, Art. 3 paragraph 6, demonstrates clearly this very dynamic claim: "The latest state of technology within the meaning of this law is the stage of development of progressive processes of facilities or of methods of operation which indicate the practical suitability of a particular measure for limiting emissions. In determining the latest state of technology, special consideration shall be given to comparable processes, facilities or methods of operation which have been successfully tried out"[7]. This requirement has been incorporated as a result of scrupulous deliberations.

These requirements are valid for new and—much more importantly—for existing facilities. Most emissions are generated by existing facilities. Therefore, it is very important to reduce the emissions of those existing sources in particular.

The most important instrument to enforce the BAT is the licensing of sources. The establishment and the operation of installations which are particularly liable to cause harmful effects on the environment or otherwise endanger or cause considerable detriment or considerable disturbance to the general public or the vicinity shall be subject to licensing[8].

For the implementation of the Federal Immission Control Act the Technical Instructions for Air Pollution Control (TA Luft) of 1974, amended in 1983 and 1986, are very important. They provide instructions for the licensing authorities in the German Federal States (Länder).

The Instructions spell out licensing requirements for installations of particular environmental significance. They include specific regulations:

1. ambient air quality standards for the most important air pollutants

2. emission standards for air pollutants and types of installations, which reflect the state of the art of emissions control

3. specific procedures for the monitoring, assessment and evaluation of immissions and emissions

4. methods to calculate the required stack height

5. authorization to establish secondary rules and regulations for improvement of existing installations according to BAT.

The Federal Water Act, notably in its Art. 7a,[9] requires also the application of the best available technology to reduce hazardous substances at their source. Administrative regulations for different industries are issued to reduce emissions from

---

[7]See also Helsinki Convention, Art. 3 paragr. 3, Regulation 3 and Annex II, in Annex A of this paper, and Paris Convention, Annex II.

[8]Federal Immission Control Act, Art. 4, paragr. 1. The respective sources are enumerated in the Ordinance on Sources that are subject to licensing (4th Ordinance on the Federal Immission Control Act).

[9]Art. 7a, Requirements to be met by waste water discharges, Federal Water Act—WHG, 27 July 1957, 23 September 1986 (Bundesgesetzblatt 1986, I, p. 1529), 26 August 1992 (Bundesgesetzblatt 1992, p. 1564).

point sources by stipulating emission standards for the pollution level or concentration.

A Waste Water Charge supports the implementation of reduction measures[10].

3.3 Best Environmental Practice (BEP)

BEP is taken to mean the application of the most appropriate combination of measures to protect the environment, such as saving resources, avoiding waste, information and education, a system of licensing of processes and environmentally relevant products, designing products to be environmentally sound during their whole lifetime, from the cradle to the grave[11].

Compared to BAT, which is mainly for the reduction of emissions from point sources, BEP shall be used for all kinds of sources[12]. BEP therefore covers a broad range of measures. It is a very new strategy for combining the acquirement of an environmentally sound product and substance policy with the strategy of the production process.

3.4 The Relation Between the Principle of Prevention and Environmental Quality Standards

A central issue of environmental strategy is the relationship between BEP and BAT and Environmental Quality Standards. If the application of BAT and BEP do not lead to environmentally acceptable results, additional measures shall be applied.[13] This means that to meet the Environmental Quality Standards it could be necessary to do more than to use only BAT and BEP; otherwise no license for operation will be granted by the authority. The reasoning behind this is that in the case of not adhering to the Quality Standards, the authority has to prevent danger to human health and the environment: the principle of prevention is superseded by the fending off of danger, which has higher priority.

In the European Union an EU Directive is under negotiation at present with nearly the same provisions.

---

[10]Waste Water Charges Act, 6 November 1990 (Bundesgesetzblatt 1990, I, p. 2431) and amended in July 1994.

[11]See Helsinki Convention, Annex II, Annex A of this paper.

[12]In detail see Annex A of this paper.

[13]Art. 3 paragr. 3 Helsinki Convention and Appendix I paragr. 4 Paris Convention.

How does the term Clean Production fit into this scheme?

The term Clean Technology is used very often to define a very advanced environmental policy to reduce pollution, but there is no broadly accepted definition. So during negotiations to modernize the Helsinki Convention and the Paris Convention, it was agreed that the use of BAT and BEP is equivalent to the provisions of Clean Technology.

4. Environmental Management in Companies—Eco-Management

4.1 A Challenge for Companies

Following environmental regulations was the only environmental action taken by industrial management for a long time. But in recent years more and more companies are establishing eco-management schemes at the top management level. After long-established cost and quality management, eco-management is becoming the third important pillar of management structure in companies today.

This is why the environment now plays an important part in company strategy as a result of high environmental standards and consumer awareness, but also because the public in general has shown a keen interest in how industry has been dealing with pollution abatement. Companies in Germany acted particularly quickly to seize this opportunity to cut costs and increase returns. On all five continents, industrial initiatives to improve the environment can be observed wherever the entrepreneurial advantages of pro-active eco-management have in the meantime been understood.[14]

Environmental protection in companies does not necessarily raise production costs.

There are many practical examples to show that it can also lead to more general improvements in cost efficiency. The main potential for cost reduction lies in the fields of energy, packaging, procurement, waste, production processes, and sewage. In the energy sphere in particular much potential remains to be exploited. In spite of the expense, studies have revealed over and over again that environmental measures often pay for themselves within a year.

Company strategies prove particularly effective and rewarding if they are designed to prevent or reduce risks. We all know from experience that unforeseen accidents and/or a neglect of pollution problems can quite suddenly impose a large financial burden on the company concerned. The bill might include anything from loss of production to rectifying the failure, and even restoring the company image. This last cost factor applies when expensive advertising campaigns are needed to convince

---

[14]Federal Environmental Agency, Umweltbundesamt (de.): Environmental Protection—an Economic Asset: Seven arguments for not selling a lead in environmental protection ... and what we think of them; Umweltbundesamt Postbox 33 00 32, D-14191 Berlin, S. 77 ff.

clients and the general public that an industrial accident or failure does not merit the severe condemnation it is receiving.

Not least, a consistent pursuit of environmental principles will bring opportunities, as long as action forms a part of a long-term corporate strategy. If company decision - makers begin early to integrate imminent regulations into their planning, they will have a head start over their rivals when it comes to exploiting their pioneer role, acquiring experience, transforming this into innovation, and ultimately consolidating the future of their enterprise (which sooner or later will no longer be possible if environmental criteria are ignored).

4.2 The European Union Eco-Audit Order

The realization that environmental protection is not a luxury, but a key factor in competition is not confined to Germany alone. The European Union, for example, attaches considerable importance to ecologically informed management. In the spring of 1993, the EU environment ministers approved a regulation which envisages the "voluntary participation by companies in the industrial sector in a Community eco-management and audit scheme". The Eco-Audit Regulation came into force on 13 July 1993[15] and will allow companies from the Member States, initially from the industrial sector, to take part. In substantiating the need for this regulation, the Council refers to "a Community programme of policy and action in relation to the environment and sustainable development". It also "underlines the role and responsibilities of companies both to reinforce the economy and to protect the environment throughout the Community".

Because of the proposed advantages for the companies economically and the protection of the environment, the expectations in the results of the EU Eco-Audit Order are high.

4.3 The European Union Eco-Management and Audit Scheme and Its Objectives[16]

The objective of the EU Eco-Audit Order scheme shall be to promote continuous improvements in the environmental performance of industrial activities by:

- the establishment and implementation of environmental policies, programs and management systems by companies, in relation to their sites;

---

[15]Council Regulation (EEC) No. 1836/93 of 29 June 1993 allowing voluntary participation by companies in the industrial sector in a community eco-management and audit scheme, 10.07.93 Official Journal of the European Communities No. L 168/1 EU ECO Audit Order

[16]See Art. 1 EU Eco-Audit Order

– the systematic, objective and periodic evaluation of the performance of such elements;

– the provision of information of environmental performance to the public.

The intention of the EU Eco-Audit is in brief to standardize the basis for assessing and improving pollution-abatement strategies in European companies.

The regulation is already legally binding. However, it cannot fully go into effect until the Member States have taken two measures towards implementation, which they are obliged to do by April, 1995. One is to establish a system for the "accreditation of independent environmental verifiers" and the other is to designate a "competent body" in their own country which will above all be responsible for registering validated sites.[17]

Such benefits as a result of an Eco-Audit could be:

– stimulating environmental protection,

– visible guarantees of compliance with regulations,

– less risk of legal disputes,

– fewer fines and penalties for non-compliance,

– improved statistics on environmental incidents,

– reduced environmental risks,

– a healthier work force,

– reduced risk of production loss due to an environmental incidents,

– lower insurance premiums based on environmental risk assessment,

– more motivated work force,

– better relations with authorities,

– indications of potential savings (e.g., by reducing energy and water consumption),

– more efficient identification of fruitful contributions to pollution abatement which deserve to be rewarded,

– greater awareness of environmental protection and responsibility among the work force,

---

[17]See Art. 6 EU Eco-Audit Order: Member States shall each establish a system for the accreditation of independent environmental verifiers and for the supervision of their activities and shall ensure that the composition of these systems is such as to guarantee their independence and neutrality in the execution of their tasks.

– more efficient identification of staff training needs,

– an important contribution to the company system of environmental information,

– greater public credibility on environmental issues.

4.4 Issues to Be Covered by the EU Eco-Audit Order

The answer to the question why these benefits could arise by applying the Eco-Audit will become clearer if we take a closer look at the issues to be addressed by the EU Regulation within the framework of environmental policy and programs and of environmental audits:[18]

1. Assessment, control, and reduction of the impact of the activity concerned on the various sectors of the environment.

2. Energy management, conservation and choice.

3. Raw-materials management, savings, choice and transportation; water management and conservation.

4. Waste avoidance, recycling, reuse, transportation and disposal.

5. Evaluation, control and reduction of noise within and outside the site.

6. Selection of new production processes and changes in production processes.

7. Product planning (design, packaging, transportation, use and disposal).

8. Environmental performance and practices of contractors, subcontractors and suppliers.

9. Prevention and limitation of environmental accidents.

10. Contingency procedures in cases of environmental accidents.

11. Staff information and training in environmental issues.

12. External information on environmental issues.

Most of the above items are very close to the criteria of the best environmental practice, BEP, which was presented above. Nevertheless, a very remarkable paragraph is number eight, because this issue not only touches on the site of the

[18] Annex I, paragr. 3. EU Eco-Audit Order.

company, it also stipulates requirements of business partners of the company. Beside national traders and contractors, international ones could also be affected by this regulation in such a way that stringent implementation by the company's business partners could be achieved by this multiplier rule.

4.5 Good Management Practices

To achieve the objectives of the Eco-Audit scheme, it is necessary to implement an excellent system of good management practices. It is noteworthy that the EU Eco-Audit Order covers such a requirement in a very advanced way:[19]

1. A sense of responsibility for the environment amongst employees at all levels shall be fostered.

2. The environmental impact of all new activities, products and processes shall be assessed in advance.

3. The impact of current activities on the local environment shall be assessed and monitored, and any significant impact of these on the environment in general shall be examined.

4. Measures necessary to prevent or eliminate pollution and, where this is not feasible, to reduce pollutant emissions and waste generation to the minimum and to conserve resources shall be taken by utilizing available clean technologies.

5. Measures necessary to prevent accidental emissions of materials or energy shall be taken.

6. Monitoring procedures shall be established and applied to check compliance with the environmental policy and, where these procedures require measurement and testing, to establish and update records of the results.

7. Procedures and action to be pursued in the event of detection of non-compliance with its environmental policy, objectives or targets shall be established and updated.

8. Cooperation with the public authorities shall be ensured to establish and update contingency procedures to minimize the impact of any accidental discharges to the environment that nevertheless occur.

9. Information necessary to understand the environmental impact of company activities shall be provided to the public, and an open dialogue with the public should be pursued.

---

[19]The company's environmental policy shall be based on the principles of action set out below; the activities of the company shall be checked regularly to see if they are consistent with these principles and that of continual improvement in environmental performance.

10. Appropriate advice shall be provided to customers on the relevant environmental aspects of the handling, use and disposal of the products made by the company.

11. Provisions shall be taken to ensure that contractors working at the site do so at the behest of the company.

5. Conclusions

To realize an ecological modernization of industrial production the Principle of Precaution is a successful strategy to minimize or to eliminate harmful or hazardous substances as a result of industrial production. In order to prevent or to eliminate pollution from industrial production, the application of the use of Best Environmental Practice (BEP) for all sources and Best Available Technology (BAT) for point sources shall be promoted.

To control whether the strategy is sufficient for the environment, Environmental Quality Objectives (EQO) should be defined. If the application of BEP and BAT does not lead to environmentally acceptable results, e.g., if the EQO are exceeded, additional measures shall be applied.

To support this comprehensive strategy, the introduction of an Eco-Management scheme is highly recommended. A combination with economic instruments, environmental impact assessment and environmental planning is possible and depends on the individual case.

Nevertheless, the implementation of this environmental strategy for the ecological modernization of industrial production could be an important step torwards achieving a sustainable development of our environment.

References

Convention for the Protection of the Marine Environment of the North-East Atlantic, Paris, 22 September 1992 (Paris Convention), Bundesgesetzblatt, 1994, II, p. 1360 ff.

Convention on the Protection of the Marine Environment of the Baltic Sea Area, 1992, Helsinki, 9 April 1992 (Helsinki Committee), Bundesgesetzblatt, 1994, II, p. 1397 ff.

Council Regulation (EEC) No.1836/93 of 29 June 1993 allowing voluntary participation by companies in the industrial sector in a community eco-management and audit scheme, 10.07.93 Official Journal of the European Communities No. L 168/1 EU Eco-Audit Order.

Federal Environmental Agency, Umweltbundesamt (ed.): Environmental Protection—an Economic Asset: Seven arguments for not selling a lead in environmental protection ... and what we think of them; Umweltbundesamt Postbox 33 00 32, D-14191 Berlin.

Federal Immission Control Act (BImSchG), 15 March 1974, 14 March 1994(Bundesgesetzblatt I, p. 880), 28 April 1993 (Bundesgesetzblatt I, p. 483).

Federal Water Act—WHG, 27 July 1957, 23 September 1986 (Bundesgesetzblatt 1986, I, p. 1529), 26 August 1992 (Bundesgesetzblatt 1992, p. 1564).

Technical Instruction on Air Quality Control (TA Luft), 27 February 1986,GMBl. p. 95.

Waste Water Charges Act, 6 November 1990 (Bundesgesetzblatt 1990, I, p. 2431) and amended in July 1994.

ANNEX A[20]

Criteria for the Use of Best Environmental Practice and Best Available Technology

Regulation 1; General Provisions

1. In accordance with the relevant parts of this Convention, the Contracting Parties shall apply the criteria for Best Environmental Practice and Best Available Technology described below.

2. In order to prevent and eliminate pollution, the Contracting Parties shall use Best Environmental Practice for all sources and Best Available Technology for point sources, minimizing or eliminating inputs to water and air from all sources by providing control strategies.

Regulation 2; Best Environmental Practice

1. The term "Best Environmental Practice" is taken to mean the application of the most appropriate combination of measures. In selecting for individual cases, at least the following graduated range of measures should be considered:

- provision of information and education to the public and to users about the environmental consequences of choosing particular activities and products, their use and final disposal;

- the development and application of Codes of Good Environmental Practice covering all aspects of activity in the product's life;

- mandatory labels informing the public and users of environmental risks related to a product, its use and final disposal;

- availability of collection and disposal systems;

- saving of resources, including energy;

- recycling, recovery and re-use;

- avoiding the use of hazardous substances and products and the generation of hazardous waste;

[20]Annex II, Convention on the Protection of the Marine Environment of the Baltic Sea Area, 1992, Bundesgesetzblatt 1994, II, p. 1413 f.

– application of economic instruments to activities, products or groups of products and emissions;

– a system of licencing involving a range of restrictions or a ban.

2. In determining in general or individual cases what combination of measures constitute Best Environmental Practice, particular consideration should be given to:

– the precautionary principle;

– the ecological risk associated with the product, its production, use and final disposal;

– avoidance or substitution by less polluting activities or substances;

– scale of use;

– potential environmental benefit or penalty of substitute materials or activities;

– advances and changes in scientific knowledge and understanding;

– time limits for implementation;

– social and economic implications.

Regulation 3; Best Available Technology

1. The term "Best Available Technology" is taken to mean the latest stage of development (state of the art) of processes, of facilities or of methods of operation which indicate the practical suitability of a particular measure for limiting discharges.

2. In determining whether a set of processes, facilities and methods of operation constitute the Best Available Technology in general or individual cases, special consideration should be given to:

– comparable processes, facilities or methods of operation which have recently been successfully tried out;

– technological advances and changes in scientific knowledge and understanding;

– the economic feasibility of such technology;

– time limits for application;

– the nature and volume of the emissions concerned;

– non-waste/low-waste technology;

– the precautionary principle.

Regulation 4; Future developments

It follows therefore that "Best Environmental Practice" and "Best Available Technology" will change with time in the light of technological advances and economic and social factors, as well as changes in scientific knowledge and understanding.

"Castling": A Possibility of Partial Reorganisation of the Economy of the Baikal Region for the Sake of Sustainable Development

M.A. Grachev, N.A. Aldokhin, A.N. Suturin,
Limnological Institute of SB RAS, Irkutsk, Russia

A.P. Sukhodolov
Institute of Economics of RAS, Moscow, Russia

One of the most difficult problems in the core of the future World Natural Heritage Site, the Baikal Region, is the problem of reprofiling the Baikalsk Pulp and Paper Plant (BPPP), with a capacity 160,000 tons per year. In 1990, the Selenginsk Pulp and Cardboard Plant (SPCP) sealed off the cycle of water consumption and ceased to discharge waste water into the Selenga River, the major tributary of Lake Baikal.

It is suggested that the possibility of increasing the capacity of SPCP to 400,000 tons be considered by partially shifting it to production of bleached cellulose (using new chlorine-free technologies) on the condition of sustaining the fully closed water-consumption cycle, an at least ten-fold decrease of air pollution compared to the present and full utilisation of solid wastes.

As for Baikalsk City, it is suggested that two major sectors be developed after BPPP is closed: the production of bottled water, beverages, beer, soft drinks and food products using the ultra-pure deep water of Lake Baikal, and tourism.

These activities conform with the principles of sustainable development, and are highly profitable. One more proposed activity is the production of prefabricated houses, furniture, and saw timber.

Problems of the Region and the Task of Sustainable Development

Sustainable development of the Baikal Region requires simultaneous progress in two directions:

– increase of the economic potential and with it the standard of living;

– improvement of the environmental situation with the purpose of protection of Lake Baikal—a future World Heritage Site.

Is the achievement of these two largely conflicting goals possible?

On what basis can the economy develop sustainably without having a negative impact on the environment?

NATO ASI Series, Partnership Sub-Series, 2. Environment – Vol. 6
Sustainable Development of the Lake Baikal Region
Edited by Valentin A. Koptyug and Martin Uppenbrink

What branches of the economy can ensure the most rapid stabilisation and growth of the economy in the region?

Finally, how to solve the problem of the Baikalsk Pulp and Paper Plant, a major source of pollution of Lake Baikal?

Ecological and economic stabilisation in the Baikal Region depends on the solution of these problems.

Our analysis and preliminary estimates show that the above problems can be solved. The region has potential for more dynamic development. This can be achieved by means of restructuring the leading branches of the economy and reorganisation of regional economical relations ("castling" of stocks, of material and financial resources).

One such branch is the wood-processing industry, which yields 25% of all the production of the region. Today, due to the economic crisis, this share has fallen to 18.8%.

A few years ago, the Siberian Branch of the Russian Academy of Sciences took part in the complete sealing of the water-consumption cycle at the Selenginsk Pulp and Cardboard Plant, which had previously been a major source of pollution of the Selenga River, the main tributary of Lake Baikal (this project won a State Award of the Russian Federation in 1994). Introduction of new, no-effluent technology in the pulp-and-paper industry opens up new opportunities for the wood-processing sector of the economy of the region.

Using the resources of the region, appropriate reorganisation ("castling") of economic relationships makes it possible to solve the problem of the Baikalsk Pulp and Paper Plant. Implementation of this plan results in a two-fold reduction of the discharge of pollutants into the Lake and leads to a radical improvement of the ecological situation due to complete elimination of the most dangerous pollutants, as well as of the risk of ecological catastrophes.

Restructuring and development of the wood-processing sector of the economy is in line with the general tendencies in pulp-and-paper industry of the world. This branch provides a good basis for sustainable economic growth.

Prerequisites of Implementation

The Russian part of the water catchment basin of Lake Baikal has an area of 300,000 sq. km on the territory of the Irkutsk (4.9%) and the Chita (18.9%) Districts, and of the Republic of Buryatia (76.9%).

A major part of the forest resources of the basin is located outside the water-protection coastal zone of the Baikal Region (the "core" zone), on the territories of the Republic of Buryatia and Chita District. Industry in the Baikal Region amounts to 78.6% and agriculture 21.4% of gross production. Wood-processing is one of the most important industries in terms of both the number of employees and the total income (18.8%).

It is quite obvious that economic welfare of the territory in the very near future will be based exclusively on the available local industrial potential, including the wood-processing industry. Sustainable development of this sector is feasible, since there are all the necessary prerequisites (raw materials, production facilities, qualified manpower). Stabilisation of the wood-processing sector is the most important condition of the development of the region as a whole. On the other hand, extensive development of the wood-processing industry in the basin, if it is based exclusively on increased felling, may lead to an impact on Lake Baikal by destroying the potential of the forests to protect the environment.

Reorganisation of the the wood-processing sector should therefore involve diversification of its final products, including increased production of cellulose, cardboard and cardboard boxes. At present, up to 50% of the wood cut in the region is dumped or burnt at the felling sites.

The mean annual volume of wood cutting is 5 to 7 million cubic meters; hence, some 2 million cubic meters of low-quality wood which is now wasted could be re-processed to produce cellulose and other articles, increasing the efficiency of the wood-processing industry as a whole.

Let us consider briefly the environmental impact of the wood-processing sector. Figs.1 and 2 show its share in the total discharge of pollutants into the water and air. It is evident that the major impact is discharge into water (59.2%); the contribution to air pollution is smaller (8.7%).

The Baikalsk Pulp and Paper Plant yields half of the total volume of waste water discharged into the bodies of water in the basin (about 80 million cubic meters per year). BPPP is the only large industrial facility sited directly on the shore of the unique lake. It does not meet ecological and aesthetic requirements and contradicts the idea of including Baikal in the list of World Heritage Sites, as well as the draft of the Law of the Russian Federation on Lake Baikal.

Another cellulose-producing facility, the Selenginsk Pulp and Cardboard Plant, is situated at a distance of 70 km off the shore of Lake Baikal, outside the central zone ("core"). In August, 1990, the plant successfully introduced a completely

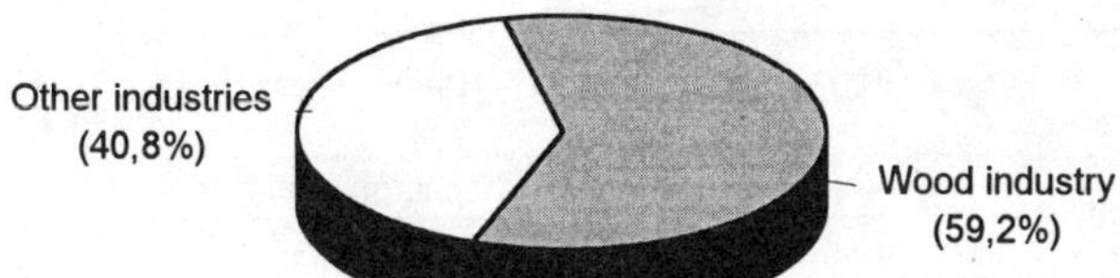

Fig.1. The contribution of wood industry into water contamination in the catchment basin of Lake Baikal.

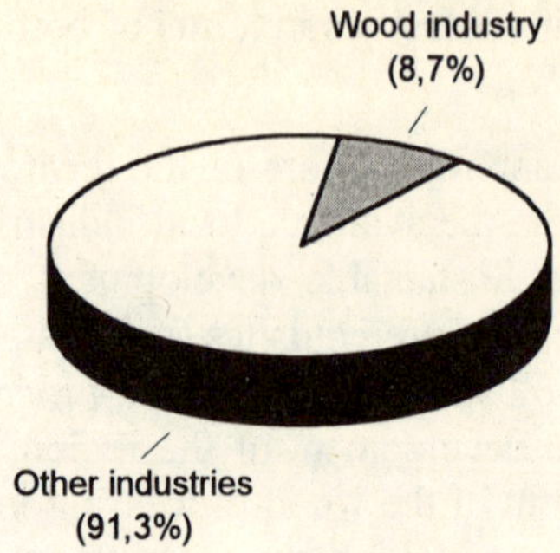

Fig.2. The contribution of wood industry into air pollution in the catchment basin of Lake Baikal.

closed water-consumption cycle, and the discharge of waste waters into the Selenga River, the major tributary of Lake Baikal, was stopped (fig.3). Discharge of pollutants into the air has also been considerably reduced due to capital investments in technology.

The general economic crisis in Russia has resulted in a considerable decrease of production. To increase the economic efficiency of the plant, and to achieve more complete processing of wastes and low-quality wood in the region, it is necessary to increase the capacity of the Selenginsk Pulp and Cardboard Plant 1.5 to 2 times. This will make it possible to improve economic efficiency and will help to overcome the recession and to satisfy the regional demands for production of cardboard containers.

The necessity of restructuring the wood-processing sector of the region is also determined by the fact that export of wood products by Russia differs from that of

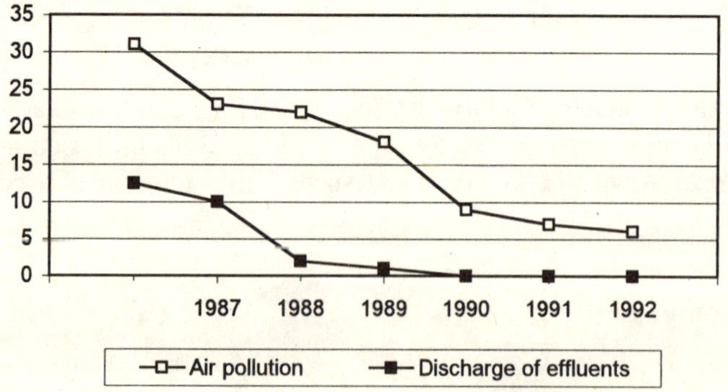

Fig.3. Dynamics of the decrease of discharge of pollutants into the atmosphere (thousand t/year) and effluents (mln cubic meters/ year) by the Selenginsk Pulp and Cardboard Plant.

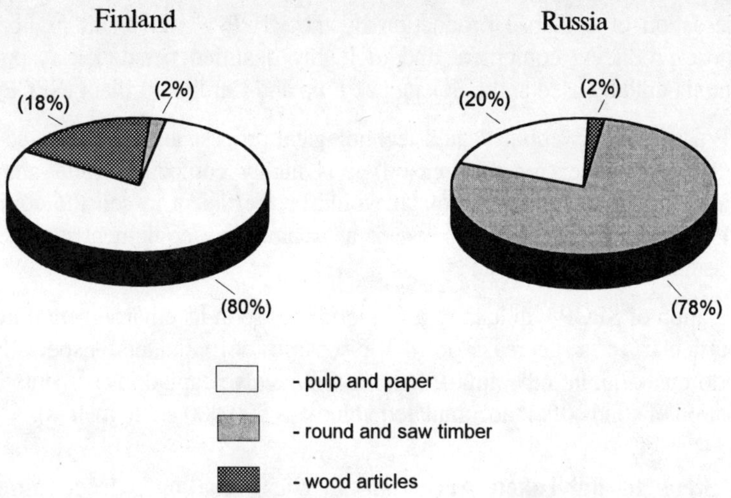

Fig.4. Structure of the export of wood products (1993).

other developed countries in its structure. For example, round timber and saw timber occupied the leading place (about 80%) in the export of the Russian wood-processing complex in 1993. The share of pulp and paper products was less than 20%. This considerably lowers the export potential, adversely influences the socio-economic situation in this economic sector in particular, and in forest-cutting territories in general. The share of pulp and paper products in the export to industrially developed countries, e.g., Finland (fig.4), is 80%, that of round and saw timber only 2%.

Hence, development of the wood-processing sector according to the proposed programme will improve the structure of the export of the region.

The Idea of "Castling"

According to this idea, it is proposed to close the Baikalsk Pulp and Paper Plant and to increase the capacity of the Selenginsk Pulp and Cardboard Plant on the condition that the water-consumption cycle of the latter will remain closed. This "castling" will bring about re-organisation of some economic relations in the Baikal Region in accordance with the concept of sustainable development.

"Castling" was promoted by the following circumstances:

1.BPPP must stop production of cellulose on the shore of Lake Baikal for the sake of the protection of the Lake. This would lead to serious socio-economic consequences if special measures are not taken to create new jobs in Baikalsk.

The problem of BPPP cannot be solved by Baikalsk alone; it requires involvement of the economic complex of the whole region.

2. Cessation of cellulose production by the BPPP would result in liberation of cellulose-producing equipment and of highly qualified personnel. A part of this potential could be used at the Selenginsk Pulp and Cardboard Plant (SPCP).

BPPP uses a highly sophisticated technological process and can produce any kind of sulfate cellulose; for this reason, it is highly competitive and adaptable to requirements of market economy. It would be inefficient to sell the equipment of BPPP outside the region or to use it as scrap. This equipment can be used to increase the facilities of SPCP.

Renovation of SPCP will lead to a further decrease in its environmental impact and in particular to a decrease in the discharge of pollutants, especially of the malodorous kind, into the atmosphere. It will involve capital investments in the full utilisation of solids (their accumulated volume is 500,000 cubic meters).

Main Steps to Be Taken According to the "Castling" Programme at the Baikalsk Pulp and Paper Plant (BPPP)

After closure of BPPP, it is proposed for Baikalsk to develop a food industry, including production of bottled Baikal drinking water (100 million bottles per year), of alcohol and alcoholic drinks, of beer and other beverages; mechanical wood-processing (100,000 cubic meters per year), including production of furniture and ready-built houses (800 houses per year); tourism (50,000 man-nights per year).

The Republic of Buryatia, not only the Irkutsk District, could be directly involved in the development of these sectors. Auxiliary activities, like production of plastic bottles from polychlorovinyl, polyethylene and polyethylene terephthalate supplied by the Angarsknefteorgsintez and Sayanskkhimprom Companies, could be also developed. Production of high-quality glass bottles with raw material obtained from the Cheremshanskoe quartzite deposit (300 million bottles per year) can take place at the Ulan-Ude Glass Plant. These auxiliary products could be used to pack water, drinks and beverages produced in Baikalsk.

Fly ash produced by the coal-burning power plant of BPPP (20,000 t/year) may be transported to the Timliui Cement Plant (200 km) and used as an additive to cement. Wastes of alcohol production can be utilised as fodder at a pig-breeding farm in Mostovka (250 km).

Environmental and economic indices before and after BPPP reprofiling are given in fig.5 and Table 1.

The necessary tentative capital investments are $120 to 150 million, the number of employees in Baikalsk will be 3,000 to 3,500 persons.

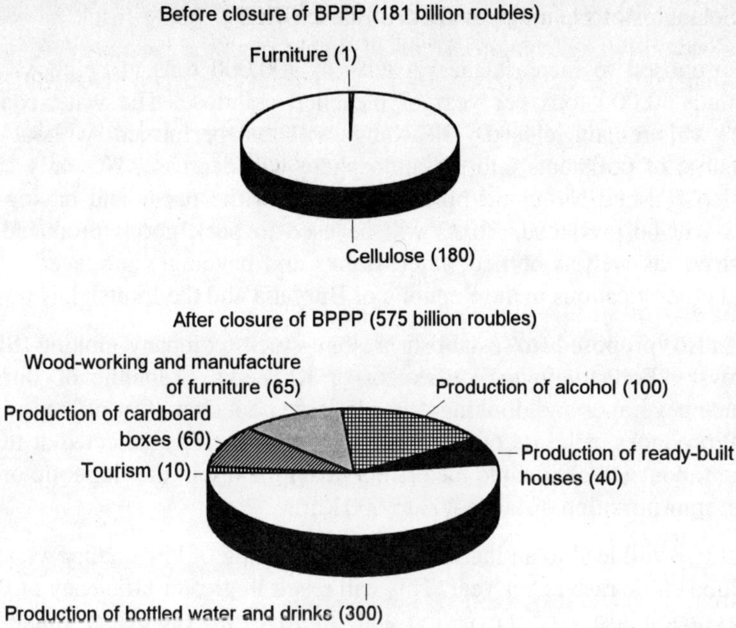

Fig.5. Production in Baikalsk.

Table 1 Ecologo-economic indices before and after closure of the Baikalsk Pulp and Paper Plant (according to the "Castling" proposal)

Indices	Before	After	+/- (times)
1. Volume of production(billion roubles/year, in prices of 1994)	200	575	+3
2. Number of employees (persons)	4000	3500	-1.1
3. Water consumption (million cubic meters per year)	80	6	-13
4. Sewage volume (million cubic meters per year)	80	5.2	-15
5. Discharge of aerial pollutants (thousand tons per year),	14	2.2	-6
6. Discharge of aerial pollutants including malodorous compounds (tons per year)	450	0	-
7. Demand for forest felling (thousand hectares per year)	12	2	-6

The Selenginsk Pulp and Cardboard Plant (SPCP)

It is proposed to increase the capacity to 400,000 tons of cellulose per year, including 60,000 tons per year of bleached cellulose. The water-consumption cycle will remain closed. Bleaching will be performed without chlorine. Discharge of pollutants into the atmosphere will decrease. Not only cardboard, but also finished cardboard boxes lined with white paper and having coloured labels will be produced. Boxes will be used to pack goods produced by local industries, as well as bottled water, drinks and beverages produced in Baikalsk and in other locations in the Republic of Buryatia and the Irkutsk District.

It is also proposed to establish a joint-stock company uniting SPCP and Kluyevsky Forestry in the settlement of Kluyevka (Republic of Buryatia), to produce beverages and drinking water in large containers (for offices and hotels). SPCP produces cellulose of low-quality wood and chips collected at the sites of felling (about 800,000 cubic meters per year) mainly in the Republic of Buryatia (mean transportation distance 400 to 600 km).

"Castling" will lead to an increase of the processing of low-quality wood equal to 2 million cubic meters per year. This will result in greater efficiency of the wood-processing industry of the region and improve the ecological situation in the forests of the basin.

Increasing the capacity of SPCP will also lead to a greater consumption of limestone of the Zaigraevskoye Deposit (Republic of Buryatia), and of chemicals produced in the Irkutsk region in the cities of Angarsk, Usolie-Sibirskoye, Sayansk (sulfuric acid, alkali, sodium sulfate, ammonia, hydrogen peroxide, polyacrilamide, superphosphate).

Solid wastes—ash of the power plant, bark, lignin—will be used to produce compost for the agriculture of the Republic of Buryatia. This will help to improve eroded soils and to avoid the planned extraction of peat near the shore of Lake Baikal. All the necessary conditions for construction works in Selenginsk are available. The terrain is flat and convenient for the siting of new buildings. All the necessary local construction materials may be obtained at a distance of less than 15 km. There is a Selengapromstroy Construction Company, with a capacity of 1 million dollars per year, in Selenginsk. The necessary capital investments are $80 to 100 million.

Plan of Action

The programme proposed here considers the situation only in general and gives an outline of the prerequisites of "castling". An extended and much more detailed specification should follow if the idea is to be accepted.

The Siberian Branch of the Russian Academy of Sciences and the Government Commission on Lake Baikal are requested to establish a working group for this purpose. It is also suggested to draft a Decree of the President of the Russian Federation for implementation of attractive conditions for investments, e.g., of a free economic zone in the Baikal Region for the sake of its sustainable development and of the protection of Lake Baikal.

References

1. Goncharov, A.I., Aldokhin, N.A. et al., Programme of Transformation of the Selenginsk Pulp and Cardboard Plant into an Environmentally Friendly One. Moscow, "Bumazhnaya Promyshlennost", no. 6, 1990, p. 6 (in Russian).

2. Aldokhin, N.A., Goncharov, A.I., Grachev, M.A., Suturin, A.N., Recycling of Waste Water and Solid Waste at the Selenginsk Pulp and Cardboard Plant. UNEP Industry and Environment. July–December, 1990, p. 21.

3. Trifonov, G.A., Aldokhin, A.N., Sugonyaev, Yu.V. et al., A Closed-water Circulation Method in Pulp and Cardboard Production. Russian Patent no. 1773270 of 18 Dec. 1989.

4. Grachev, M.A., Aldokhin, N.A., Goncharov, A.I., Suturin, A.N., Closure of the Water-consumption Cycle and Recycling of Solid Wastes at the Selenginsk Pulp and Cardboard Plant. Abstracts of the Conference: CHEMRAN VIII—Chemistry and Sustainable Development (IUPAC). Moscow, 1992, p. 85.

5. Sukhodolov, A.P., Samarukha, V.I., Ecology and Economy in the Water-catchment Area of Lake Baikal. Institute of Economics and Industrial Engineering of SB RAS, Irkutsk, 1992, 96 pp. (in Russian).

[illegible]

References

1. [illegible]

2. [illegible]

3. [illegible]

4. [illegible]

5. [illegible]

Problems of Sustainable Development of the Agricultural Complex in the Lake Baikal Region

R.K. Salyaev
Siberian Institute of Plant Physiology and Biochemistry of SB RAS, Irkutsk, Russia

V.M. Korsunov
Buryat Institute of Biology of SB RAS, Ulan-Ude, Russia

The state of the agricultural complex of the Lake Baikal Region is examined and a number of anticrisis measures contributing to its transfer to the regime of sustainable development are given. Recommendations for an ecologization of agriculture are developed based on ecologically optimum land use.

The sustainable development of the region implies sustainable use of natural resources on the condition of optimum supply of various goods to the population. It is evident that the sustainable development of any region is impossible without environmental protection.

As applied to our topic, such development of the agricultural complex (ACC) of the Lake Baikal Region could provide a sufficient supply of diverse foodstuffs to the population, with maximum preservation of non-renewable natural resources, of unique natural complexes, and of biological diversity.

It is certainly impossible to present here a comprehensive plan of the system of sustainable development of the agricultural complex; this task requires the long-term work of various specialists.

An attempt is made in the paper to analyze and interpret possible ways for stabilizing the ACC under new economic conditions and of its transition to sustainable development based on the ecologization of land use.

Analysis of the ACC activity under the new economic conditions has shown that the agriculture of the region is in a critical situation caused by disruption of the former links of the ACC with industry, banks, trade networks, and the population. For instance, in the main agricultural area of the Lake Baikal Region (Buryatia), grain yield has decreased to 500–700 kg/ha, a large part of pastures and hay fields is heavily degraded, idle land area has increased and cattle livestock reduced by almost half because of forage shortage. The percentage of eroded land is up to 30% in the Irkutsk district, and more than 60% in Buryatia. In Buryatia the total length of gullies is about 9,000 km. In 1993 alone, the number of cattle decreased by 18%, sheep by 17%, pigs by 24%, and horses by 7%; the production of milk fell by 30%, of eggs by 14%, of wool by 22%, etc. The gross yield of agricultural production decreased by

NATO ASI Series, Partnership Sub-Series, 2. Environment – Vol. 6
Sustainable Development of the Lake Baikal Region
Edited by Valentin A. Koptyug and Martin Uppenbrink

35% on an average in the Lake Baikal Region. Profitability of all kinds of agricultural production has decreased. Most agricultural lands are experiencing loss of humus reserves and biophylic elements.

The agricultural complex is part of the national economy. Therefore it should be considered together with other spheres of economic activity. The analysis of the situation shows that the ACC continues to lose normal financial and economic interactions with industry, including the fuel- and energy-producing complex, production of fertilizers and pesticides, machine manufacturing, production of building materials, etc. all of which directly supported its viability. Inadequate prices for agricultural produce make it impossible for agricultural workers to purchase necessary fertilizers and agricultural machines, to construct roads and to improve social and living conditions.

Federal support of the agricultural complex still remains insufficient. The private sector of the economy, including banks and trade, is not encouraged to invest in agriculture. On the contrary, high interest rates do not allow farmers to take out loans in private banks. Private trade companies often refuse to sell local agricultural products or even to purchase them at prices far below the net cost. They are more eager to get involved in export-import operations. Selling raw materials and some industrial goods abroad, they bring a great number of foodstuffs (often of doubtful quality) into Russia, thus decreasing the demand for domestic agricultural products and complicating the situation of agriculture as a whole. A great portion of the capital gained from these operations is not added to the financial system of the country, but rather remains abroad as real estate and deposits in banks. According to preliminary calculations, the flow of currency from Russia abroad amounts to tens of billions per year.

From the above, it may be concluded that the agricultural complex of the country is operating under conditions of total economic crisis and that without elimination of the crisis phenomena it cannot develop sustainably, which implies significant investments both in agricultural production itself and in environmental protection. Therefore, the problem of sustainable development of the ACC should be classified into two sections:

- development and efficient accomplishment of anticrisis measures in the national economy as a whole and in the ACC in particular, to provide a favorable background for the sustainable development of the ACC;
- making provisions for the transition of the ACC towards sustainable development.

Considering the processes mentioned, anticrisis measures should, in our opinion, comprise a sensible governmental regulation of the market, a more efficient price and tax policy, introduction of a government monopoly on foreign trade, provision of measures to prevent the flow of capital to foreign banks, abolishment of inadequate prices for agricultural produce, as well as federal and regional subsidies of the ACC in view of the necessity to compensate for the difference in the net cost of agricultural production and its purchase price. The result may be the financial stability of farms of all types of ownership and a foundation for reaching sustainable development of the ACC.

And finally, in our opinion it is reasonable to change priorities from the fuel- and energy-producing complex, which is of prime importance today to the economy and price formation, towards the agricultural complex. This complex is supposed to provide the nation's welfare, especially since during the last 75 years the agricultural sector has never been given due attention. Life made us create a powerful military-industrial complex. The best resources, the best brains, highly qualified engineering personnel and skilled hands were concentrated there. And the result is at hand: a competitive military- industrial complex, one of the most powerful in the world. The same can be applied to agriculture: to develop a powerful ACC with ecologically and scientifically sound highly efficient technologies, which will supply the population with diverse high-quality foodstuffs, it is necessary that the government provide important financial and legal support, as has long been the case in all developed countries.

For instance, the USA invests 30% more in agriculture than in any other sector. 12.7% of the gross output is invested in agriculture in the FRG, 13.7% in Austria, 17.7% in Denmark, 26.2% in Great Britain and as high as 32.5% in Norway. In some countries governmental support of agriculture amounts to 40–80% of the cost of all the agricultural production (up to 80% in Finland and Japan). In Russia this support is 3.5% of the budget and is not constant.

In the second part, the paper emphasizes that the Lake Baikal Region possesses unique natural complexes. It includes the World Natural Heritage site, the unique fresh water Lake Baikal, inhabitated by many relict and endemic plant and animal species and coastal areas. On the other hand, it is a region in which economic activity has been, to our regret, destructive to nature for many decades.

The ecologically unsafe situation in the Lake Baikal Region was caused, in many respects, by expansion of the ACC itself, which was accompanied by violation of engineering and nature protection regulations. For instance, the virgin-soils campaign of 1954–1978 resulted in scientifically unjustified and hence uncontrolled plowing of large areas of meadows, pasture and forest lands (564,000 ha), where soils are easily eroded. As a result, in the valleys of the Selenga and Barguzin rivers, the largest in the Lake Baikal basin, we have areas of man-made deserts, over 100,000 ha of loose sands. In addition, thousands of hectares of croplands, natural pastures and hayfields have degraded. The long-term intense and irregular forest cutting (over 10 million cubic meters per year) caused tremendous damage through natural erosion of forest soils and transformation of vast areas of former forests into steppe. The forest area decreased by more than 20% and changed the hydrological balance in the Lake Baikal basin (shoaling or drying out of small rivers and rivulets). This resulted in ecologically endangered regions.

Hence, ecologization of the agricultural complex should be of prime importance for the sustainable development of the region, provided the following approaches are taken:

1. Management of agriculture on a landscape-ecological basis, i.e., land use with regard to climate, relief, exposition and steepness of slopes, agro-ecological zoning of agricultural lands, adaptive land classification with preservation of natural components of the environment and ecologically optimum ratio of agricultural lands (croplands, hay lands, pastures, etc.).

2. Ecological certification of agricultural enterprises of all types of ownership.

3. Differentiated land use in accordance with zonal systems of agriculture.

4. Maintenance and improvement of soil fertility by field- management methods, restoration of fertility of degraded soils.

5. Application of ecologically sound methods of plant protection against diseases and pests (mainly by biomethods).

6. Designing of new types of agricultural machines providing use of soil-protective techniques and new methods of crop cultivation.

7. Development and application of the system of economic measures for optimization of nature use: encouragement of soil-improving environmental activity, economic incentives for the cultivation of ecologically clean products, as well as systems of fines for compensation of the damage by pollution, degradation and depletion of soils and other damage and disorders.

8. Organization of environmental monitoring of natural resources in the region.

9. Application of biotechnological methods for recycling of agricultural and industrial wastes to manufacture fertilizers, forage for domestic animals, as well as chemical reagents, medical and biological preparations, etc.

10. Large-scale phyto-melioration of heavily degraded soils: sowing to grass, takeover by bushes, planting of groves and parks.

11. Mapping of refuges and sites within the whole area of the Lake Baikal Region, extending the network of national parks, reserves, sanctuaries and other natural sites. This system of natural standards would serve not only to preserve biological diversity, but also could help permanently control the degree of anthropogenic impact on agro-ecosystems.

12. Intensification of the efficiency of the application of results of agro-ecological investigations of the research centers of the Siberian Branch of the Russian Academy of Sciences, of the Russian Academy of Agricultural Sciences, of the institutions of higher learning and department services, and the use of this potential for the solution of many problems of nature preservation and sustainable development of the ACC.

The above approaches have been partially realized and are being implemented in the Lake Baikal Region today. For example, "Norms of Maximum Permissible Impact on the Ecosystem of Lake Baikal" (1987) suggested by the Academy of Sciences have stimulated a number of nature-protection measures.

As for the optimization of the structure of agricultural production, particular attention should be given mainly to sustainable crop production. Of particular importance here should be an efficient system of selection and seed growing of zoned varieties and substitution by new varieties, providing higher yields of more resistant crops. An important aspect in the sustainable development of agriculture is the efficient use of fertilizers based on optimization of plant nutrition, which can help save up to 20–25% of fertilizers, preserve the environment from agrochemical pollution and obtain ecologically pure production.

In crop farming, it is important to use new cereal varieties with high crop capacity adapted to local conditions and resistant to pests and diseases. In some areas, especially in the Irkutsk region and partially in Buryatia and the Chita region, the introduction of winter wheats producing high yields and high-quality grain used for baking bread and in the pasta industry is very promising. Barley, rye, oats and legume-oat mixtures as high-caloric forage for animal husbandry should take up larger areas.

In providing fodder resources for stock-raising, the main branch of agricultural production in the Lake Baikal Region, particular attention should be paid to the restoration of productivity of degraded meadow-pasture lands (grass cultivation, irrigation, agrochemical melioration, etc.), as well as to new methods of preparation, preservation and storage of forage. Forage balance with regard to protein-amino-acid composition is considered very important; thus, it is possible to significantly increase the number of forage units and nutrient value of forage production. As for cattle breeding, the following management and structural changes should be followed. According to the recommendation of the All-Union Academy of Agricultural Sciences (1987), it is necessary, first of all, to develop meat-wool production in sheep raising, and to reduce livestock by 20 to 30% in order to conserve pasture lands. The cattle livestock would increase to between 600,000 to 650,000, and it could be re-oriented to meat production except for suburban farms. Beef production can be increased by 75,000 tons through intense feeding of young milk super-replacement stock as well as an increase in beef cattle livestock capita (up to 2,450,000).

Milk production should be raised by increasing annual milk yield per cow to between 3,000 to 3,500 kg (from 1,500 kg obtained at present). With this new productivity, 130,000 to 170,000 milk cows will be necessary to meet the demand of the population for dairy products.

In steppe zones, it is necessary to increase the stock of horses to 100,000 head and horse-meat production to 3,000 tons of live weight. Furthermore, it is desirable to increase the total amount of meat of all stocks by 1.7 to 1.8 times in all types of farms by the year 2000 as compared to 1985.

A few words about the development of highly productive vegetable farming. It is very necessary because of the constant vegetable deficiency in the diet of the population of the Lake Baikal Region. Using sheltered ground, it is possible to satisfy the needs for vegetable varieties. At the same time, the cultivation of vegetables and especially of potatoes on the watered open ground should be expanded, which can give high yields every year. Besides, it can strengthen the fodder resources for stock-raising.

Vegetable factories and construction of modern vegetable and potato storehouses have to be developed.

Attention should also be paid to the large-scale collecting and preserving of wild fruit, berries and mushrooms. Supplemented by production grown in farms, orchards and garden plots, it may largely satisfy the demand.

When organizing the sustainable development of the agricultural complex in the Lake Baikal Region, it is important to determine the most efficient forms of farming. By its natural conditions, the region may be characterized as a zone of risky agriculture, and most promising should be considered management on an industrial basis. It could be associations of farmers and peasants, as well as state farms having large areas of agricultural lands.

Small farms appear to be most effective, in our opinion, in the field of apiculture, traditional cattle breeding, growing of vegetables and berries, and in establishing facilities for the processing and preservation of agricultural products (smoked products, sausages, cheese, butter, kumyss, etc.).

We believe that successful realization of the measures described above could provide conditions for the sustainable development of the agricultural complex of the Lake Baikal Region. These measures will also help to increase the number of jobs for the local population.

Ecological Modernization of Agriculture

Christina von Haaren
Fachbereich 13: Stadtplanung, Landschaftsplanung
Universität Gesamthochschule Kassel
Henschelstrasse 2
D-34109 Kassel
Germany

This paper presents a general strategy for the planning and implementation of an environmentally compatible agriculture that should be adapted to the unique situation of each country and region. Agricultural policy has to be adapted to environmental demands as a precondition for the implementation on the regional and local scale. Nevertheless, in areas with high-priority problems there will always be a demand for area-specific regulations or incentives for agriculture. Two examples from Germany illustrate how a concept for a regional transition process might be developed. Some proposals for the Baikal region are made.

The general goal for a reform of agriculture in the Russian Federation should be the production of high-quality products and at the same time sustaining underlying natural resources. This goal is faced with a tight economic situation of agriculture that is very likely to lead to increasing pressure to exploit natural resources beyond their carrying capacity. This paper presents a general strategy for the planning and implementation of an environmentally compatible agriculture that should be adapted to the unique situation of each country and region. Agricultural policy has to be adapted to environmental demands as a precondition for the implementation on a regional and local scale. Nevertheless, in areas with high-priority problems there will always be a demand for area-specific regulations or incentives for agriculture. In particular, the implementation in a regional or local context should try to combine the goals of environmental protection with economic and/or cultural prospects for the region. Two examples from Germany illustrate how a concept for a regional transition process might be developed. Some proposals for the Baikal region are made.

1. General Problems of Agriculture and Environment

The transition towards a market economy in the new independent states of the former Soviet Union (NIS) is characterized by several effects of the macro-economic change (OECD 1994: 7ff). Examples for such effects and difficulties are:

NATO ASI Series, Partnership Sub-Series, 2. Environment – Vol. 6
Sustainable Development of the Lake Baikal Region
Edited by Valentin A. Koptyug and Martin Uppenbrink

– farm enterprises are caught in a "cost-price squeeze" (prices of input and farm equipment rise; at the same time, state subsidies have been terminated, and the demand for domestic food and feedstuff is decreasing);

– lack of capital and credit facilities to purchase farm inputs and improve farm operations;

– poor access to information about the overall economic-reform process;

– lack of long-term objectives because of intense political debate about its pace and direction;

– shortage of farm-management expertise;

– during the transition period there is a lack of stable and sufficiently large tax bases for the protection of natural resources (ibid.).

In former times industrialized large-scale agriculture caused serious ecological problems such as pollution of water and decreasing soil fertility. The present economic crisis is leading to lower use of agrochemical inputs. But there is a high risk that in the future, after the present initial decline, the quantity of inputs is expected to rise in keeping with increased agricultural production. Even now there is an increased presence of western agro-chemical companies in the Russian Federation.

In this situation, the modern concept of agriculture is faced with being more than food production, especially in ecologically sensitive sites like the Baikal Region, where different functions of agriculture and problems caused by agriculture are concentrated. The situation of agriculture between those different and partly contradictory demands and tasks may be characterized as follows:

– Agriculture is improving the productivity and efficiency of a natural system for human purposes. Examples are food production, enhancing the production of ground water in watershed areas or the enlargement of species diversity.

– Agriculture is entrusted with the conservation of rural landscapes. These landscapes are generally very important to people and regarded as a part of the national heritage (Nishimura 1994: 20). Much of the agricultural countryside of the Russian Federation still contains more extensive traditionally farmed landscapes than in Western Europe, with a high amenity value and potential for eco-tourism (OECD 1994).

– On the other hand, the way agriculture is practiced is usually aimed only at food production. The highly sectorized agricultural policy usually lacks an integration component with environmental or other needs of society. In most countries, environmental benefits and costs of environmental degradation are not adequately reflected in input or product prices. External costs, caused by the deterioration of the environment, must not be calculated in the agricultural enterprise. This is causing a lot of hidden costs for the public economy.

– There is also often no appropriate regulatory structure to protect those environmental assets without a market value such as biodiversity.

– These conditions lead to farming that does not take into account its environmental implications and the needs of society as ecological functions and recreation. Deterioration of natural resources caused by agriculture include pollution of air, water, soil and elimination of endangered species, habitats and attractive landscape features.

– At the same time agriculture itself may be affected by industrial pollution.

The consequence of this situation is not that we want to expel agriculture from sensitive sites, but that we have to conserve the positive roles of agriculture in the landscape and minimize environmental damage.

The task is to do this by taking the political and social environment into account. Planning must be integrated to develop a suitable strategy for planning, implementation and management under these circumstances.

2. Problems and Principles for the Implementation of an Environmentally Regulated Agriculture

2.1 Range of Possible Instruments

We must distinguish three basic approaches in order to influence the development by specific instruments:

– national agricultural policies such as price policies, laws (for instance threshold values of emissions), property policies, enhancing certain structures of agriculture;

– area-related incentive programs for environmentally friendly agriculture without limitation to special areas or cases demanding protection;

– area-specific, direct regulation of farming or habitat management.

The first approach usually has an indirect effect on the method of farming and its environmental implications. The effect is general, affecting a large area. The instruments, however, fail to solve special local problems. The demand for administrative efforts is usually rather low.

The second approach sometimes has a stronger effect on the farming method than the first, but this effect is limited to special farm properties that are not chosen under environmental aspects. The demand for administrative effort is high, and therefore the efficiency may be considered rather low. Area-specific, direct regulation of farming also places high demands on administration, but the adequacy to specific problems is also very high.

Therefore the ecological modernization of agriculture has to start on the first level, trying to integrate agricultural and environmental policy and by that creating rather high requirements on environmental quality for "regular agriculture" (in general we may assume that the conditions for regular agriculture also determine the environmental precautions agriculture has to practice without receiving any compensation or subsidies). These policies may start with a regulation of product standards and product use, such as testing agro-chemicals before sale. Regulations like that are relatively easy to implement (OECD 1994: 17). A lot of agricultural economists, as well as environmentalists, also demand taxes on fertilizer, pesticides and energy. Also, measures must be developed to remove easily identifiable sources of major pollution. This includes, for instance, dismantling large livestock units and developing an appropriate waste management of the smaller units. An environmental legacy may develop restrictions for banning certain agricultural techniques. Other important goals are threshold values fixed by national law or state subsidies based on ecological criteria and the development of farmer-advisory and extension services (ARUM 1990). The main task is to choose the best option in terms of effectively solving the problem in minimizing the amount of money and administration needed.

However, even an optimized agricultural policy is of itself incapable of solving all the problems. The reason may be either that the problem is very urgent, very unique to a particular area or situation or that it demands a high degree of agricultural adaptation, which farmers normally cannot cope with economically. Examples for areas with such specific requirements are the water supply (especially in sites where the ground water is highly susceptible to pollution) and sites which are important for the protection of natural landscapes, habitats and species or land contaminated by toxic and radioactive substances. Therefore, we need concepts for different scales of planning that indicate the areas to concentrate on with specific efforts and the areas where it is sufficient to use general instruments as characterized by the current definition of regular agriculture. Of course the areas needing a specific treatment grow larger if we have low standard requirements on environmental quality for "regular agriculture".

The economic conditions and laws set by national policy play a key role in the ecological modernization of agriculture. If we do not succeed in changing agricultural policy, the framework of agricultural economy counteracts many of our measures. However, effective changes are extremely complicated not only because of G.A.T.T. and the conflict between agricultural, industrial and other national interests, but also because of strong thinking along sectorial lines in the administration on a national scale.

Very often our activities to control the problems are restricted to the local or regional scale.

For this task, we may use different categories of instruments:

– existing legal compulsory instruments

– incentive programs

– persuasion intruments

– marketing efforts.

The use of existing legal compulsory means (as for instance prohibiting certain actions or the establishment of nature-conservation areas) is usually effective and adequate to certain problems, but acceptance is low and control is often difficult. Incentive programs usually have good acceptance, but their drawback is a high demand for administrative effort and high costs. They do not solve area-specific problems. However, persuasion instruments, such as consulting, may be very effective, but they work only within the framework of regular agriculture and not against the economic interests of the farmers. In general, there is no ideal instrument that could solve all the problems. Instead, it is nessesary to use the whole set of possible instruments and choose the best according to the specific problem.

2.2 The Social and Cultural Dimensions

In the process of implementation, one must consider both the human factor and "human capital". That means that we have to take the social, cultural and political structures in the area into account and try to organize an optimum communication between people of different professions such as scientists, farmers, planners and other affected people (Luz 1994). Insufficient communication is a very common problem between representatives of agriculture and the environmental disciplines in most countries and a major cause of unintegrated policies. To improve acceptance and implementation, it is nessesary to take into account the following aspects:

– The ability to analyse and accept roles of the different parties involved in the implementation processes (e.g., lobbyists, officials of farmers' organisations, administrative officials, environmentalists) is important because it helps us to find out how willing participants are to negotiate, where the margin for negotiations is, and it explains the way people act as economic entities and as representatives of an organization.

– Techniques of communication and mediation may help to organize the process of implementation with more success.

– Environmental goals for agriculture should as far as possible be integrated in a practicable concept for farming.

– These goals should be combined with agro-economic goals, as well, to try to open new economic, social and cultural perspectives for the region. The task of making economic prospects, social equity and environmental sustainability complementary may be a field for rural development policy.

2.3 Principles of Planning in a Regional and Local Context

Considering the described background, there may be some consequences suggested for planning and implementation in a regional and local context:

– The comprehensive study of the landscape functions is the first step in recognizing hidden costs of environmental degradation. Their recognition encourages a process of adaptive trade-offs which move toward sustainable conditions.

– If we based our performance only on control-oriented land-use regulations, it is unlikely that we would inspire any creative development (see Ahern 1991: 135, 140).

– As a consequence, we need a dynamic management that is more oriented to landscape context and functions than to the goal of a specific ideal model landscape.

– In addition, landscape analysis must rely not only on a comprehensive study of the agricultural situation in the region, but also the economic and cultural information.

– Flexible objectives and measurements, for instance in target plans, should show various stages of priority (indispensable measures, advisable measures). In this way, the plan prepares and encourages democratic decision-making.

– Rural-development planning has to strive for acceptance by using a language and pictures that people understand.

A concept of sustainable agriculture heeding these principles could also be an investment for opening new horizons for the economic development of the region. It might, for instance, develop a potential for eco-tourism, organic agriculture and agro-forestry. This may contribute to a greater social acceptance of the paradigm of sustainability.

3. Examples for Planning and Implementation in a Regional and Local Context

Considering the above-mentioned conditions and principles, each region and locality has to develop its own concept of sustainable development adapted to the unique local situation. The following case studies may give a concrete impression of such processes and their results.

3.1 Example on a Regional Scale: Integrated Agricultural and Environmental Planning in the Hannover Region

The goal of the study (ARUM 1992) was to define new planning categories for agriculture which could be integrated in the subregional plan. The regional planning board also wanted to establish closer cooperation with the agricultural authorities.

Method

The available survey data about the environment was evaluated in a comprehensive study to determine the productivity of the natural system for certain needs of society (such as drinking water, habitat conservation and recreation) and the vulnerability of the system to agricultural impacts. The second step involves an appraisal of the actual positive and negative impact of agriculture on the environment, using statistical data and other information.

Results

The central result of the planning is four categories of goals for the development of agriculture according to the needs of the environment. These categories are supplemented by several more specific measures. Each measure is combined with

category of subregional plan	measures (examples)	realization (examples) (1992)
regular agricultural production	farming according to the needs of soil protection (erosion by wind/water; concretion)	- state program for distribution of milk contingents (only for ley farming in areas endangered by erosion)
	special farming along river/creek banks	- water legislation (state law)
agriculture with special ecological obligations	farming compatible to the needs of ground waterprotection	- extensification program (EEC) - plant protection legislation (FRG) - water supply companies (in protected areas)
	farming according to the goals of nature conservation	
food production restricted/excluded	afforestation	- set aside program (EEC)
	no cultivation, except primary production for industrial use	
	habitat management	
agriculture with special functions in the urban fringe	ley farming for recreational use	- reduced rent for farmland in public property

(from ARUM 1992)

Figure 1 Improvement Scheme for Agricultural Area in the Hannover Region (environmentally favorable farming)

possible instruments to finance and implement the goal. A main criterion for separating the goal categories from each other was the state budget's obligation to compensate most of the farming restrictions which are not fixed by law. These categories are shown in a final map with the following features (cf. Fig. 1):

1. Areas of so-called regular agriculture, with basic environmental requirements but without any further needs or obligations. The minimum requirements cover the use of fertilizers according to the need of the crops as well as measures to preserve the natural basis of agricultural production itself, such as erosion control. Compensation must not be paid to farmers working according to these rules.

2. Areas of agriculture with special environmental requirements; these are highly sensitive and/or contaminated sites where the farmer has to fulfill environmental regulations, which are combined with more severe economic sacrifices. These should be compensated by the public budget. An efficient policy should concentrate financial means for compensation in these areas.

3. Areas where food production should be restricted or excluded. In these areas, agriculture itself is afflicted by effects, especially emissions, caused by industry and other sources causing soil or air pollution. The required restrictions for agricultural land use should be compensated, in this case, by the polluter. Hitherto there has usually been no satisfactory regulation of such compensation, as it is very often not possible to identify the polluter.

4. Areas of agriculture in the urban fringe include the direct vicinity of towns and agglomerations where agriculture has to fulfill other special demands. Requirements beyond those of "regular agriculture" should be compensated.

(cf. Nishimura 1994: 275–292).

This concept enables the regional planning board and the authorities of agriculture to guide agriculture in environmental concerns and to choose among several possible measures the most efficient for implementation. The method itself is considered to be transferable to other situations, such as the Lake Baikal region, having different ecological and political conditions.

The planning process

The entire planning process was supplemented by a participation process involving authorities of the agricultural administration and consulting offices as well as local representatives of the farmers' political organizations. The regional planning board was coordinator and chairman of the discussion meetings. Here, the authorities who must decide, for instance, about land-consolidation schemes or priorities of consulting became familiar with its goals at a very early stage. The process is an example for a type of participation involving mainly representatives. A major advantage of such a level of participation is that people are not personally affected and therefore may be more willing to compromise on difficult questions. A disadvantage may be that the representatives of political interest organizations in particular (in this case the farmers' organizations) have to reduce their agreements to the lowest common denominator of all the members they represent. Nevertheless, the process in general was very effective. Not only was the concept frequently scrutinized by the experts, even the discussion of the conflicts contributed to a regional cooperation between environmental conservation and agriculture instead of the former rigid controversy. In particular the results were:

– agreement of the officials participating to the integration of the proposed planning categories into the subregional plan;

– better acceptance of the paradigm of sustainability among those participants;

– the beginning of a widespread political discussion involving farmers' organizations and political parties in the region;

– cooperation with water-supply companies in the region (initiative for marketing efforts);

– cooperation with the city of Hannover (several implementation efforts; in particular, the organization of a farmer's market in the city).

These results may not be transferred elsewhere, where very different social and political conditions exist. But the paradigm of democracy and the prospect of an improved implementation may encourage such processes even in a very different context.

3.2 Example on a Local Scale: Local Conservation Master Plan for the "Wedemark" District

—a concept study about an integrated implementation of enviromentally friendly agriculture and achievements for the income of farmers—

Premises

The Wedemark district is situated about 15 kilometres north of the city of Hannover. The decision to work out the study was motivated by many conflicts between agriculture and environment in this area. The premises for planning and implementation here may be characterized as follows:

– The natural productivity for agricultural production is rather low in most sites of the district. Dry and sandy soils with a high demand for artificial irrigation are dominant. The need for irrigation is partly caused by the pumping of drinking water.

– Besides its significance for the production of raw materials, farming contributes to the conservation of species and habitats and other natural resources. The replacement of farmland by forests and woodlands does not necessarily achieve the required results.

– In many areas, even proper sustainable farming has to be combined with further measurements to achieve amenity benefits for the environment. Here farmers have to be paid for further conservation work.

– The need for action in the environmental sector is faced with a tight agro-economic situation that will force many full-time farmers out of business. The size of farms run full-time varies between 30 and about 100 ha.

– Young farmers and other managers of farms which have been economically viable up to now are therefore interested in additional sources of income, as their perspective for the future is insecure.

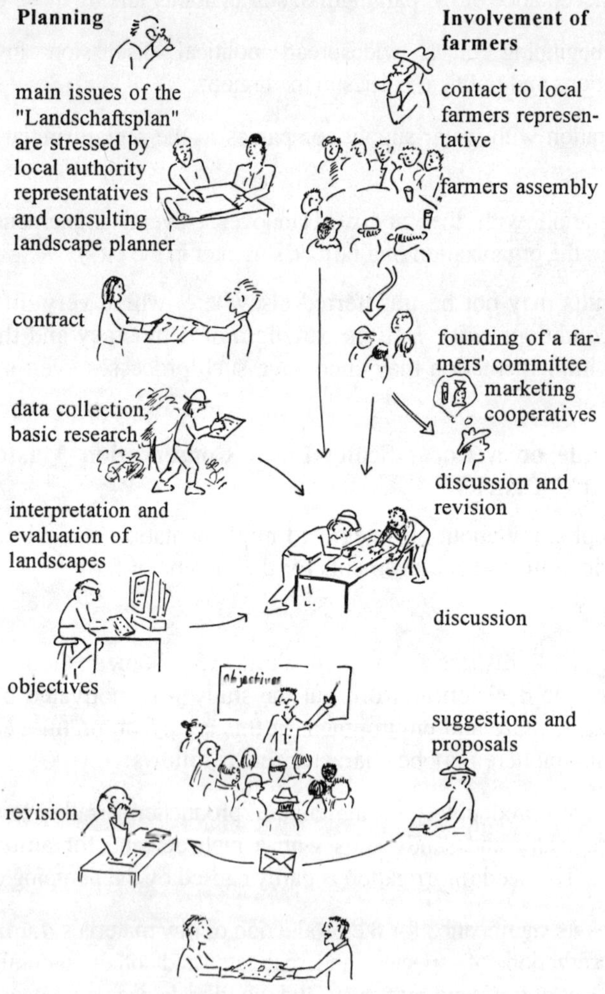

Figure 2 The Drawing Up of a Local Conservation Master Plan - Interaction of Planning and Implementation (Wedemark example)

"Local conservation master plan"

The information basis to develop goals for an integrated development of agriculture and the environment was worked out in a local conservation master plan that contained a comprehensive study of landscape functions. Data on the present situation of farming was taken from statistical sources and by questioning farmers. The concept is to combine conservation issues with other local economic, cultural and social interests, such as new additional sources of income for farmers.

The planning process was accompanied by an intense involvement of the farmers (see Fig. 2). The process involved common discussion and analysis of local problems, discussion and revision of basic data collections and topical maps (e.g., soil erosion) by the farmers as well as an intense discussion about all objectives, measurements and restrictions made by the conservation plan before the final draft. The objectives and measurements proposed by the conservation master plan considered the implementation and realization. The presentation was adapted to psychologically important aspects. It was very much avoided to pillory farmers as the ecological evil and by that to put them on the defensive against new developments. To encourage a discussion about the goals and measurements, they were given different priorities.

Implementation

These priorities also play an important role for an implementation concept that is based on two different approaches, according to two different types of initiators dealing with the implementation of environmentally friendly agriculturc (Fig. 3):

– In areas with a high priority for environmental action, the Wedemark District Council and the local waterworks develop activities with particular aims. The parish council, for instance, involves farmers who possess fields beside rivers and brooks turning their fields into grassland of high conservation value. It also raises funds for the realization of these conservation projects. In other areas which are important for ground-water conservation, the Hannover waterworks is funding an advisory service and several projects for the improvement of proper farming techniques. The funding of proper farm management is less expensive than technical treatment of groundwater polluted with fertilizers and agro-chemicals.

– The second approach is not restricted to special prioritized areas and works via cooperation with all farmers interested in environmentally friendly or organic farming. Several farmers would take up environmentally friendly or even organic farming on the condition that there is a good promotion and marketing strategy to gain new clients and market shares for the more expensive organic products. On the short term, sufficient grants must be provided to facilitate the conversion of farming techniques.

The farmers willing to cooperate organized themselves in the "Farmers' Cooperative for Environmentally Friendly Farming in the District of Wede-

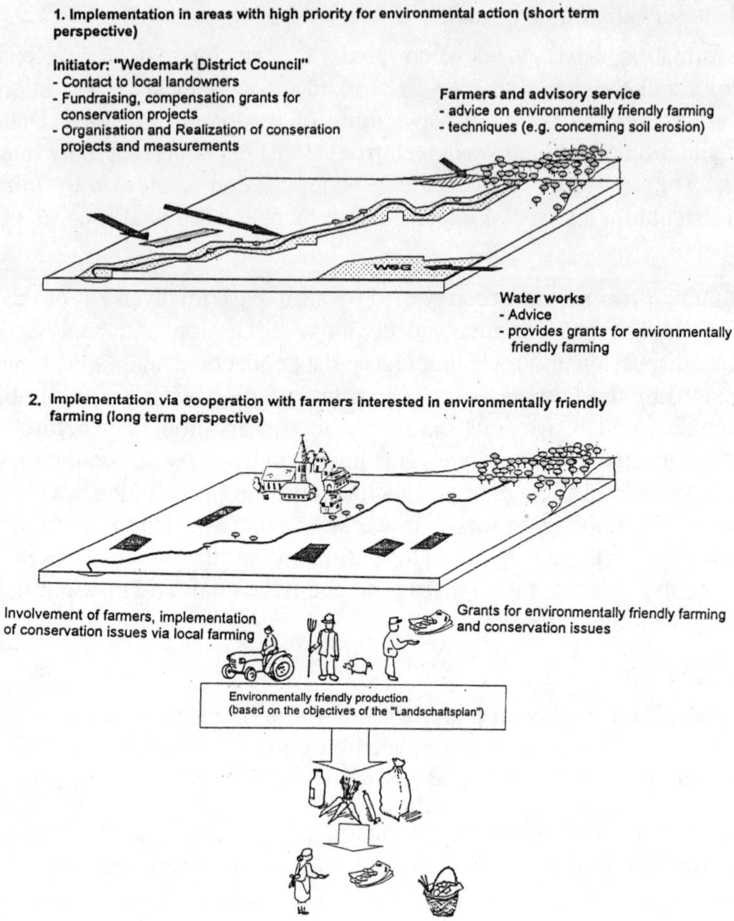

Figure 3 The Implementation of a Concept Concerning Environmentally Friendly Farming in a Local Context

mark".The objectives and principles of this group are: conservation work as an additional source of income and part of the farmers' livelihood and gradual conversion towards organic farming, depending on economic success.

These goals are approached by four main steps:

– Organization: the association consists of 11 farmers, the chairman of the local farming community and a representative of the farming women's association. They work together with the local authorities and the landscape consultant. The long-term perspective is to cooperate with other existing production cooperatives, setting up an office with a full-time staff dealing with organization and marketing. A regional cooperation would enable the farmers to offer more easily marketable quantities.

– Adjustment of agricultural production towards more environmentally friendly techniques.

– Advertising by developing a specific regional product identity and quality.

– Marketing (direct marketing, cooperation with bulk purchaser and other marketing strategies).

Previous problems of the cooperative have been:

– Available public grants concerning organic farming do not always fit the local situation.

– Farmers are often overloaded with work and inexperienced in cooperation with public bodies. Their involvement in fundraising and marketing is therefore limited.

– The organization of new marketing strategies can be very difficult and time-consuming.

4. Conclusion

The general objective for an ecological modernization of agriculture in the Russian Federation should be the development of an agricultural sector which produces high- quality products and which sustains underlying land, water, and air resources and protects habitats and species. Implementing such a sustainable agriculture presupposes the adaptation of agricultural policy to ecological demands. This is a task of both the Russian Federation and the republics. It includes price politics, the refocussing of subsidies, taking account of environmental protection and rural development objectives (OECD 1994: 8), and may also include a framework for land reform.

In addition, a need for specific planning and implementation will remain in certain areas and special cases requiring special management or which are affected by high-priority problems. For this purpose it is nessesary to identify highly sensitive or polluted sites by means of comprehensive ecological studies. The study of the landscape has to be combined with economic studies such as analysis of farm structure, potentials of alternative production, analysis of local and regional markets. Implementation on the local scale has to work on two tracks:

– Short-term implementation of high-priority objectives in selected areas (initiated, for example, by the state or local authorities); within these areas there might be a need for regulations governing farm practices.

– Implementation by farm businesses which are willing to change the structure of their farms or the management of certain fields and want to combine their economic interest with results for ecology.

The development of farm advisory and extension services to establish an information transfer about better farm management skills is very important in this context. In addition, a new type of agricultural consultant would be helpful. This consultant will work as "site supervisor" of the implementation of environmental protection objectives, combining them with economic benefits.

Planning and implementation concepts should always be worked out in cooperation with agriculture and environmental planning. This offers opportunities for:

– a better coordination of sectional branches of administration and cooperation among government ministries (possibilities are interministerial tasks and commitees)

– political discussions about sustainable agriculture involving officials and farm business

– initiatives of local governments and private persons for the implementation of sustainable agriculture according to the needs of environmental functions

Within this framework, the Baikal Region has to find its own ways matching the special conditions of nature and agriculture in the region. Appropriate policy measures are a function of conditions unique to particular areas, land suitability, farming practices and political and public opinion. Therefore, the goals may be achieved only by a policy mix. Possible steps in the process of reform of agriculture in the Baikal Region might be the comprehensive investigation of the situation concerning present harmful effects as well as amenity benefits caused by agriculture, sensitivities of the ecosystem against effects of farming, agrarian structure, economic situation of farm enterprises, structure of property, and market conditions. The connections and dependencies between these different sectors have to be considered. According to the ecological conditions, a set of fairly simple ecological guidelines for farming should be worked out for the different areas. Farming should adapt to these regulations especially in areas with high-priority problems. Restructuring must start here. If land reform takes place, the new property rights could be combined with such obligations. These might be reflected in the price of the land to avoid competitive disadvantages for the rural enterprises starting business in such areas. Another way to compensate farming restrictions, which are not established by national law, is to use special subsidies to stimulate farmers to take better care of the environment. Other challenges of the market for individual farmers may be encountered through such cooperative approaches as production and marketing cooperatives, information networks and other alliances. The development of educational and extension services and targeted programs for priority areas and problems are also important. Although European models cannot be copied without changes specific to the Baikal Region, new ways of linking economy, environmental and social objectives deserve scrutiny.

Acknowledgments

I would like to thank S. Losch (Bundesforschungsanstalt für Landeskunde und Raumordnung), E. Güldenberg and J. Niebuhr-Ette (Kommunalverband Großraum Hannover) and U. Schwertmann (administration of Wedemark district) for the support and stimulating dialogues while working on the different projects that contributed to this paper. I also cite the committed work of U. Kriese (ARUM) on the "Landschaftsplan Wedemark". Thanks go to Prof. M. Uppenbrink for the opportunity to come to know the Baikal Region.

References

Ahern J.(1991) Planning for an extensive open space system: Linking landscape structure and function. Landscape and Urban Planning 21: 131–145, Amsterdam

ARUM (1994) Landschaftsplan Gemeinde Wedemark. Unpublished, Hannover

ARUM (1992) Umweltschonende Landwirtschaft im Großraum Hannover. Beiträge zur regionalen Entwicklung, Kommunalverband Großraum Hannover (ed.) 27, Hannover

ARUM (1990) Raumordnung und Extensivierung. On behalf of the Bundesminister für Raumordnung, Bauwesen und Städtebau. Bonn

Luz F.(1994) Zur Akzeptanz landschaftsplanerischer Projekte. European University Studies, Series XLII Ecology, Environmental Studies Vol.11. Peter Lang, Frankfurt /Main

Nishimura H.(ed.)(1994) Farmland Use in Suburban Areas in the Developed Societies: Based on Studies in U.K., Germany, U.S.A. and Japan. Fumin Press, Tokyo

OECD (1994) Agriculture and the Environment in the Transition to a Market Economy. OECD Documents, Paris

The Role of Scientists in Developing Sustainable Land Use in the Lake Baikal Region[1]

George D.Davis
President Ecologically Sustainable Development, Inc.,
Elizabethtown, NY 12932 USA

It is a pleasure and a privilege to be here today with my distinguished colleagues from around the world. This workshop is potentially the most important international gathering since the 1992 UNCED Conference in Rio. I say that because we are gathered not just to talk about the Rio Conference and how to implement its many recommendations, but to review, evaluate, and determine how to refine a program that actually implements the most important recommendation from Rio: developing and implementing ecologically sustainable economic development programs that preserve our global environment and biodiversity while improving the quality of life of the region's people.

Between 1991 and 1993 a joint Russian-American team of scientists and public-policy specialists developed an ecologically sustainable development program for the Lake Baikal Region. More importantly, later that year the governments of the Buryat Republic and Chita Oblast adopted this program as official policy. These two regional governments control 95% of the entire Baikal watershed. Only the Irkutsk Oblast, with authority over only 5% of the watershed but 40% of its shoreline, has yet to adopt the program.

This program, entitled "A Comprehensive Program of Land-use Policies for the Russian Portion of the Lake Baikal Region", stands today as the first region-wide sustainable development program actually being implemented in the entire world. It is administered by two regional departments, established by the Buryatia and Chita regional governments solely for the purpose of putting the program in effect. Financial assistance to implement this program has come from the regional governments and U.S. Department of State. In addition, the U.S. Agency for International Development is investing $3 million in 12 sustainable development projects in the region to demonstrate the program's ecologic and economic feasibility.

The role of scientists in developing the program was critical. Their role also marked a turning point in the former Soviet Union, an example of how scientists

---

[1] Presented at the NATO/RAS-SB sponsored International Workshop "Sustainable Development of the Lake Baikal Region as a Model Territory for the World" held in Ulan-Ude, Buryatia, Russia, September 11-17, 1994.

NATO ASI Series, Partnership Sub-Series, 2. Environment – Vol. 6
Sustainable Development of the Lake Baikal Region
Edited by Valentin A. Koptyug and Martin Uppenbrink

can influence, and even drive, public policy. I believe that the role of scientists, here and elsewhere, must increasingly be one of insuring that scientific knowledge be more effectively used in determining public policies.

It must be remembered that the "Comprehensive Program" is not, and was not intended to be, a scientific document; rather, it is a social-political-economic document built on a foundation of good science and the integration of numerous scientific disciplines. Indeed, scientists who chose to be involved in developing the "Comprehensive Program" took a professional risk. Some were scoffed at by fellow scientists who studiously avoided the project because they did not want pure science to be diluted by other public considerations, or because they wanted more research, more certainty before making recommendations, more individual control of the project, or more involvement of scientific institutions. But those willing to be involved persisted and their persistence, coupled with a willingness to venture into the public-policy arena, paid off. As a result of the work of Russian scientists such as Gregori Galazy, Valerian Vikulov, Vladimir Korsonov, Valery Mikheev, Tatiana Strizhova, Vladimir Molozhnikov, Oleg Votakh, Alexander Cherkachin, Arnold Tulokhonov, Vladimir Bizyukin, Galina Chimitdorzhieva, Tsyden Tsybzhitov, Irina Urbanaeva, Olga Goryunova, Balzhan Zhimbiev, Roman Zlotin, the late Eduard Zhbanov, and numerous others, 95% of the Baikal watershed is now at the global forefront of sustainable development. It is receiving millions of dollars of foreign investment and will receive millions more, to protect the waters of the fabled Lake Baikal and create jobs for the region's residents.

Of course these scientists did not accomplish this undertaking alone. They relied on numerous colleagues and had the strong support of Vladimir Saganov, then Chairman of the Buryatia Council of Ministers, and Valery Lomako, Deputy Governor of the Chita Oblast. And now, two Russian scientists who were involved in developing the program, Sergei Shapkhaev and Irina Glazyrina, have been chosen by their governments to oversee implementation of the program in Buryatia and Chita.

What did these scientists do that made a difference? First, they agreed to set aside institutional jealousies and the common reluctance to work with other scientific and policy disciplines, to work together as a team. Second, they were willing to make recommendations on existing scientific knowledge or even professional hunches. They understood that we scientists too often seek more and more research on a subject, until, too often, we have an environmental or social disaster that could have been avoided. Third, they recognized that although they might know more about Lake Baikal than any outsiders, they could always learn more, particularly about public-policy-making and market economics; so they reached out and listened to others from around the world. They recognized that no individual nor country knows everything about anything. We can all learn from one another.

These Russian scientists, men and women, believed that the opportunity to save Lake Baikal now, before it is too late, was worth a sacrifice, even worth risking their reputations. They deserve our heartfelt thanks, and the thanks of our grandchildren's grandchildren.

No one is foolish enough to say that the "Comprehensive Program", the world's first regional program of ecologically sustainable development, has saved Lake Baikal. Or even will save Lake Baikal, without further refinement based on more scientific research, without further financial investment, or without the active involvement of the Irkutsk Oblast, which although only 5% of the watershed, controls 40% of the Baikal shoreline and most of the Baikal airshed at a time when air pollution may be the single greatest threat to the Lake's future.

The "Comprehensive Program" is admittedly only a first step. But the first step determines the path and in any worthwhile effort that is usually the most difficult step. Russian and American scientists have now been joined by our German colleagues working on the western shore of Baikal. They have the opportunity to join hands and help show us the second step. Others are certain to join in, for Lake Baikal is truly a global treasure and a potential model of sustainable development that all peoples can learn from.

But what precisely did these people do that was able to so quickly turn the vision of Rio into a workable program to preserve Lake Baikal? And what is the proper role of the scientific community in developing sustainable land-use policies?

Let me summarize the history and work that has resulted in "A Comprehensive Program of Land Use Policies for the Russian Portion of the Lake Baikal Region".

Collective awareness and concern for the environment is probably higher today than ever before. We concern ourselves with air pollution, water pollution, hazardous waste disposal, toxic metals, and endangered species. But how often do we step back for a holistic view of our environment and its ailments? Policy-makers must be taught that the word environment is singular and all-encompassing. We must teach them that the land and its communities of life are the barometer of environmental health. They must understand the importance of naturally functioning ecosystems and biological diversity.

Sustainable development—or, more accurately, ecologically sustainable economic development—has become a popular buzzword since its use in the 1987 report of the United Nations' Commission on Environment and Development, the Brundtland Commission. Although specifically defined by the Commission, the term is often said to be vague and is frequently used to mask the true nature of both environmental and economic impacts of ill-advised development.

In releasing her Commission's report, Norwegian Prime Minister Gro Harlan Brundtland dispelled the nonsense coming from the then American administration, and elsewhere, that we must make choices between environmental protection and economic prosperity. Not only did she debunk that philosophy, she convincingly argued that if we do not soon learn how to seek both goals simultaneously, our future as a species is bleak indeed.

The Brundtland Commission defined sustainable development as "development that meets the needs of the present without compromising the ability of future generations to meet their own needs". I would add to that definition that sustainable development should improve the quality of human life while remaining within the carrying capacity of supporting ecosystems.

At the request of a coalition of government officials, scientists, and non-governmental organizations in Russia's Buryat Republic and Chita and Irkutsk Oblasts, a team of Russian and American scientists and public-policy specialists developed an ecologically sustainable economic development program for the thirty-two-million (32,000,000) hectare Russian portion of the Lake Baikal watershed. The theme of the process we chose was "listen to the land and listen to the people". It is an approach that borrows heavily from regional planner Ian McHarg's "design with nature" approach and E. F. Schumacher's "small is beautiful" economic philosophy.

Any such program must be tailored to the particular country involved to reflect the culture of that country. However, we have found that the best process for creating sustainable development programs is amazingly similar for all countries in which we have worked, regardless of political process, economic system, or bioregion.

The first requirement at Baikal or elsewhere is to develop support for the project among scientists, non-governmental organizations, and various levels and branches of government. Without such broad-based support from the beginning, there is far less likelihood of successful implementation of any program developed. In addition to initial support, these three sectors must also remain involved throughout the development of a program. This insures the final program, rooted in good science, will reflect the desires, needs, and political reality of the country, rather than those of the outside consultants providing the technical assistance.

The second step was to assemble an interdisciplinary team of scientists and public-policy specialists. Scientists and public-policy specialists, including government representatives, are equally important to the process because the final report, as stated earlier, must be a social-economic-political document based on sound science. The inclusion of a minority of foreign specialists added new perspective and valuable experience to the team. Policy people must be involved at the inception of a sustainable development project and throughout its evolution, not simply handed a scientific report at the conclusion. Communications between the scientists and the policy people must be frequent and in a spirit of true collaboration.

The scientists included experts from disciplines such as agronomy, resource economics, forestry, terrestrial ecology, environmental engineering, anthropology, limnology, protected-areas management, land-use planning and remote sensing. But equally important as depth of knowledge in such fields is a commitment to working together coöperatively, a commitment to sustainability, a commitment to bettering the lives of people, a commitment to working within existing cultural norms, and a commitment to future generations.

Public policy specialists included professionals from academia, government officials from economic and environmental positions in both the executive and legislative branches, land-use lawyers, and leaders of influential non-governmental organizations.

The assembling of a broad but compatible team is critical, but a partnership and a close working relationship is far from automatic. Scientists and bureaucrats are too often not used to working together and neither are necessarily comfortable working with public-interest-group representatives. Nor are foreign specialists automatically accepted as experts until they prove themselves in the new situation; but this is as it should be. In addition, cultural differences abound and continually surprise us, even when we have diligently tried to identify potential cultural obstacles. At Lake Baikal there was great pride in our final document and map among the entire team. Unfortunately, we soon encountered a near total lack of understanding about what it really meant, in terms of implementation, from scientists brought up under the Soviet system, scientists who had never had the opportunity or reason to think about implementing their recommendations.

The first work of the multi-national scientific and public policy team, once all members understood the process to be used, was to prepare specific objectives for an ecologically sustainable economic-development program for the region. These vary with the region, culture, and political and economic realities. For the Lake Baikal region eight primary objectives were set:

1. Preserve the purity of Lake Baikal's water for all time;

2. Inextricably link sustainable economic development and environmental protection;

3. Preserve the natural ecological processes and biological diversity of the Lake Baikal watershed;

4. Preserve cultural traditions and diversity found in the Baikal Region;

5. Insure that present and future generations can live in dignity and will enjoy an improved quality of life;

6. Involve the public in land-use-policy decisions;

7. Expand intergovernmental collaboration between Russia and Mongolia and among all levels of government within the region; and,

8. Achieve UNESCO World Heritage Site designation to reflect the global significance of the Baikal Region and the determination of the Russian people to protect it.

As previously stated, the process used to create the Lake Baikal Region sustainable-development program can be summarized as "listening to the land and listening to the people". Listening to the land simply means evaluating, based on existing data and knowledge, the inherent capabilities and limitations of the land and water resources for various potential uses.

All lands have inherent capabilities and limitations reflected by their soils, slopes, biological communities, visual amenities, and similar characteristics. These characteristics were individually evaluated and rated for various potential land uses, and then aggregated so a land-capability and limitation rating could be applied. This aggregated rating reflects situations where, although the soils and slopes may indicate suitability for development, an area should remain undeveloped because of other factors: an endangered species or visual sensitivity for example.

Of course, many parcels of land are suitable for several uses, while others are clearly inappropriate for all but one or a few uses. The latter speak for themselves but the former require a priority hierarchy, one in which the program's objectives must be accurately reflected.

Although priorities vary with each country, and often within each region of the country, it was decided to give special priority to the creation of a comprehensive protected-areas system (designed to preserve biological diversity) and to the preservation of productive agricultural soils.

A comprehensive protected-areas system should insure that the maximum biological diversity and examples of all functioning natural ecosystems are preserved for future generations. By comprehensive, I mean serious efforts to include all of the World Conservation Union (IUCN) protected-area classifications, such as national parks, wildlife refuges, natural anthropological reserves, scenic rivers, and so on, as appropriate. But no matter how comprehensive, we must never fool ourselves into believing that any protected-areas system relieves us of our responsibilities to practice state-of-the-art environmental stewardship on all lands we impact.

The Earth's productive agricultural soils are perhaps the most important natural resource from an anthropogenic viewpoint. To foreclose their use for either present or future crop production is short-sighted and generationally selfish to the extreme. In the long run it is usually economic folly as well. Of course, productive agricultural soils are not necessarily synonymous with tillable land. Americans discovered this in our dust bowl. Russians discovered it with the now infamous "virgin lands" policy, where productive grazing lands were plowed to grow grain. The Buryat people of Zabaikalsk, who, as nomadic grazers, had been excellent stewards of their lands for over two thousand years, knew this new policy would result in disaster, but no weight was given their concerns. The same was undoubtedly true of soil scientists. The result, in Buryatia and elsewhere, was the loss of immeasurable quantities of rich soil unsuited to tillage.

In evaluating the land's capabilities and limitations, our project methodology required the use of existing data. Frequently scientists on the team wished, justifiably, for additional or more recent data. We have found that it is invariably a mistake to wait for more data before developing regional comprehensive land-use policies for at least two reasons. First, if we have been asked to develop such policies and there is broad support at the time, it is important to follow through quickly before the enthusiasm wanes and implementation becomes more difficult or impossible. And secondly, our land resources have been abused or ignored for far too long— it is time to act. Additional research and the resultant data will always remain desirable and in any logical program will be encouraged as the basis for providing refinement and amendment. But in the meantime we must stop wringing our hands about how terrible our stewardship of the earth is and start doing something about it. I believe scientists can make substantial improvements in the way land is viewed and used, even in areas where little hard data exist.

Listening to the people means public involvement early in the process to learn the cultural traditions of the people in the region and to learn their needs and desires, both for themselves and for future generations. This was done by meeting with individuals, local officials, and organizational leaders. Public meetings were held throughout the region to review the project objectives and the first draft of a proposed sustainable use program. As part of these meetings scientists explained to the people our first attempt at mapping future land use zones—such as proposed arable areas, commercial forest areas, and protected areas—and asked for their comment. During this process, preliminary objectives were revised to better reflect the desires of the residents for their future.

Before being completed, the draft program and policies were reviewed by a broad group of international natural resource, development, and public policy experts, not only for the sake of peer review, but also to encourage the incorporation of appropriate lessons learned elsewhere.

As sustainable development policies emerge and are applied to land use and economic development, it is imperative that we recognize that while each region of our planet is distinctive, it is also part of a larger whole. "Think Globally, Act Locally"— a sound concept, but as with all concepts it must trigger a larger thought process. The community, the region, the state, the nation, the continent, the earth; somehow we must intuitively realize that all levels are important, all are interconnected, all are dependent on one another, and none is superior to the others.

In summary, our global society simply must find a way to make ecologically sustainable development a reality rather than an academic exercise. The scientific community can and should be at the forefront of this effort. To do so we must first break down barriers among ourselves and then learn to work with both government policy-makers and the general public. We must influence rather than dictate. We are all on the same global spaceship and we do have the ability to determine its course. Let's get on with it.

Thank you.

Forest Resources of the Lake Baikal Region and Perspectives of Their Industrial Exploitation

E.A. Vaganov, A.I. Buzykin, M.D. Evdokimenko
V.N. Sukachev Institute of Forests of SB RAS, Krasnoyarsk, Russia

Reasons and ecological impacts of damaged forests are analyzed. The main aspects of ecological stabilization and steady development of the forest complex are discussed. Alternatives for change of forest utilization structure as well as a reasonable course for improving logging technologies and wood processing are considered. The decision to adopt an ecosystem forest management will promote forest-complex stability.

The forests of the Lake Baikal basin occupy an area of 20.6 million ha within Russia. It is difficult to overestimate their biospheric role: they control the state of other very important components in the region: water, soil and the atmosphere. This is their main function. However, the forests of the Lake Baikal basin are very important for industrial exploitation, as well. The total stock totals 2010 million cubic meters.

Their composition and structural diversity is significantly determined by relief. Woodland changes from 10–25% in the mountain forest steppe, up to 50–90% in mountain taiga, and on average it makes up 60% in the regions adjacent to the lake. The forests are not of high biological productivity (fig.1). The total annual increment amounts to 21million cubic meters. Of more than 400,000 ha of

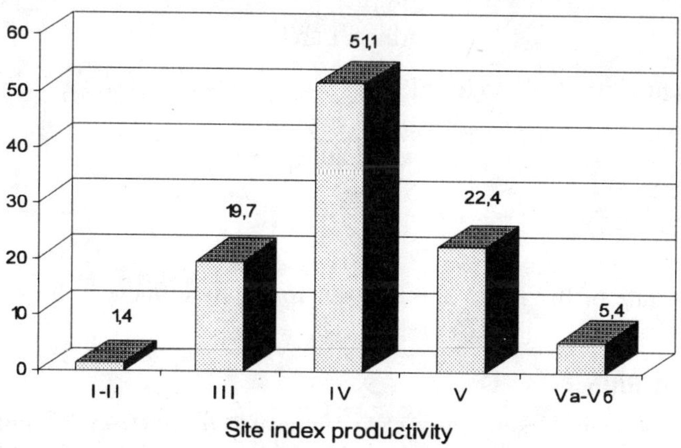

Fig.1. Distribution of wooded area according to the site indexes (According to Protopopov, 1990)

NATO ASI Series, Partnership Sub-Series, 2. Environment – Vol. 6
Sustainable Development of the Lake Baikal Region
Edited by Valentin A. Koptyug and Martin Uppenbrink

unwooded area, half is occupied by burnt areas and by not yet afforested felled areas.

The forests of the Lake Baikal basin are generally characterized by large differences of altitude-zonal complexes as determined by orography, climate and soil diversity (Table 1). There is much to be said for the observation of regular dynamic processes of the Baikal mountain forests and their strong dependence on the violations of the environmental balance.

Table 1 Characteristics of altitudinal and zonal complexes of Lake Baikal basin

Altitudinal and zonal complexes	Areal (%)	Prevaling forest types	Sum of positive temperatures (Grad)	Sum precipitation (mm/year)
1. Meadow-steppe	17	-	1720-2140	240-600
2. Subtaiga-forest-steppe	19,5	Light-coniferous (pine,larch)	1430-2170	380-1100
3. Larch taiga (Larix sp.)	37,5	Larch	1080-2170	380-1170
4. Pinus sibirica-taiga	4,0	Pinus sibirica, Larch	1040-1820	560-1550
5. Abies-Pinus sibirica taiga	7,0	Pinus sibirica, Abies, Larch	840-1600	670-1210
6. Mountain tundra	15,0	Pinus sibirica open woodlands, Shrubs	490-1200	570-1570

The present state of the Lake Baikal basin forests is influenced by a number of factors:

1. Industrial Cuttings

About 90 million cubic meters were cut from 1970 to 1981, or 7.5 million cubic meters per year on average. Annual cutting decreased to approximately 6.5 million cubic meters by the late 80's. The negative feature of the last decades is extremely irregular distribution of cutting areas in the basin. Felling areas make up approximately a quarter of the total forest-fund area in regions with intensive logging. The structure and composition of forest associations are severely damaged; primary coniferous stands are substituted by birch and aspen forests. The replacement of mature and seasoning stands by young stands has reduced the water-protective potential of the forests. The area's heterogeneity demands a differentiated approach to reforestation. So in many felled areas, natural forest regeneration is not observed. As a result, surface run-off accelerates erosion processes.

2. Forest Fires

Some circumstances reveal the significance of the pyrogenic factor for the basin. First, take orography. Mountain ridges are mainly situated perpendicular to the direction of prevailing winds. Second, light coniferous stands dominate, which constitute a high fire hazard. Third, the low precipitation characteristic of the forest-steppe zone (230–250 mm per year) as well as its extremely irregular distribution during the vegetation period make for a long season with a high risk of fire. In the fourth place, emergency fire situations occur in some years. For example, in 1969, 1976, 1979 and 1987, the number of forest fires totalled 1,500, and burnt areas exceeded the annual forest felling areas by several times. According to the data of the recent forest inventory (1993), the forest area which was not regenerated after cuttings and fires amounted to 230,000 ha. Forest plantings on eroded lands (approximately 1,000 ha per year) are virtually unsuccessful.

3. Forest Decline

The total area of damaged forests near the Baikalsk Pulp-and-Paper Plant (BPPM) amounted to 145,000 ha by 1986. By now, major forest damage and shrinkage has been observed in an area of 50,000 ha and weakened forests in an area of 250,000 ha. The dry season, air and water pollution caused by BPPM and subsequent insect propagation in the weakened stands should be considered as possible reasons for shrinkage.

4. The Impact of Recreation

The increased recreational pressure mainly effects forests of the first group (forests near settlements, cities, coastal forests and, also, forests in the areas having well developed transportation networks).

The results are the decrease of class quality (productivity) of tree stands, deterioration of other inventory parameters, disturbance of natural forest regeneration (the rate of viable regrowth is over 3 times lower; there is a decrease of grass phytomass by 2 to 2.5 times, a reduction of protective cover by 1.4 to 8 times, as well as of forest litter (mass) and of soil saturation.

At present, there is an obvious discrepancy of forest management and logging with regard to the main goal of a water-protective regime in the Lake Baikal basin, i.e., with the regeneration of clean water.

A problem of this complexity also demands a complex approach to its solution. What forest-management regime can be suggested as strategic which could be aimed at the sustainable forest management and sustainable development of the Lake Baikal Region?

It must be an ecologically balanced regime which takes into account regular dynamic processes and the large differentiation of the forest area which is based not on the theoretical model, but on the recent state of the forest and on the existing forest-management structure. The ecosystem forest management realized in the United States National Forests can serve as an example of practical realization of sustainable forest-management regime.

Its ideology, goals and realization can be a good model in managing the Lake Baikal basin forests. However, some measures are needed to follow such a strategy based on the specific state of forest management.

First, there are no clear formulations of objectives in the Forest Legislation of the Russian Forest Service to control and to manage the forests of the Lake Baikal basin.

A reasonable and coördinated approach between the federal and local forest legislation must be introduced which will take into account, as much as possible, this region's specificity. Since 1970, all the forests of the basin were assigned to the first and second groups of forest management. The raw-material potential of the forest complex significantly decreased, according to this zoning. Also, the structure of the exploitational stock changed (for example, the share of stands located on slopes 16 to 25' steep increased), transport communications decreased, taxes (initial wood price) increased, coniferous forest share decreased in comparison with leafy species. These changes must be taken into account in the local forest legislation, in establishing new economic relations between enterprises and forest owners. As long ago as 1987, the Institute of Forestry of SB RAS recommended a realistic method of determining (correcting) cutting volume, taking into account the ecological role of the forests, their productivity, transport and technological accessibility.

Second, the necessity of reorientation from round timber export to the deep processing of wood raw material is obvious. The net cost of production is such that the greatest portion must be processed locally. And as the Lake Baikal Region is surrounded by forest regions, the utilization of wood output must be mainly oriented to this region.

Therefore, the forest complex and many other related industries could not be developed without investments in the local enterprises. Small business involved in deep wood processing is a primary task for the near future here.

Third, it is necessary to increase the share of leafy wood in forest management because it makes up large reserves even in the most exploited areas. Probably local legislation and the complex of tax breaks as well as the establishing of competitive conditions for enterprises providing leases for forest areas may stimulate the utilization of leafy species.

Fourth, the increase of wood-felling volume in cleaning cuttings performed with machines of the "Harvester" or "Forwarder" types, which are ecologically safer, will provide an impetus for forest-utilization development.

The cleaning-cuttings potential covers the timber needs of the regional market. Cleaning cuttings themselves, however, provide even and sustainable forest utilization. Within the ecosystem forest management, the problem of utilization of wood and raw material even in the protective riverside zone is solved where final cuttings are forbidden because of the special status of these forests. It should be noted that methodological and technological aspects of sanitary reconstructive cuttings were elaborated for the Lake Baikal Region by the Institute of Forests of SB RAS in the late 80's, and the technical support (machinery) was provided by the Ministry of Forestry of the Republic of Buryatia on the occasion of the establishing of joint Russian-Swiss and Russian-Italian ventures for wood felling and processing. Annual logging (wood felling) is considered to be 700,000 cubic meters.

It is to be regretted that the projects of the Institute of Forests of SB RAS regarding selection and gradual cuttings on steep slopes were not accorded financial support.

As for selection and gradual cuttings, they are ecologically safer, but they are less profitable. This is the price of ecology: the wood will cost more than that produced by other methods. Therefore, the increase of the cost and expenditures must be compensated in direct or indirect ways. And if we are to speak about Lake Baikal's status as a World Heritage Site, the federal measures alone are insufficient. International decisions and measures are needed here (for example, decrease of tariffs on wood, etc.)

Fifth, improvement to an optimum and well thought-out system of forest-fire protection is needed in the region which would take into account its extreme differentiations.

Finally, the sustainable development of the Lake Baikal forest complex is possible only with an environmentally balanced and sustainable forest management. Even if the local forest legislation is strictly followed, it is not possible to regulate in detail the ecosystem forest management for such a complex as the forests of the Lake Baikal basin.

Ecologically safe forest-utilization projects for specific areas can be prepared only by specialists of the National Forest Service. So the training of specialists for ecologically safe forest management can be considered an important task of today.

Summary

The most acceptable forest management in the Lake Baikal basin as a possible World Heritage Site must be based on the ecosystem forest management (fig.2).

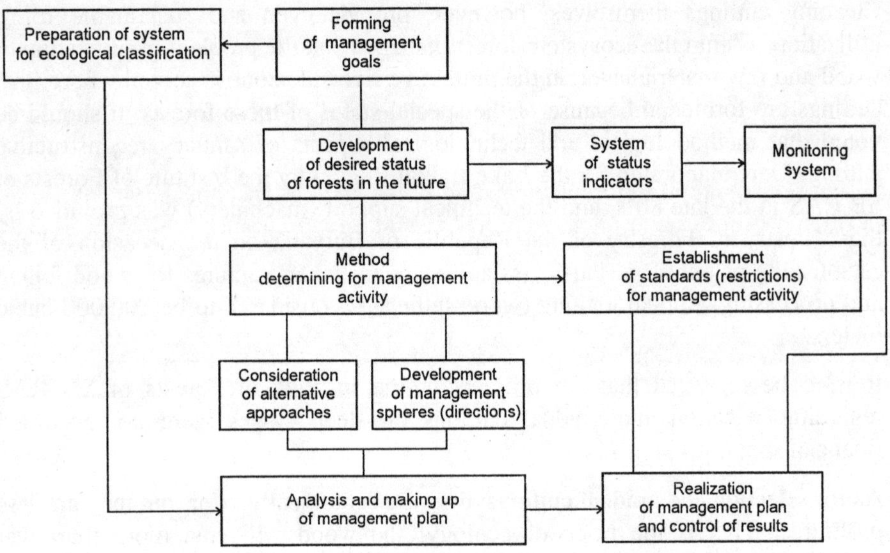

Fig.2. Scheme of ecosystem management for forest area

The strategy of the ecosystem management determines relations of users and forest owners. The final aims of management are maximum conservation of ecological and biospheric Baikal forest functions effectively satisfying the needs of the people who live there.

Informational and Cartographic Support of Sustainable Development in the Lake Baikal Region

V.V. Vorobyev, A.R. Batuev
Institute of Geography of SB RAS, Irkutsk, Russia

The paper analyzes the achievements of thematic cartography and geo-information in solving the problems of natural-resources management and environmental protection in the Baikal Region. A plan of cartographic support is suggested for sustainable regional-development management on the basis of specialized information systems containing the data on natural resources and the state of the environment. The unique nature and efficiency of such information systems are achieved by a cartographic form of information and efficient opportunities of up-to-date information technologies. Thematic maps in these systems are the main means of geographical data interpretation and organization. The necessity of establishing a regional cartographic geo-information center is recognized.

An important role in the implementation of a policy of sustainable development must be taken up by informational cartographic support, which makes it possible to perform operative analysis of the condition and possible variants of development of the region and to arrive at a synthesis of administrative decisions for future developments in cases where several scenarios are available. Moreover, it facilitates working out proposals for eliminating deviations when the technico-economical development of the region exceeds the scope defined by ecological regulations and limitations.

It appears that the core of such support must be an automated informational cartographic system (ICS). ICS is a system of scientific and technical methods and means of automated collection, storage, transformation and operative representation of manifold and, primarily, ecological and economic information about the region in cartographical and other form convenient for the user.

The main demands on ICS are of a systematic nature, entailing purposiveness, high efficiency and multivariance in generating new computer maps.

This combination of properties is achieved by a high scientific intensity of the system, encompassing geographical, cartographical and other data bases as well as original models and methods of applied mathematics and artificial intelligence.

Until recently, the main efforts of cartographers were focussed on creating a series of thematic maps and atlases, i.e., major cartographical publications which present stabilized and thoroughly verified scientific knowledge. Therefore the atlases are intended for repeated, multipurpose use. They still retain their

NATO ASI Series, Partnership Sub-Series, 2. Environment – Vol. 6
Sustainable Development of the Lake Baikal Region
Edited by Valentin A. Koptyug and Martin Uppenbrink

importance as a basis of information. At present there are a sufficient number of cartographic works on the Lake Baikal Region. They provide a solid foundation for developing measures for the sustainable development of the Lake Baikal Region.

First of all, the general purpose of cartographic material should be noted. There are general geographical and topographical materials on the scales 1:1,000,000, 1:500,000, 1:300,000, 1:200,000 and 1:100,000. For some regions, there are maps on the 1:50,000 and 1:25,000 scales and even larger. All large- and medium-scale topographical maps are revised regularly by Roscartographia. Lately, Roscartographia has begun digitization of topographical maps on the 1:200,000 scale for the Irkutsk region.

Apart from topographical maps, a large group of thematic maps and atlases of territories adjoining Lake Baikal has been compiled. Among them are the "Atlas of the Irkutsk Region" and "Atlas of Transbaikal".

The "Atlas of the Irkutsk Region" was published in 1962. It contains plentiful factual material on nature, population, economy and culture. The Atlas, as a handbook, is intended for a wide group of users. At present, the Atlas is of historical significance and may contribute to the solution of economic and cultural problems, provide sceintific information about nature and the history of the development of productive forces and culture of this large region of Russia. The general scale of maps of the Atlas is 1:4,000,000; the south of the region adjacent to Lake Baikal is on the scale 1:2,000,000. The Atlas contains 115 maps, with a circulation of 15,000 copies, and is a bibliographic rarity today.

The "Atlas of Transbaikal" (Buryatia and the Chita region) was published in 1967. This atlas was planned to be a part of the integrated cartographic work based on the newest data on nature, the economy and population of the Transbaikalian area at that time. Most of the thematic maps are made on the basis of selected and systematized factual material.

At the time of compilation, original cartographic and bibliographical sources reflecting fundamental scientific ideas of cartographic objects were used. That is why many maps of the Atlas (there are over 200 of them, their general scale being 1:3,500,000) are fit for investigation by the cartographical method of various regional regularities and peculiarities of the Transbaikalian area.

Comparison of maps of the above two atlases with the maps of the atlas of the Asian part of Russia (as regards the Lake Baikal Region) published in 1914 provides fundamental information about the changes in the economic and cultural development. In the atlases of such large areas as the Lake Baikal Region, map scales cannot be very large (1:4,000,000, 1:3,500,000). However, even an insignificant scale enlargement in thematic cartography will considerably improve the scientific and practical potential of maps. The unified series of maps of the nature, population and economy of the south of Eastern Siberia on the scale of

1:1,500,000 has made it possible to obtain an integral representation of the Russian part of the Lake Baikal Region. The first map of this series, the "Economic Map of the South of Eastern Siberia", was published in 1969. Thereafter, the maps "Vegetation of the South of Eastern Siberia" (1972), "Landscapes of the South of Eastern Siberia" (1979), "Land Use in the South of Eastern Siberia" (1988) were compiled and published. Unfortunately, the map "Soils of the South of Eastern Siberia" was omitted from the series. In 1988, a map of soils of the Irkutsk region was published on the same scale.

Together with the general geographical series of maps, a specialized series "Agroclimatic Resources of the South of Eastern Siberia", including maps on the scale 1:2,500,000, was published in 1969. The series was drawn up in response to the demands of the administration. The scales of the maps are somewhat smaller than those of the previous series, yet they make it possible to give information necessary for practical specialists.

The Lake Baikal Region and some of its districts are represented in other thematic maps: that of the Baikal–Amur Railway, scale 1:2,500,000, a series of thematic maps of North Baikal at 1:500,000, etc. In the recently published handbook atlas "Baikal" (1993), a compendium on natural conditions and resources of the lake and its basin is given.

Other series of thematic maps—geological, hydrogeological, geochemical, mineral, etc.— have been published, including the maps published by the Buryat Scientific Center ("Especially Protected Territories of the Lake Baikal Basin", "Landscape-Ecological Map of Buryatia", etc.).

A series of large cartographical publications has also been made for the Mongolian territory of the Lake Baikal Region. The largest of them are the "National Atlas of Mongolia" (1990) and "Regional Atlas of Lake Hubsugul" (1989).

Examination of cartographic funds of science centers, large libraries, design and industrial organizations and universities, and, also, treatment of large data files of bibliographical and information sources provide an integral picture of the state of cartographic support in the Lake Baikal Region. For example, the bulk of maps of ecological content (72%) was prepared already after 1985, and, among them, hand-drawn maps prevail (the number of printed maps considerably diminished). The series of large-scale ecological maps of urbanized sites of the region were drawn most intensively, and among thematic directions snow and soil geochemical maps predominated.

At the Institute of Geography of SB RAS, a series of 105 ecologo-geographical maps of the city of Irkutsk and its suburban zone was prepared (map scales from 1:50,000 to 1:500,000), and at the Institute of Geochemistry and the Sosnovgeologiya Association a large-scale series of snow and soil geochemical maps was compiled.

Compilation of a new large cartographic work, "Ecological Atlas of the Irkutsk Region", has been completed. In this atlas there are 212 maps, the general scale of which is 1:2,500,000. Some southern districts of the region directly adjoining Lake Baikal are shown on a larger scale, 1:1,000,000 and 1:500,000. The Atlas is of a comprehensive character, since it gives an all-round characteristic of the environmental condition in the Irkutsk region, and all the environmental components are shown in their independence, with anthropogenic impact, forecast changes of nature components and planned measures for nature protection. Interrelation of nature components and branches of the economy, scientific and technological progress and changes in the development level of the region, all this determines the dynamic character of the processes of interaction between nature and society and requires special modes of representation. This complexity is solved in the Atlas by means of special maps showing the dynamics of phenomena, and various materials showing the environmental situation in their development.

The Atlas is a part of projected works to determine the strategic trends in nature management and optimization of the environmental situation in the region. The spatial-informational arrangement of cartographic material is intended for the basic regional structures of planning and administration, of regional and district levels, thus increasing the usefulness of the Atlas by providing a united informational basis.

The program of the ecological atlas envisages the use of materials by appropriate organizations and services of local authorities in working out socio-ecological, ecologo-economic and economic solutions.

At present, the Atlas is being prepared for publication.

Apart from fundamental tasks, an important special task has become operating cartography, i.e., production of cartographic materials to represent frequently occurring or rapidly changing situations. Such maps are mainly intended for solving a simple, often very specific problem, e.g., to obtain ecological indications. They show, as a rule, a number of probable variants rather than some indisputable finding.

Operating cartography based on Geo-informational System (GIS) technologies called geo-informational cartography has deep roots in comprehensive and systemic thematic cartography as a method of versatile and integral representation of reality. The program-coördinated creation of a series of thematic maps and atlases of natural, social and economic geosystems that ensures their coördination, compatibility and mutual complementarity can be regarded as a birth stage of GIS technologies and a basis for creation of computer atlases representing a system of video maps.

Lately, the prospect of sustainable socio-economic development of territories has made people more and more aware of the necessity of solving these environmental problems within the framework of optimization of relationships in the "environment-society" system (Vorobyev et al., 1989). In such a system, thematic cartography as a specific information-communication system with a language of its own (Lutyi, 1988) is used as one of the most important and effective elements in the control of this development.

At present, environmental maps are characterized by a bilateral direction. A part of them are anthropocentic, i.e., oriented to problems of human ecology in a single region. The environment in these cases is estimated from the positions of living conditions and human economic activities, taking into account the consequences both for man and for nature. In others, objects of study and cartography are concrete elements of nature—relief, soil cover, vegetation, and animal populations. Other natural and anthropogenic components characterize parameters of development or habitat.

The level of ecologo-geographic cartography reflects to some extent the degree of maturity of the society (state or its subordinate entity) from the viewpoint of awareness of the importance of the ecological imperative in the socioeconomic life of the territory. This is fully confirmed by the world ecologo-cartographic theory. At early stages, resource and nature-protecting trends dominated. In Europe, Canada, the USA, and Japan, single maps and specialized atlases were drawn up which disclosed the structure of the regions' resource potential. Considerable attention was paid to biotic, mineral, water and climatic resources. A large number of maps and atlases was dedicated to characteristics of especially protected territories—reserves, national parks of the USA, Germany, France, and Africa.

The necessity of the most rational use of land resources has determined the appearance of many land-use maps at various scales. Such maps have been drawn up in various states of the USA, France, for some prefectures of Japan, even for the Cook Islands. However, already by the end of the 60's, maps of the anthropogenic impact on the environment began to appear. In Great Britain, maps of environmental pollution (average concentrations of smoke and sulfurous gas) and of surface water flows were created in 1957–1958. In 1971, a map of surface-waters contamination of France was published. In the early 70's, maps of technogenous environment pollution in the GDR and Poland were drawn up (Leman, 1975; Leszczycki, 1975). In 1971, a map entitled "Use of the Great Lakes Waters" was prepared in Canada, where sources and parameters of contamination of surface waters with technogenous and household sewages were shown. Preparation of comprehensive maps to characterize the environmental condition was started. In 1973, a medium-scale map for the Czech Socialist Republic was published and followed in 1974 by a nature-protection map of Slovakia. In 1978, a map of the state and dynamics of environment in France appeared. At that time in Germany (FRG), large-scale landscape maps were published for the planning and improvement of the condition of lands (Salishchev, 1976).

Apart from single maps, specialized atlases of the environment are appearing which disclose the natural potential of territories and the character of anthropogenic changes that have occurred by the present time. Such atlases were drawn up for Alaska, the State of Washington, and for the city of Anchorage. These trends include large cartographic publications in Canada, such as the "Economic Atlas of Ontario" and the "Atlas of British Columbia". They show the resource potential of these regions, give their economic estimation, taking into account the changes in the environment.

It is noteworthy that the above mentioned cartographic works were not defined as environmental. Meanwhile, in France, an independent cartographic school was founded in the 70's under the guidance of P. Osend (1977). Since 1973, selections of environmental cartographic materials (Documents, 1973–1983) began to be published with appendices in the form of environmental maps of the most interesting and important regions of France.

Lately, increased attention has been devoted to the monitoring of accumulation of technogenous elements in the environment. This problem has become an object of discussion at international symposia in Rova (1989) and Uppsala (1991). Ecologo-geochemical cartography is beginning to develop as an independent trend. It is being realized through the projects "International Geochemical Cartography", "Global Reconaissance Testing". The idea is being discussed of setting up a world center of geochemical data, standard specimens of composition and a system of continuous monitoring of information thanks to the introduction of new technology and equipment.

By now, in world cartography, a rapid development of new methods of ecologo-geographical cartography is being observed. It is based on aerospace material and powerful geographic-informational system (data bases) that make it possible to start operating computerized maps and atlases of large scale. This affords more efficient use of the cartographic-informational systems in controlling the environmental quality of territories.

In many countries, the process of creating modern cartographic geo-informational complexes to rapidly solve a wide range of problems of ecological mapping has been completed (the USA, France, Germany, Great Britain, etc.). Widely developed are new cartographic "paperless" technologies. The Electronic Atlas is quite a new type of cartographic work with unlimited geomodelling possibilities.

In this country, ecological geographic cartography also has a rich history associated with cartographic support in the solution of practical nature-protection and management problems.

This trend of thematic cartography was most actively developed in the period when cartography faced the problems of supporting the processes of developing the new regions of Siberia and the Far East and later in solving the problems of formation and development of territorial industrial and agroindustrial complexes (TIC and AIC).

The current period of ecological geographic cartography in Siberia is characterized by a rapid transition to regional and local levels of environmental study and development of principles and methods of cartographic analysis and synthesis based on new technologies, i.e., ecological geo-informational cartography.

Geo-informational cartography leads to a complete integration of methods of creating and using maps when the models of analysis and synthesis form integral chains.

In Russia, these tasks are given the status of priority problems for the nearest years, which are provided with a certain financial support on the part of Roscartographia, even under the conditions of difficult economic situation of the country.

Development of GIS and of geo-informational cartography is an important and necessary stage of the "informatization" of society and the transformation of geographical information into a social resource.

The absence, for example, of a comprehensive automated system of informational-cartographic support of cadastres of natural resources makes it impossible to control rapidly and effectively nature management in the region. This considerably delays the solution of the problem of property of these resources (at regional and local levels), economic assessment and levying of taxes for their use.

A good cartographic support of sustainable development of the region, of rational utilization of its natural resources, environmental protection, of implementation of new regional policy in the conditions of transition to the market economy is determined not only by the existence of modern material and equipment by global standards, but mainly by the degree of investigation of fundamental questions of estimation and expertise of natural resources and environmental condition.

The high dynamic nature of geo-ecological situations, factors and processes of nature and resource management dictate the necessity of their mapping in various temporal sections, at various territorial hierarchical levels, and at least in three basic directions: creation of basic cartographic works (atlases and map services), development of comprehensive cartography systems and drawing up express-information cartographic documents. That is why in 1990 a Commission of the Siberian Branch of the Russian Academy of Sciences worked out a program "Ecological Cartography of Siberia". Its main purpose is the creation of a system of operational cartographic support for solving regional problems of inventory, monitoring, expertise and forecasting of the state of natural resources, their rational management and environment protection on the basis of formation of a Cartographic Geo-informational Center (CGC) equipped with modern scientific technology.

The CGC's goal is to create conditions for operating ecological cartography based on GIS technologies. These conditions consist of the elaboration and observation of necessary requirements in order to specify, analyse, draw up and edit maps; putting all subject and combined maps on a single scale so that they could be included on this basis in the international cartographic network; efficient analysis and editing of subject maps which come to the CGC in the course of exchange and collaboration, thus making it possible to raise the level of map preparation for publication and to avoid duplication of subjects for concrete regions; orientation toward more complete and meaningful ecologo-cartographic support of the development of the national economy complex.

The experience of countries that have long since convincingly embarked on the information journey indicates that geo-informational cartography should be concentrated in a few major regional centers and supported by an integrated GIS. For example, in the North American countries, where by the early 90's there were over 4,000 GIS's at various levels, the main task has become consolidation of forces, overcoming the "numeric chaos", organization of exchange and interaction between GIS (Berlant, 1990). Instead of a single problem-oriented GIS, regional integrated GIS's are formed, and supplemented with a network of industry-oriented GIS's. The most important principle of setting up a highly developed geo-information structure is the modularity principle, which ensures exchange of masses of cartographic information, data bases and even concrete models.

In the process of comprehensive and operative ecological cartography of the Siberian region, information systems must be set up that contain data on the condition of natural resources and environment which are suited to conducting rational discussions on the control of nature management, devising of economic systems, and implementation of territorial ecological expertise at different levels.

Comprehensive and operative cartographic support can serve as an efficient means of controlling the processes of nature management and regional development on the whole. For each universal function of nature management control, special groups of thematic maps can be drawn up, and on their basis mathematico-cartographic and other models of sustainable development of the region can be elaborated.

In the Institute of Geography of SB RAS, a schematic plan of cartographic support for controlling sustainable regional development has been worked out in which maps are united into series, groups and blocks. A group of maps on the modern state of nature management contains a series of maps reflecting the present situation of environmental management and its changes under the influence of economy and population. These and resource-evaluative maps, as well, constitute an inventory-evaluation block. The block of optimization maps consists of two map groups: prospective production development and state of the environment (forecast) and nature-management organization. The next block comprises two groups of maps: nature-management condition after optimization (control), and nature-management administration (regulation). The control-map group includes series of environment monitoring and control of production processes; the group of regulation maps includes regulation in accordance with changes in the environment and deviation in the production process.

The main function of maps for regulation is representation of such parameters or methods by which it is possible to eliminate various deviations in the nature-management system in a concrete territory or to improve this territorial system for achieving the main goals of sustainable development. This, in turn, has made it necessary to evaluate the degree of vulnerability of the environment to negative anthropogenic influences, which has become a stimulus to elaboration of methods of drawing up maps of critical loads and critical concentrations, resistance and vulnerability of natural media, etc.

Combination of the available experience of creation and use of comprehensive mathematical models and methods of analysis of complicated social and ecologo-economic objects makes it possible to achieve a new quality in setting up informational cartographic systems (ICS) for support of the development of the Baikal Region as a modular, gradually integrable distributed multilevel and multifunctional system (information collection, forecast, planning, administration).

Typical blocks of such ICS are data bases, development models, expert systems and information-cartographic blocks.

In data bases, the special and thematic information about the existing and expected situation in the region is concentrated and can be obtained also by means of development models. Expert systems in application to cartography make it possible, using discrete, fragmentary, heterogeneous, incomplete information, to formulate algorithms for the complex painstaking processes of preparation and drawing up of maps, using the knowledge of the geographical situation, heuristic approaches and methods, and direct cartography. This leads to a new level the technology of creating maps on the whole and computerized cartography in particular.

Information-cartographic blocks containing complexes of basic digital thematic maps of varying scale and purpose summarize the results of modeling. This type of block, on the one hand, makes it possible to "see" the information not observed directly by man, since it is obtained on the basis of system-model forecast calculations of expected conditions in the spatial and thematic aspect. On the other hand, they can play the role of a computerized experimental testing ground on which, e.g., optimization of structure and functioning of the ecologo-economic system of the region will be followed up.

With the aid of ICS, indices and goals of development of the Baikal Region, essential factors and limitations, the existing administration mechanisms, initial prerequisites for drawing up scripts of the functioning of the region can be elucidated and analyzed. Socio-ecologo-economic models of development of the region and methods of their investigation will be created, information technology of modeling and preparation of administrative solutions will be developed. It will become possible to analyze the scenarios of sustainable development with the help of juridical and economic mechanisms, using different assumptions (of availability of nature-sparing and resource-sparing technologies, informational support of ecological education) and under conditions of multicriteriability, incompleteness of information, conflict between economic and ecological criteria, possibility of occurrence of emergency situations due to technogenous and natural catastrophes.

For setting up a regional ICS, solution of a number of problems is envisaged by means of their stage-wise realization. Among the primary tasks it will include will be the following:

- creation of a geo-information model of the territory containing the initial numerical and cartographic material (on various scales);
- development of structure, principles, methods and information support of ICS;
- working out methods of mathematical modeling of cartographical knowledge.

At the next stage is envisaged:

- summary of concrete experience of incorporating the primary model into the technology of elaboration of documentarily reliable and regularly renewable problem-oriented geo-information model of the territory for the problems of sustainable development;
- development of methods of coupling the mathematical modeling with data and knowledge in cartographic representation;
- elucidation of conditions of coding, coupling and representation of results of modeling in the cartographic geo-information model of the territory.

At the third stage is envisaged:

– operative generation of computerized maps in various applications, their normative-juridical estimation; – formation of a communication system for exchange of information with a wide circle of users, taking into account their perception level and spatial-problematic situations.

The problems of creation and utilization of ICS for analytical support of territorial development permit the planning of a number of international projects for the Lake Baikal Region with the purpose of solving some tasks of sustainable development.

Ecologo-Economic Models of the Lake Baikal Region Development

V.A.Baturin, S.N.Vassilyev, A.V.Lakeyev, A.I.Moskalenko
Irkutsk Computing Centre of SB RAS, Irkutsk, Russia

I.I.Dumova,
Baikal Institute of Rational Nature Management of SB RAS,
Ulan-Ude, Russia

ABSTRACT. Within the frames of problems of sustainable development more particular mathematical models for the scenario analysis of variants of the Lake Bailal Region's development are proposed.

Results of the scenario analysis of evolution of the region's nature-economic criteria and, also, results of solving problems of rational nature management (resouce consumption), including examples with wood resources and with resources of the unique population of omul (Baikal salmon), are given. A constructive solution for the problem of harmonizing interests of an enterprise and the region is suggested with examples of payments for the environment impact (pollution).

1. Introduction

The role of mathematical aids in solving problems of sustainable development of socio-ecologo-economic systems consists in providing: 1) descriptive analysis within the frames of some model of development processes (under definite assumptions on exogenc factors) and 2) constructive synthesis of parameters and controls providing a desired character of development within the frames of the model.

In this case, even very rough models may appear to be useful if when applying such models a man becomes careful of possible negative results of his activity.

The wish to have realistic optimistic solutions necessitates refining the model, and hence, as a rule makes it more complex from the viewpoint of both description and investigation. An attempt to overcome the model's complexity with the aid of computers is probably the deadlock way. The natural compromise is to decompose the system and to develop a system (hierarchy) of interconnected mathematical models, since furthermore a man may introduce and introduces elements of hierarchy into the structure of socio-ecologo-economic systems themselves. Existence of structure in the system of models allows to deal with the models finite in the description and more movable in the aspect of their investigation.

Sustainable development of socio-ecologo-economic systems will be considered below for the Lake Baikal Region (LBR) and defined as a complex dynamic property of controllability class, which includes the requirements of: 1) occurrence of the development trajectory in the target set of states in a definite time; 2) the requirement that the trajectory does not leave some set of "dangerless" states during the predicted time interval; 3) almost monotone increase of some development criteria (living standards, etc.); 4) asymptotic stability (stabilizability) of the trajectory.

NATO ASI Series, Partnership Sub-Series, 2. Environment – Vol. 6
Sustainable Development of the Lake Baikal Region
Edited by Valentin A. Koptyug and Martin Uppenbrink

The practical development of the regional development strategy on the basis of a system of models is based on the analysis of different scenarios from the viewpoint of above criteria of sustainable development.

But condition 4) may turn out to be insufficient to compensate the inevitable roughness of the model. Therefore, the formal scenario-type analysis must be complemented with the constructive analysis of the problem of scenario realizability. The latter problem goes beyond the frames of mathematics and science, but any other applied problem goes beyond these frames as well. The frequently observed discrepancy between the competence of governments and the complexity levels of management (control) problems in the world and its regions, the problem of concordance of criteria of justice and efficiency for a socio-economic system, the problem of increasing the feedback from diverse social groups, and other problems [1] do not exclude but require their modelling and elaboration of appropriate theories of harmonizing the parties' interests directed to providing the convergence of negotiation processes.

The present paper does not claim to solve the whole set of abovementioned problems. We describe the models within the frames of the problem of regional sustainable development and analyze the following problems: 1) structure of the general system of different level models of the region; a scenario approach and its application to analysis of LBR development (sect.2); 2) technology of solving problems of rational consumption for the forest resources (as an example) and for the endemic population of Baikal salmon (omul) (sect.3,4); 3) harmonization of interests of an enterprise and the region considered on an example of payments for polluting the natural environment (sect.5).

2. A System of Development Models for the LBR and its Subsystems

The contemporary state of control theory gives a possibility to investigate sufficiently complex nonlinear interconnected models of dynamic systems. In case of high dimensions, the analytical methods are combined with numerical modelling, and decomposition of models (decomposition of motions) is also employed, etc. The class of properties assuming its analytical investigation (for appropriate dimensions of the model) is practically unlimited (different definitions of these properties admit precise description in the language of positively constructed formulas for applying methods of nonlinear analysis and mathematical logic [2,3]). Effective methods for finding optimal controls have been developed [4,5]. In numerical multivariant constructing the development trajectories and searching for most preferable alternatives, powerful computers and methods of decision making may be effectively used [6].

Constructing the model itself, identification of the model's symbolic specification, forecasting the technological progress, the account of social, political, psychological problems in the region, etc. are more complicated problems.

The technology of simulation and finding solutions of problems of overcoming the crisis for such a complex socio-ecologo-economic object of analysis and control as the LBR is based on unique methods of mathematical modelling, systems analysis, operation research, control theory, computer science and Artificial Intelligence under the conditions of multiple criteria, conflict of interests, incompleteness and fuzziness of information, and large structural perturbations.

Besides the models of LBR proper, in order to take account of exogenous perturbations, there is need to have sufficiently aggregated models of Russia development and world

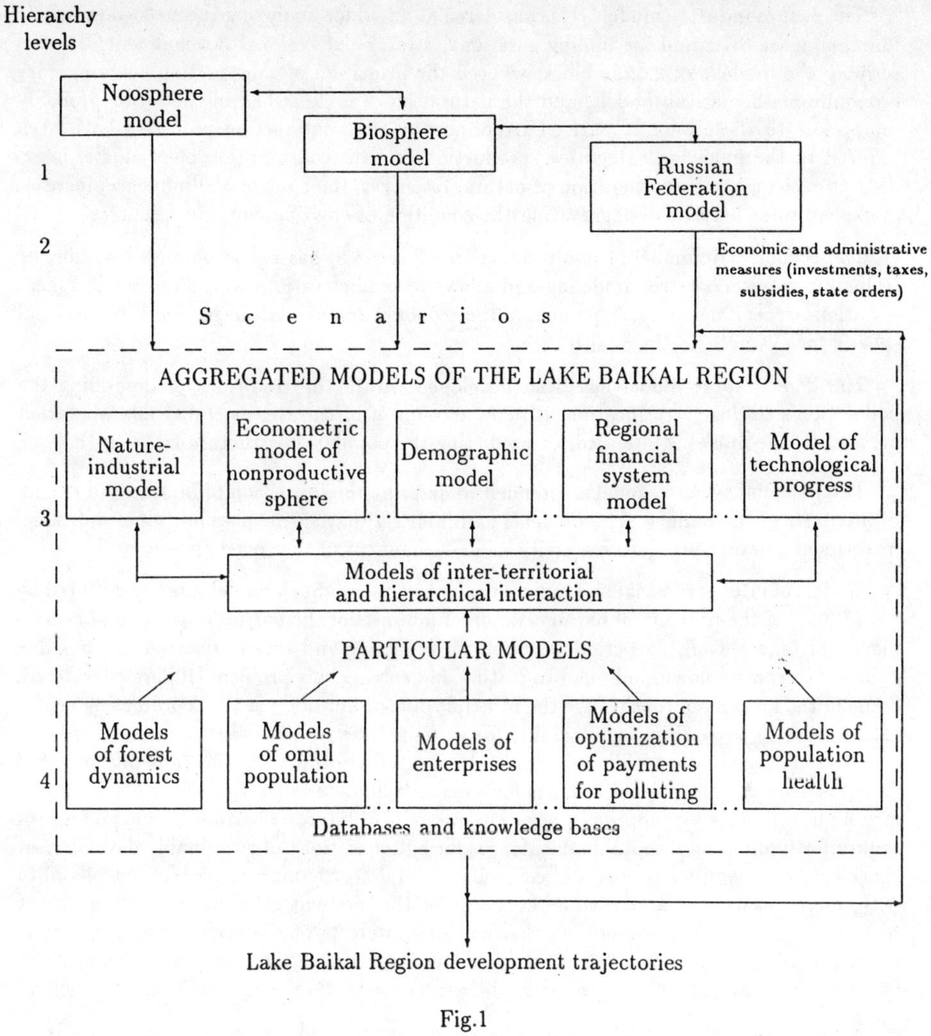

Fig.1

development (Fig.1). The global models known (such as World, LINK, GLOBUS, FUGI, GEYA, etc.) describe planetary ecologo-economic processes, global variations in the biosphere due to socio-cultural, political, military and other factors, which add to the factors of industrial impact and are considered as those of reason sphere (noosphere) which will further determine more and more the evolution of the planet. The information obtained from these models gives a possibility to formulate general restrictions on ecological and economic processes in the region, and determine social, political and ecologo-economic norms characteristic of the whole world. As far as the present work is concerned, in the process of constructing regional development scenarios, such information has not yet been used, particularly due to the fact that the objective of modelling has been finding the ways of bringing the LBR from crisis to the sustainable development trajectory rather than futurologic predicting. So the interval of consideration has been limited by 2005 (the time period for which biospheric and many other global processes have not been taken into account).

The nature-industry model [7] is considered as a tool for analyzing the ecologo-economic functioning of LBR and for finding a rational strategy of regional development. For describing the model's economic block we used the principles of constructing contemporary economic-mathematical models, and the natural block is described in the terms probably more close to the economic part. Furthermore, various interaction processes have been reduced to the following categories: production growth, consumption of products, inter-district exchange, self-regeneration of natural resources, their reciprocal influence, increase of expenditures because of aggravating the condition of environment and resources.

The econometric model of non-production sphere [8,9] has been completed within the technology of econometric modeling and allows to predict various aspects of social sphere variations depending on developement of the economy and capital investments in the social sphere, for example, in the health care.

The demographic model has been developed within the traditions of describing the population's sex/age stratification; it takes account of birth-rate, mortality and migration processes. The model is intended for predicting the population structure in the LBR.

The financial system model is intended to describe the formation of income and expenditure parts of the budget. The model is useful for estimating the possibilities of industrial investments, nature protecting activity and development of the social sphere.

Models of inter-territorial and hierarchical interactions. Such models are necessitated by importance of the problem of harmonization of interests of the parties to the end of removing conflicts, especially in periods of political, economic and social crises. A few possible approaches (both known and new) to stating and solving this problem [10] are considered. Some of the known approaches are the principle of coordination via the coordinating centre (M.Mesarovic); providing Nash stability for system states in noncoalition games; introduction of coalitions and providing dynamic stability of a compromise in cooperative games (L.A.Petrosyan); concordance of plans for a region and state industries (K.A.Bagrinovskii), etc. As new models we propose: a negotiation model of interests harmonization and a procedure for finding an acceptable decision; sufficient conditions of sustainable development considered as a complex property of controllability (almost monotone, stable controllability with phase constraints and with an estimate for the moment of occurrence in the target set), an ecologo-economic model for harmonizing interests of the territory and its enterprises [11,12]. These approaches to the problem of harmonization of interests of the parties (subsystems, players) allow us to model different aspects of realistic situations in practice.

A model of technological progress has been constructed on the basis of analysis of numerous information sources about technological solutions and give forecasts of the rate of perspective changes of technological indices and relationships.

Fig.1 also represents some more particular 4th-level models providing 3d-level models with the necessary information. Note, there are models for repulsion of enterprises to the possible changing tax rates.

There are problems with information filling the models [13] since :

1. It is difficult to determine the character and rates of further development of Russia and the former Soviet Union. Processes of privatization and demonopolization of industries, fixing free prices and tariffs impact both industrial and social goals of any region and the motivation of its economic links. All regions of Russia are characterized by industrial decrease, strong inflation, limited investments. We need additional information on both law and economic aspects of the market tendences in Russian territorial-administrative regions.

2. It is difficult to predict the situation of development in Siberia on the whole. Setting up such regional economic associations as "Siberian Agreement" and, also, the so called free economic zones may both reinforce and weaken the positions of LBR territories.

3. Perspectives of the LBR industrial complex development are uncertain. The present specialization of industrial branches of LBR does not give any possibility to reorient enterprises in a short time for producing the consumer goods essential for intraregional needs.

4. The territory of the LBR is too large. Interests of the LBR districts are quite different as well as the initial levels of socio-economic development.

All above aspects ground the usage of the scenario-type approach to analysis of LBR development.

Characteristics of some scenarios are given in Fig.2.

The scenario of retaining the industrial+raw-material orientation of the regional economic complex is the "basic" one. According to the "normative scenario", the state of the water and air environment may be improved with respect to the "basic" scenario at the expense of resource-saving technologies beginning from 1995. But it is obvious that direct nature-restoring activity is needed to compensate the supernormative load on the natural environment. The comparative analysis of scenarios shows that the "normative" scenario

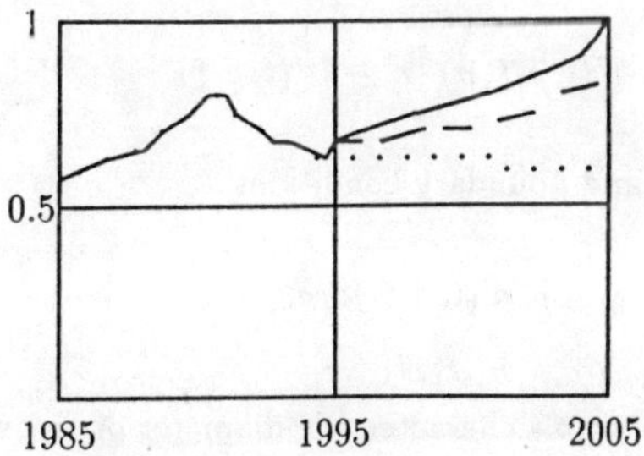

Total regional produce.

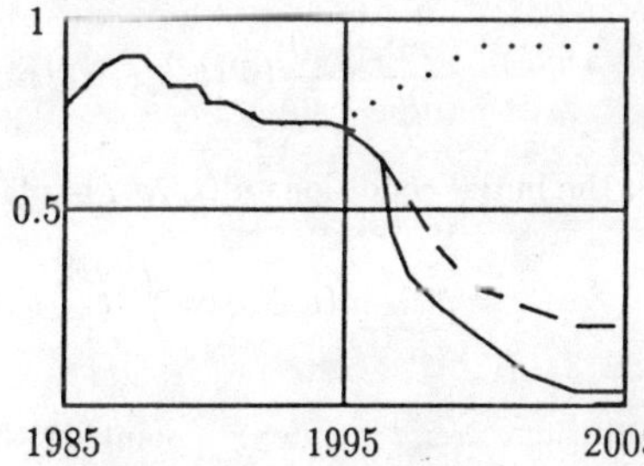

Average annual concentrations of scaled pollutions in river mouths vers. the background value.

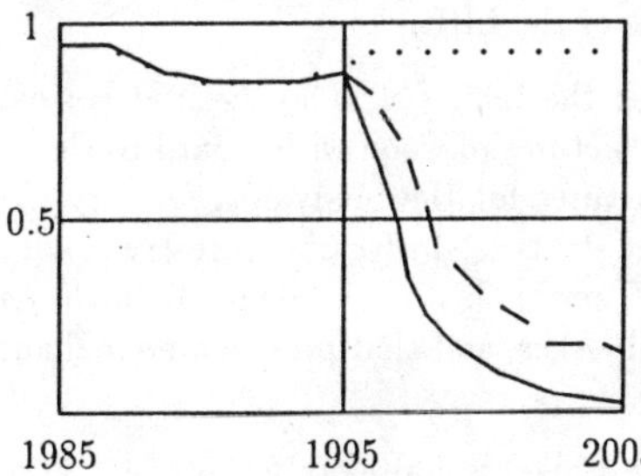

Excess of air pollution vers. the background.

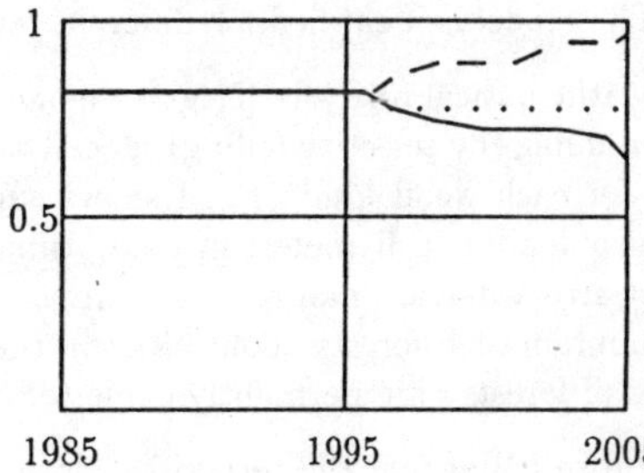

Scaled biological resource.

Several scenarios are listed:
· · · - basic; — — - recreative; ——— - normative.

Fig.2

allows to pass to the acceptable development trajectory. But this will require substantial capital investments for reconstructing the economy (about 6 billion roubles for 10 years in the prices of 1985). Analysis of both the tax policy and the taxation system shows insufficiency of LBR financial resources.

The following two sections discuss possible solutions of more particular problems of sustainable development via examples with rational forest management and payments for environment pollution.

3. A Model of Rational Forest Management

The problem of sustainable development in the aspect of consumption of natural resources, which are capable of self-regeneration, consists in finding strategies of permanent and nonexhaustive (nondestructive) usage of resources. Since the needs in natural resources grow steadily, the problem of determining the maximum possible nonexhaustive usage of resources is the principal one in rational resource management.

A variant of the model of spatial forest structure dynamics, which takes account of both the natural content and the diameter distribution of wood trunks has been developed [5,7,14,15]. The model may be written as a system of partial differential equations

$$\frac{\partial n^i}{\partial t} + \frac{\partial}{\partial \rho}(\vartheta^i(t,\rho,\xi,n)\cdot n^i) = -d^i(t,\rho,\xi,n)\cdot n^i - U^i(t,\rho,\xi)$$

with the initial conditions $n^i(t_н,\rho,\xi){=}n^i_н(\rho,\xi)$ and boundary conditions

$$n^i(t,\Delta,\xi){=}\int_A\int_\Delta^{\rho^i_m} b^i(t,\rho,\xi,\xi_1,n)\cdot n^i(t,\rho,\xi_1)d\rho d\xi_1.$$

Denotations are: t - time; ξ - point on the plane; ρ - characteristic diameter of the wood trunk; Δ - diameter of young growth; i - wood kind number; $n^i{=}\ n^i(t,\rho,\xi)$ - forest density ($n = (n^1, n^2, n^3$, where 1 - dark conifer, 2 - light conifer, 3 - soft deciduous forest); ϑ^i - tree growth rate in width; d^i - dying off coefficient; b^i - coefficent of resumption and territorial distribution of forests; $U^i(t,\rho,\xi)$ - rate of wood felling.

The model is identified for different territories of the LBR.

Mathematical analysis [5] gives evidence that the felling strategy desired consists in maintaining (by selective felling) special target structures of wood with regard to the thickness for each wood kind. Fig.3 shows such structures for the Sludyanka forestry. There are two dominant diameters in the optimal target struture. In Russian forestry slang such forest structures are named 'two-humpbacked'. There is historic evidence that they used to maintain such forests about cities in the Middle Ages, and that provided nonexhaustive usage of forests on a permantly exploited territory.

Entire fellings on the territories close to Lake Baikal are already forbidden. So introducing the forest target structures in the forest exploitation practice could be opportune.

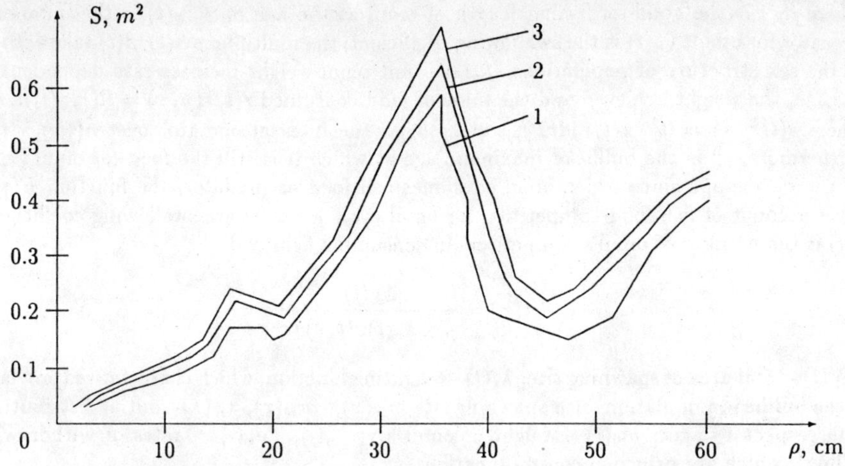

Fig.3. Distribution of the sum(S) of tree section areas with respect to diameter (ρ). for Badan kind forest: 1 - dark conifers; 2 - light conifers; 3 - leaf-bearing forest.

4. Modelling and Control of the Omul Population with Account of Fish Nutrition Dynamics

We developed a refined model for solving problems of forecasting, optimal exploitation and reproduction of Baikal omul with account of the most important component of its nutrition - the food represented by the population of bullhead. The first version of the model was described in [16].

Let t be time, τ - age. The age density of omul quantity $n(t,\tau)$, the average weight of one fish $m(t,\tau)$, and the age density of bullhead biomass $\pi(t,\tau)$ are described by the equations:

$$\frac{\partial n}{\partial t}+\frac{\partial n}{\partial \tau}=d_n(t,\tau,Q(\tau,m))n-\chi_n(\tau,n)\chi(t)-u(t,\tau),$$

$$n(t,0)=\alpha(t)\int_0^{\tau^m} b_n(t,\tau,Q(\tau,m(t,\tau)))n(t,\tau)\,d\tau,$$

$$\frac{\partial m}{\partial t}+\frac{\partial m}{\partial \tau}=f_m(\tau,m,r(t,\tau,n,\pi)),\quad m(t,0)=m_0(t),$$

$$n(t_H,\tau)=n_H(\tau),\quad m(t_H,\tau)=m_H(\tau),\ \tau>0,$$

$$\frac{\partial \pi}{\partial t}+\frac{\partial \pi}{\partial \tau}=d_\pi(t,\tau)\pi-\kappa_\pi(\tau)\int_{\hat{\tau}}^{\tau^m} r(t,\tau;n,\pi)n(t,\tau)dt-$$

$$-\kappa_\gamma(\tau,\pi)\gamma(t)-u_\pi(t,\tau),\ \pi(t_H,\tau)=\pi_H(\tau),\quad \tau>0,$$

$$\pi(t,0)=\beta(t)\int_0^{\tau_\pi^m} b_\pi(t,\tau,S_\pi(t,\tau))\pi(t,\tau)d\tau.$$

Here: $d_n(\cdot)$ (resp., $d_\pi(\cdot)$) and $b_n(\cdot)$ (resp., $b(\cdot)$) are coefficients of natural dying and birth; and τ^m (resp., τ_π^m) is the maximum omul (resp., bullhead) age; $Q(\tau,m)$ - Fulton's well-feeding quality,

$$Q(\tau,m)=\frac{m}{l^3(\tau)}100\%,$$

where $l(\tau)$ is the good-for-fishing length of omul at the age of τ; $\chi(t)$ is the number of predator for omul; $\chi_n(t)$ is the swallowing coefficient; the multipliers $\alpha(t), \beta(t)$ take account of the sex structure of populations; $f_m(\cdot)$ - unit omul weight increase rate dependent on the age, the weight achieved and the mass of food consumed $r(t,\tau;n,\pi) = R(\tau, z(t;n,\pi))$, where $z(t;n,\pi) = \int_0^{\tau_\pi^n} \pi(t,\tau)d\tau / \int_{\hat\tau}^{\tau^m} n(t,\tau)d\tau$ is the integral operator over $n(t,\cdot), \pi(t,\cdot)$, furthermore, τ_π^n is the bullhead maximum age at which it is still the food for omul ($\tau_\pi^n < \tau_\pi^m$), $\hat\tau$ - the age since which omul consumes fish food as predator; the function $R(\tau,z)$ takes account of fish-food competition for omul ; $\kappa_\pi(\cdot)$, $\kappa_\gamma(\cdot)$ are swallowing coefficients; $\gamma(t)$ is the number of omul's competitors in consuming fish-food;

$$S(t;\pi) = \frac{S_\Sigma(t)}{\int_0^{\tau_\pi^m} \lambda_\pi(t)\pi(t,\tau)d\tau}$$

($S_\Sigma(t)$ - total area of spawning site; $\lambda_\pi(t)$ - weighting function, which characterizes provision of the bullhead population with spawning site; $n_H(\tau)$, $m_H(\tau)$, $\pi_H(\tau)$ - initial distributions with respect to ages; $m_0(t)$ - weight of omul sry; $u(\cdot)$, $u_\pi(\cdot)$ - rates of withdrawing (filling), which are principal control functions.

Planned tasks of fishing may be represented by the equalities

$$\int_0^{\tau^m} r(t_k,\tau)d\tau = R, \quad \int_0^{\tau_\pi} r_\pi(t_k,\tau)d\tau = R_\pi,$$

where $r(t,\tau)$, $r_\pi(t,\tau)$ - are auxiliary phase coordinates,

$$\frac{\partial r}{\partial t} = m(t,\tau)u(t,\tau), \quad r(t_H,\tau) = 0, \quad \frac{\partial r}{\partial \tau} = u_\pi(t,\tau), \quad r_\pi(t_H,\tau) = 0.$$

Age densities of quantity undergo natural constraints $0 \le n$, $0 \le m$. As the optimality criterion we consider the functional $I = \int_0^{\tau_\pi^m} m(t_k,\tau)n(t_k,\tau)d\tau$, which has the sense of the total omul population biomass at the final time moment. The optimal control problem consists in maximization of the functional I on the set of functions $n, m, \pi, r, r_\pi, u, u_\pi$ and consists in obtaining such a fishing strategy that would allow to fulfil planned tasks with the best value of the criterion I , which characterizes well-being of the omul population at the final time moment.

The practical identification of the model has been made on the background of literature data and experimental materials of the ichthyology laboratory of the Institute of Biology in Irkutsk State University.

The initial state of the system "omul - bullhead" corresponds to that of 1978, when due to both devastating fishing and abrupt rise of the water level in Lake Baikal (erection of Irkutsk hydroelectric power station) the system was in the pathological state. The optimal solution obtained determined an optimal fishing strategy, which ensures the transit of the system from the pathological to the "well-being" mode. Furthermore, besides biomass (quantity) which is of paramount interest for us, the optimal solution contains also the optimal age structure.

Results of computations for this variant are presented in Fig.4. Let us analyze the results. Firstly, note $N(t_H) > N_H$, $M(t_H) > M_H$, where

$$N_H = \int_0^{\tau^m} n_H(\tau)d\tau, \quad M_H = \int_0^{\tau^m} m_H(\tau)n_H(\tau)d\tau,$$

$$N_H = 150,376mln., \; M_H = 8344.8t$$

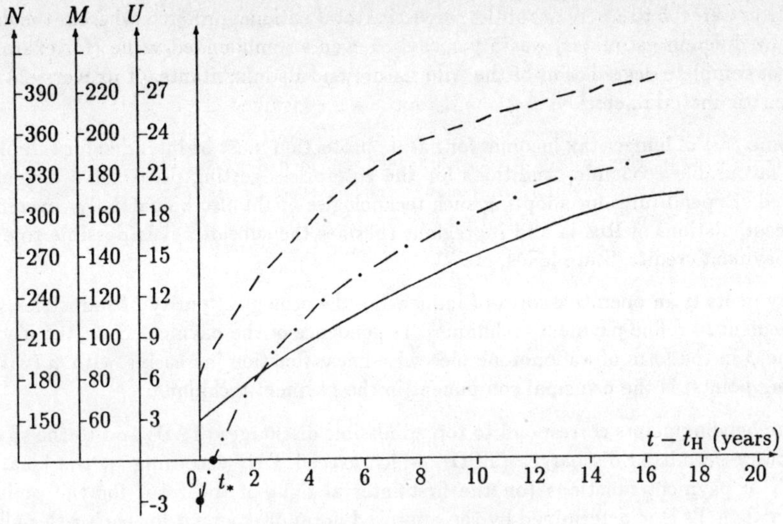

Fig.4. Dynamics of principal characteristic of the omul population:
—— - total quantity (N, mln.);
— — - total biomass (M, hundreds tons);
—·— - total rate of fishing (U, hundreds tons).

Formally, this gives evidence of the necessity of immediate replenishment of omul population. Since it is practically impossible, the values of $(N(t_*) - N_H)$, $(M(t_*) - M_H)$ may serve only for computing the required time $(t_0 - t_H)$ of industrial fishing prohibition needed to bring the system into the state $N(t_0) = N(t_*)$, $M(t_0) = M(t_*)$ by measures for improving inhabitance conditions and may be also correcting fishing. The industrial fishing starts from the moment $t = t_0$ (this corresponds to the graph's shifts right by $(t_0 - t_*)$ in Fig.4) and increases from zero to some quasi-stationary value. In this case, the total biomass and omul population quantity also take quasi-stationary values.

5. Harmonization of Interests of the Parties in Cases when Enterprises are Obliged to Pay for Polluting the Environment

There are different ways to stimulate nature protection. Let us consider compulsory ways, i.e. payments for polluting the environment. As far as 'limiting of discharges' mechanism (with storing the "bank of permissions") is concerned, its shortcoming consists in the necessity to have a large number of agents in the supply and demand market of discharge limits. The form of direct management, which would require the obligatory usage of definite technologies, is also less attractive.

Furthermore, note, when speaking about payments, that the technique of payments accepted in Russia and approved (in the main part) by law cannot solve the problems of nature protection since ecological funds made by such payments in the centre and the regions are too poor. As a more effective way we consider combining such payments with with appropriate taxes, which could form a fixed fraction of total national (or regional) product specially intended to be used for nature protection. For example, according to the published data, in recent years, expenditures for the purposes of pollution level control in the

EC states were 0.5 to 2 per cent of the respective total national product, whereas the damage cost (by different estimates) was 5 per cent or even an unbounded value (for example, in cases of complete degradation of the wild nature and its inhabitants, of irreversible loss of the genetic material, etc.).

Some part of budget tax incomes for nature protection must be intended for establishing more favourable economic conditions for the enterprises getting progressive technologies adopted. Expenditures for adopting such technologies might also be partially covered from tax accumulations of Russia and its regions (besides the subsidies it is possible to employ easy payment credits, State loans, etc.).

Payments is an operative form of influencing the ecological purity of industries. So, it is expedient to refine payment techniques. Dependence of the payment S on the discharge volume φ in the form of a monotone piecewise-linear function increasing with φ (with two bending points) is the principal component in the payment technique.

The bending points correspond to top admissible discharges (TAD) and to the so called temporary permitted discharges (TPD), which exceed TAD according to the idea. The growth of payment functions (on the first interval right of the zero) for the pollutions smaller than TAD is determined by the approved norm of payment for each i-th pollution type; on the next interval, between TAD and TPD, the rate of growth increases by 5 times; finally, beyond TPD, the growth rate is 25 times as large as that of the initial growth.

The dependence $S(\varphi)$ itself is not correlated with the mass and cost of the produce. To concord interests of the territory and an enterprise we need a more flexible form of payments. Particularly, it is expedient to connect selection of parameters of $S(\varphi)$, as of the principal component of the payment function, with the enterprise's manufacturing activity.

A manufacturing model of an enterprise is used. By control of the manufacturing $v(t)$ and of the purification degree $z(t)$ an enterprise may maximize its profit. By varying the parameters of the payment function under consideration, the territorial (and central) administration may regulate the increase of produce output and decrease of pollution discharges.

The problem of harmonization of interests of the parties implies selecting such payment function's parameters that an enterprise in the course of maximizing its profit would simultaneously guarantee the Pareto optimum for territory's criterial functions. It is natural to consider this problem in a wider class of functions, allowing for breaks at earlier bend points of $S(\varphi)$ (see Fig.5).

It has been proved [11] that the harmonization problem can be solved if and only if the enterprise is profitable. In the latter case, parameters of the payment function must satisfy such found, easily verifiable, algebraic relations as equalities and inequalities with respect to indicated parameters and parameters of enterprise's operation. For example, if the linear convolution of the territory's criterial functions has the form $\eta_1 v - \eta_2\varphi(v,z)$, where $\varphi = \xi v - z$ is the residual pollution (ξ - const) then for $\eta_1 - \xi\eta_2 > 0$ satisfaction of the following estimates is one of possible variants of sufficient conditions of harmonizability

$$k_1 < \xi v_0 - z_0 < k_2, \quad b < \mathrm{tg}\beta < \frac{1-a}{\xi}, \quad \xi R_1 < (1 - a - \xi \mathrm{tg}\beta)(\xi v_0 - z_0 - k_1),$$

where v_0, z_0 are the maximum possible output and the maximum possible degree of purification; a and b are some derived parameters of the enterprise's manufacturing process.

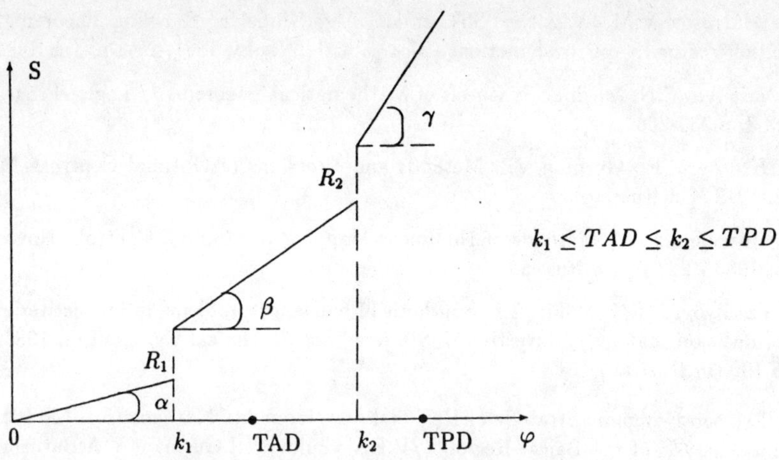

Fig.5. The function of pollution payments $S(\varphi)$

6. Conclusion

In sect.1 we noted that it is important to develop the theory of parties' harmonization. The game theory and the theory of economic behavior may form its basis. Problems of conservation of "the green lungs" in the LBR, designing, erection and exploitation of large-scale enterprises, which are of inter-territorial interest (e.g. Irkutsk hydroelectric power station), and some other problems require, from the theoretical viewpoint, the transition from the coalitionless descriptions to the co-operative one, as well as necessitate investigation of problems of existence of dynamically stable compromises. Classical examples from the game theory show that for an obvious absense of antagonism of interests of the parties, but when co-operative tendences are insufficient (when the co-operative game is nontransferrable), the players have a gain smaller than they might have.

As other bases for harmonizing the interests of the parties we may consider the theory of multicriteria decision making, and also the theory of negotiation processes, which are actively developed presently as well as the theory of small groups, coalitions, and the theory of leadership [18].

Furhtermore, since it is possible to solve regional problems strategically only in the context of solving global problems, it is important to be able to construct and anaiyze global and multiaspect world models [19]. In particular, using these models, it is possible, to our mind, to find comparative estimates of (i) justifiability and (ii) realizability of known conceptions of world development: "Agenda 21", "Golden Billion" and "Multipolar World" (according to the last conception, there must be at least two superpowers or supercoalitions).

References

1. Koptyug V.A. Siberia - problems of ecological policy and sustainable development // Proc. of the All-Russia Conf. on Siberia Ecological Development.- Novosibirsk, 1993, pp.88-109. (in Russian)

2. Matrosov V.M., Vassilyev S.N. et al. Algorithms for Deriving Theorems of the Method of Vector Lyapunov Functions.- Novosibirsk: Nauka, 1981.- 283 p. (in Russian)

3. Vassilyev S.N. Machine Synthesis of Mathematical Theorems // Logic. Progr., 1990, 9, No.2-3, p.235-266.

4. Krotov V.F., Gurman V.I. Methods and Problems of Optimal Control.- Moscow: Nauka, 1973. (in Russian)

5. Moskalenko A.I. Methods of Nonlinear Mapping in Optimal Control.- Novosibirsk: Nauka, 1983.- 222 p. (in Russian)

6. Vassilyev S.N., Seledkin A.P. Synthesis of efficiency problems in multicriterial problems of decision making // Izvestiya USSR Acad.Sci., Technical Cybernetics, 1980, No.3, pp.186-190.(in Russian)

7. Ecologo-Economic Strategy of Regional Development: Mathematical Modeling and Systems Analysis of the Baikal Region. /V.E.Vickulov, V.I.Gurman, V.A.Baturin et al.- Novosibirsk: Nauka, 1990.- 184 p. (in Russian)

8. Potorochenko N.A. Mathematical model of social development for the Baikal Region // Proc. of the All-Russia Symp. "Problems of Analysis of Sustainable Development and Strategic Stability".- Irkutsk, 1994, Vol.2, p.144-152. (in Russian)

9. Matrosov V.M., Golovchenko V.B., Noskov S.I. Modeling and Forecasting Indicators of Regional Socio-Economic Development // Novosibirsk: Nauka, 1991.- 144 p. (in Russian)

10. Abdullin R.Z., Vassilyev S.N., Lakeyev A.V., Moskalenko A.I. Scenario analysis of state industry degradation, and harmonization of interests for providing sustainable development// Proc. of the Conf. "Regularities of Social Development - Orientations and Criteria of Future Models".- Novosibirsk, 1994. (in Russian)

11. Vassilyev S.N., Baturin V.A., Lakeyev A.V. Ecologo-economic Model and Solvability of Harmonization Problem // Proc. of IEEE Int. Conf. on Systems, Man and Cybernetics, France, 1993, Vol.5, pp. 339-343.

12. Vassilyev S.N., Lakeyev A.V., Moskalenko A.I. Aids of mathematical modeling for analysing the problem of Baikal Region sustainable development // Proc. of the All-Russia Symp. "Problems of Analysis of Sustainable Development and Strategic Stability".- Irkutsk, 1994, Vol.2, p.102-121. (in Russian)

13. Baturin V.A., Dumova I.I., Tulokhonov A.K., Vassilyev S.N. Modelling of Processes of the Region Development//In: The Present and Future State of the Baikal Region (Possibilities for Sustainable Development). Part III: Perspectives on Sustainable Development, Novosibirsk, SB of RAS, pp.48-51.

14. Models for Control of Natural Resources./ Baturin V.A.,Gurman V.I., Moscalenko A.I. et al.- Moscow: Nauka, 1981.- 264 p. (in Russian)

15. Optimal Control of Natural-Economic Systems./ Belchanskii G.I., Gurman V.I., Moskalenko A.I. et al.- Moscow: Nauka, 1980.- 294 p. (in Russian)

16. Butin A.A., Moskalenko A.I., Toporkov I.G. Mathematical model of a fish population under limits on the food // Modeling Processes in Nature-Economic Systems.- Novosibirsk, 1981, p. 221-232. (in Russian)

17. Moskalenko A.I. Mapping Methods in Optimal Control of Distributed Parameters. Doctoral Thesis (mathematics). Irkutsk, 1993.- 253 p. (in Russian)

18. International Multilateral Negotiation: Approaches to the Management of Complexity / I.W.Zartman, ed., San Francisco: Jossey-Bass, 1994.

19. Matrosov V.M. Models, methods and informational technology of unstable development analysis // 2d Intern. Workshop "Analysis of Development Instability on the Basis of Math. Modeling".- Moscow, 1993, pp. 5-34.

Sustainable Development and Tourism. Ecotourism: A Popular Slogan or a Powerful Tool?

Bernhard Müller
Lehrstuhl Raumordnung
Institut für Geographie
Technische Universität Dresden
Mommsenstrasse 13
D-01062 Dresden
Germany

Sustainable development has become a popular catchword in the international development debate. Criticised as ambiguous and open to a wide range of interpretations, the meaning of the term was specified in Agenda 21 of the United Nations Conference on Environment and Development in Rio de Janeiro and in successive proceedings. Although tourism is hardly mentioned in Agenda 21, it is closely related to sustainable development at least in two ways: first, as one of the most important economic sectors worldwide, and second as a tool to actively support sustainable development by raising public awareness of environmental matters.

The connection between tourism and sustainable development is of special relevance to concepts of nature protection. Such concepts are shifting from preservation to conservation. This is based on the assumption that nature reserves may more easily be protected in the long run if they generate economic benefits for national governments and for the local population living in natural areas. Ecotourism, though a vaguely defined and variably used term thus far, is widely considered to be an appropriate instrument for increasing the attention to nature-protection projects.

Against this background, ecotourism is defined and analysed in this article as an instrument for facilitating the realisation of sustainable development. It is argued that ecotourism activities should be developed cautiously and only be encouraged if a number of requirements are met. Fields of activities for the promotion of ecotourism are presented. Finally, a concept of Concerted Tourism Management (CTM) is introduced and discussed. It is argued that the Lake Baikal Region could not only become a world model for sustainable development, but also a forerunner in the creation of new concerted management practices in endangered areas if this concept was applied.

Tourism: A Threat or an Opportunity for Sustainable Development?

Sustainable development has become one of the most popular terms in the international development debate of the nineties. According to the World Commission on Environment and Development (WCED), the term refers to

NATO ASI Series, Partnership Sub-Series, 2. Environment – Vol. 6
Sustainable Development of the Lake Baikal Region
Edited by Valentin A. Koptyug and Martin Uppenbrink

development that meets the needs of the present without compromising the ability of future generations to meet their own needs. Criticised as ambiguous and open to a wide range of interpretations, the meaning of the term was specified in Agenda 21 of the United Nations Conference on Environment and Development in Rio de Janeiro and in subsequent proceedings.

Although tourism is hardly mentioned in Agenda 21, it is closely related to sustainable development in at least two ways:

- Worldwide, tourism is one of the most important industries. In many countries it is an important source of income, and it provides a great variety and number of job opportunities. If tourism is not carried out in an environmentally friendly and sustainable way, it may not only severely affect the natural and the socio-cultural environment, but also destroy its own economic base.

- Tourism has a potential to actively support sustainable development. For example, tourism may raise public awareness of environmental matters, and it may also become an instrument for increasing the attractiveness of conservation strategies.

Various initiatives have recently proven that decision makers at all levels, the local population as well as international donors, are becoming more and more interested in tapping the potentials of tourism in natural areas. For example, they argue that ecotourism could become an integral part of the development strategies for the Lake Baikal Region and support the efforts to make Lake Baikal a world model for sustainable development (cf. The Lake Baikal Region in the Twenty-first Century 1993, The Russian Academy of Sciences (Siberian Branch), Irkutsk and Buryat Scientific Centres 1994).

However, the results of the two conferences on environmentally friendly tourism development in Russia—held in Berlin in 1992 and in Listwjanka in 1993—have illustrated that tourism may also conflict with sustainable development (cf. BMU 1993, Resolution der 2. russisch-deutschen Konferenz 1993). Tourism may

- become a financial burden for the public sector because high public investments are usually required before tourism starts,

- create unfavourable labour market structures,

- contribute to cultural disruption and speed up the erosion of social structures as has been experienced in many tourist places around the world,

- lead to a decrease in autonomy of regions or nations due to growing dependence on external investments, and

- have a negative impact on the environment, resulting in air and water pollution, land "consumption", degradation of flora and fauna.

Moreover, one should bear in mind that action potentials are unequally distributed among those involved in tourism. The population living in a tourist area fully depends on the sustainable utilisation of the potentials. If tourism is not carried out in a sustainable way and if, as a consequence, tourism deteriorates after some time, people living in a tourist area may be deprived of the ability to meet their own needs. In contrast, those who promote and organise tourism on national or international levels have a wider range of possibilities to react to adverse trends endangering the continuation of tourist operations in a specific area. They are also affected by the individual undertaking. However, tourist entrepreneurs or investors still may have the choice to develop and market similar but different tourist destinations worldwide.

Finally, one should pay attention to the fact that even if tourism is carried out cautiously, it tends to grow into a big business if a tourist site is attractive or has been marketed successfully as an attractive destination. If limits of acceptable use are neglected, this may have harmful effects on the environment and endanger sustainable development. This is of special relevance for natural areas in which touristic capacities are usually very low.

Ecotourism: A New Magic Formula for Sustainable Development Through Tourism?

Recently ecotourism has been discussed with increasing interest. Many politicians and experts regard it as a new magic formula for harmonising economic development and environmental concerns. Travel agents, tour operators, local dwellers and non-governmental organisations unanimously regard it as a suitable tool for combining business interests with nature protection. Moreover, ecotourism usually coincides with the "high income-low impact" dreams of its promoters worldwide. Many experts and decision makers expect that the natural potential of an area can be easily tapped and protected at the same time if an ecotourism strategy is implemented.

However, what has become a popular slogan is in fact a highly disputed and very vague term. Ecotourism is related to responsible travel to natural areas that conserves the environment and sustains the well-being of local people (cf. The Ecotourism Society 1993). The term stands for new environmentally friendly and socially acceptable forms of travelling as a part of a culture and society that is oriented towards ecological principles (cf. Grenier et al. 1993). It aims at providing a programmatic orientation rather than a scientifically correct differentiation.

Ecotourism has to be seen in the light of integrated ecologically oriented development approaches. These are characterised by principles of regional autonomy and based on concepts of self-reliance as formulated by Sachs (1980) in his concept of ecodevelopment in the late seventies.

However, the term ecotourism has meanwhile been used very ambiguously and in an inflationary way. A recent review of tourist literature has shown more than 30 similar terms, most of them no less problematic than the term ecotourism itself (cf. Scace et al 1992):

– Terms such as alternative tourism, ecological tourism, ecotravel, ecotripping, ecoventures, environmental tourism, ethical travel, green tourism, resource-based tourism, sustainable tourism are used as generic terms.

– Terms such as natural-areas travel, wilderness tourism, rural tourism, nature-oriented tourism, nature tourism, nature vacations, tourism to protected areas, jungle tourism refer to destinations of tourists.

– Terms such as appropriate tourism, low-impact tourism and socially responsible tourism reflect the intended socio-cultural impact of tourism.

– Terms such as adventure travel, non-consumption tourism, consumption tourism (e.g., controlled hunting and fishing for sport), cultural tourism, ethnic tourism, anthropological tourism, primitive and remote travel, safari tourism, science tourism, soft-adventure tourism or special-interest tourism refer to interests and activities of tourists.

In order to clarify the concept of ecotourism here, we may understand it as a form of tourism to natural—often protected—areas which is conducted in a responsible way, and which aims at minimising any negative impact on the environment, avoiding adverse socio-cultural effects, mobilising funds for environmental activities and generating income for the population living in the tourist area. Thus, ecotourism is to be distinguished from nature tourism which indiscriminately comprises all forms of tourism to natural areas.

What we know about ecotourism in general is very limited. For example, we know that ecotourism is a growing segment of tourism. However, we know very little about how fast it is growing, whether there is stable growth, in which direction ecotourism is developing and how it is structured.

If one looks at some of the results of so-called ecotourism projects worldwide, one may find a number of undertakings which are at a standstill (Arbeitsgruppe Ecotourism 1994, Boo 1990). In many cases the revenue potentials have turned out to be far less considerable than expected. This is because in many cases projects are started as rather unprofessional undertakings.

A major drawback is that policy statements and project documents usually don't specify which type of tourist the tourism industry is addressing. For example, the "hard core" nature tourist and the "casual" nature tourist may be distinguished as two types with extremely different demands. The hard-core nature tourist has a very specific interest in nature and relatively low demands on infrastructure. In contrast, the casual nature tourist has a superficial interest in nature; he is interested in spectacular sites and his demands regarding infrastructure are comparatively high. In between these two groups, "dedicated" nature tourists and "mainstream" nature tourists may be distinguished (cf. Lindberg 1991).

In many cases, projects are primarily based on vague ideas and initiatives of local groups who are interested in tapping assumed tourist potentials of their home area. Expectations are usually very high. However, the local groups often lack technical know-how, business experience and connections with the national and international markets. This makes it difficult to succeed. Moreover, proper market analyses are lacking. Planning and management is weak. Consequently, tourist undertakings fail. Most of these projects may be understood as nature-tourism efforts, but none of them should be called an ecotourism undertaking in the true sense.

Ecotourism as a Strategic Choice—What Should Be Observed?

Against this background, it becomes clear that ecotourism should be understood as a rather sophisticated undertaking. Ecotourism is a strategic choice. However, it is advisable to make this choice, i.e., to promote or develop ecotourism projects, only after determining whether the conditions for ecotourism are favourable. Besides the tourism-related evaluation of the specific conditions of an area, the in-depth analysis of its tourist potential and a proper market analysis, one should be aware of five major aspects that need to be considered.

(1) One should find out whether tourism activities are apt to contribute to the improvement of the living conditions of the people living in a tourist area. This means that profitable undertakings are needed. Only feasible projects provide possibilities for deriving revenues that can be channelled to the population and the local communities. It makes no big difference whether benefits are collective benefits or individual benefits. Collective benefits are, for example, tax revenues or income from land leases which may be added to the local budgets and spent for local projects. In contrast, individual benefits provide direct income for the population. The demand for profitable projects gives ecotourism a certain bias. Ecotourism should not be left to local initiatives only. There is some evidence that in many cases these initiatives are rather idealistic, but also rather ill-prepared. Ecotourism is more likely to succeed if it is developed and managed with the assistance of experienced promoters and tourism professionals with strong relations to the national and international markets.

(2) One should ascertain whether tourism activities can be expected to introduce innovations with regard to attempts to solve regional and sectorial problems. Environmentally friendly tourism in a true sense has to adhere to high environmental standards, and it requires refined procedures. Thus, certain innovations, for example with regard to building technology, energy, water consumption, waste-water treatment, should be expected from ecotourism development. However, one should bear in mind that the breakthrough of innovations is always very much dependent on a social and a politico-administrative climate that encourages innovations and social communication. Consequently, a favourable socio-cultural and politico-administrative environment is highly relevant for the development of ecotourism.

(3) It is advisable to analyse whether and to what extent politico-administrative structures allow for a fair sharing of revenues between private investors, the state

budget, environmental authorities or agencies and local communities. Environmental protection may be weakened, and the co-operation of the population in the protection of nature becomes uncertain, if financial benefits are mainly "eaten up" by the general budget of the central government or if they are fully used to carry out projects in other sectors or regions.

(4) There should also be an evaluation of whether tourism activities are or will be carried out by responsible investors or operators. This is a crucial question for ecotourism development based on principles of sustainability. It is certainly difficult to ensure that only responsible tourist agencies operate in an area. Moreover, it is difficult to decide whether this is achieved more easily by control, competition, persuasion or co-operation. There is, however, a first step that can be made by decision makers on local, regional or national levels: to define limits of acceptable use and environmental standards for tourist operations. This may contribute significantly to explaining the expectations of the authorities and the local population to investors and to specifying the meaning of the term "responsible" with regard to ecotourism development in a certain area. Consequently, environmental-management planning for tourist areas by the respective authorities is of utmost importance. It should comprise land-use aspects, e.g., zoning of accessible and restricted areas, as well as standards with regard to operational issues like defining types and quality of accommodation and transport, the quality of guided tours and the required qualification of tour guides.

(5) Finally, it is advisable to ascertain whether control mechanisms for ecotourism undertakings are in place or can be implemented. Control instruments are crucial to assure that operations stay within the defined limits and adhere to environmental standards. They provide a means to identify the "black sheep" and separate them from the herd of responsible investors. Control mechanisms should preferably be in place before ecotourism is started in a specific area. Although this is generally agreed upon in principle, the implementation of control mechanisms is very much in question in the individual case.

Promoting and Managing Ecotourism—What Can Be Done?

The promotion and management of tourism in natural areas increasingly attracts the attention of decision makers on local, regional and national levels of the respective countries as well as of international organisations. For example, the World Heritage Centre of UNESCO and the Industry and Environment Division of UNEP organised an experts' workshop in 1993 that dealt with the management of tourism in natural world-heritage sites.

The participants addressed management issues and elaborated the following recommendations (UNEP/UNESCO 1993): management should ensure that tourism development considers and respects ecological and socio-cultural values of the site. A management plan should be established and regularly updated. Environmental-impact assessments should be carried out before granting approval. Monitoring programmes should be implemented and their outcomes should be taken into account in decision making.

Local populations should be involved in tourism development. Co-operation with different stakeholders should be sought. Staff should be aware of the values of the site and well trained in visitor management. Relevant information and educational programmes should be in place, and a substantial proportion of the income generated through entrance fees should be directly allocated to the site for its improvement and management.

Referring to such principles and summarising recommendations of different authors, five major fields of activities can be distinguished that may contribute to ensure that ecotourism is adequately planned and managed (cf. Arbeitsgruppe Ecotourism 1994).

(1) Ecotourism should be integrated in national policies and planning. The general tourism policy and ecotourism objectives often conflict with each other. Whereas the general tourism policy may be oriented towards quantitative goals, ecotourism places emphasis on qualitative aspects. Moreover, there may be conflicts of interests between tourism and other sectors. Thus, it is advisable to define the specific objectives and goals of ecotourism, to define priorities and to co-ordinate different sectorial policies on a higher level of decision taking in order to avoid or to solve conflicts of interests.

(2) A second field of activities is related to the consideration of ecotourism in the management of protected areas and vice versa. This means that, for example, methods for regulating the visitor access and flow should be developed and applied. Moreover, land-use planning and zoning could become a powerful tool in this context. The impact of tourism on the ecological conditions should be constantly monitored by the responsible authorities. And finally, adequate systems of entrance fees should be developed and applied in protected areas in order to assure that the financial potentials of protected areas are not under-utilised.

(3) The active participation of the population of an area in all respects of planning, decision making, implementation and monitoring should be a third field of attention. A tourism development strategy should be elaborated, and it should reflect the ideas of the population. Moreover, ways and means should be sought for increasing the rate of local employment in tourist operations. Marketing of local products should be actively encouraged. Joint ventures between external tourist investors or companies and local entrepreneurs could be promoted.

(4) The co-operation with the private sector is a fourth field of attention. On the one hand, the private sector should be made aware of the specific potentials for ecotourism of a certain area and attracted to invest in ecotourism facilities by the concerned authorities. On the other hand, the private sector must at the same time also be made aware of the requirements and restrictions of ecotourism. This task can be fulfilled more easily if associations of hotel and tourism managers exist which

could be asked to become involved in defining appropriate environmental standards for tourist operations.

(5) Finally, co-operation with non-governmental organisations and scientific institutions should be given special consideration. On the one hand, environmental organisations could become active in the promotion of specific natural areas among their clientele. However, they will only become involved in a supportive way if they are fully convinced that tourism in the respective area is carried out in an environmentally friendly way and adheres to the limits of acceptable use. This is one of the reasons why authorities should seek the co-operation and involvement of environmental organisations instead of keeping their distance. On the other hand, non-governmental organisations and scientific institutions provide ample potential to assist the authorities in monitoring and controlling tourist operations as well as in assessing the environmental impact of ecotourist undertakings.

Concerted Tourism Management: The Institutional Solution?

Ecotourism comprises conflicting components. On the one hand, there are development components that are oriented towards attracting investors to a tourist area. On the other hand, there are protection components that are in need of setting limitations and restrictions and require regulatory measures and control.

Regulations and control mechanisms, however, may—for at least two reasons—pose severe problems to ecotourism development: on the one hand, potential investors may hesitate to invest in ecotourist undertakings if they perceive ecotourism as restrictive rather than development oriented. On the other hand, the respective authorities may not have the capacities for effective control. Moreover, they may not be too keen on conflicts of interest with the tourism sector. The reason for this is that environmental authorities usually do not have a strong position in the politico-administrative set-up of a country. Therefore, the probability is very high that economic interests prevail in cases of conflict of interest between environmental authorities and private businesses.

Thus, a way out of the dilemma between promotion and control should be sought. A solution may be found in the concept of Concerted Tourism Management (CTM) that has recently been introduced into the international development debate (cf. Arbeitsgruppe Ecotourism 1994). The basic idea of this concept is to involve all important players in the field of ecotourism, i.e., the national and regional authorities with special consideration to tourist and environmental agencies, representatives of the population living in an ecotourism area, tourist investors and operators as well as non-governmental organisations and scientific institutions.

CTM is based on the assumption that each partner assumes specific functions that complement each other:

- The national and regional authorities, for instance, could have the function to set up the necessary legal and regulatory framework.

– The authorities responsible for nature protection could be mainly responsible for technical aspects of environmental management, zoning, limiting and directing visitor flow.

– The representatives of the population living in the tourist area would have the chance to introduce the local perspective in decision making and management.

– Tourist investors and businesspeople would be able to establish a link between local management and the national and international markets. Moreover they would represent the touristic interest in area management.

– Last but not least, the non-governmental organisations and scientific institutions could take on functions with regard to providing know-how in nature-protection issues, promotion of an area through their relations with the international environmental community and tourists' awareness of environmental concerns.

Concerted Tourism Management (CTM) could be established in a more formal way, e.g., in the form of committees. It also could be institutionalised in a rather informal way by networking the different parties concerned. Formal committees have the advantages of being able to take firm decisions and formally decide about actions to be carried out. However, they are relatively inflexible and have the disadvantage that participants are obliged to give up part of their autonomy and therefore may be reluctant to commit themselves. Informal networks are flexible and have the advantages of being governed by the interest of their members and the joint efforts. However, they are not very well suited for solving conflicts in a sustainable way and for making binding decisions. It depends very much on the individual situation which institutional solution should be chosen.

It may not be easy to implement the CTM approach and to convince decision makers about the necessity of co-operation for achieving sustainable development. However, if co-operation is weak or reduced to mere formalities, ecotourism—owing to politico-administrative weaknesses—may stay a mere popular slogan. In contrast, ecotourism may become a powerful tool for achieving sustainable development if co-operation works and if development potentials are tapped only according to environmentally and socio-culturally friendly standards.

If ecotourism was implemented as a powerful tool to achieve sustainable development and was based on co-operation, the Lake Baikal Region could not only become a world model for sustainable development, but also a forerunner in the creation of new concerted- management practises of endangered areas integrating conservation and protection of nature, sustainable utilisation of natural resources, as well as economic and social development of an environmentally sensitive region.

References

Arbeitsgruppe Ecotourism (1994): Ökotourismus als Instrument des Naturschutzes? Möglichkeiten zur Erhöhung der Attraktivität von Naturschutzvorhaben. Berlin, Dresden, Göttingen, Hannover. (Final report to be published soon).

BMU (Bundesministerium für Umwelt, Naturschutz und Reaktorsicherheit) (1993): Deutsch-russische Konferenz. Umweltverträgliche Tourismusentwicklung in Rußland. Grundlage, Perspektiven, Ansatzpunkte, Maßnahmen. Berlin 13.–15. Oktober 1992 Schloß Niederschönhausen. Bonn.

BMZ (Bundesministerium für wirtschaftliche Zusammenarbeit und Entwicklung) (ed.) (1993): Tourismus in Entwicklungsländer. Bonn. (Materialien Nr. 88).

Boo, E. (1990): Ecotourism: The Potentials and Pitfalls. Washington D.C.

Eagles, P., Buse, S., Hvenegaard, G. (eds.) (1993): Ecotourism—an annotated bibliography for planners and managers. The Ecotourism Society. N. Bennington, USA.

Gormsen, E. (ed.) (1985): The impact of tourism on regional development and cultural change. Mainz. Mainzer Geographische Studien, 26.

Grenier, D., Kaabe, B., Miller, M., Mobley, R. (1993): Ecotourism, Landscape Architecture and Urban Planning. In: Landscape and Urban Planning 25 (1993): 1–16.

Lindberg, K. (1991): Policies for Maximizing Nature Tourism's Ecological and Economic Benefits. International Conservation Financing Project Working Paper. World Resources Institute, USA.

Lindberg, K. (1991): Mixing Environment, Economics and Tourism. In: The Adventure Travel Society 1991.

Müller. B. (1983): Fremdenverkehr und Entwicklungspolitik zwischen Wachstum und Ausgleich: Folgen für die Stadt- und Regionalentwicklung in peripheren Räumen. Mainz. Mainzer Geographische Studien, 25.

Müller, B. (1985): Tourism paving the way for an integrated development in peripheral regions? A case study of the Pacific coast in Mexico. In: Gormsen (1985) 37–46.

Müller, B. (1994): Ökotourismus in Entwicklungsländern: Umweltpolitische Leerformel oder wirksame Regionalentwicklungsstrategie? In: Festschrift für Erdmann Gormsen zum 65.Geburtstag. Mainz. Mainzer Geographische Studien, 40, 361–374.

Resolution der 2. russisch-deutschen Konferenz "Umweltverträgliche Tourismusentwicklung in Rußland—allgemeine Tendenzen und Probleme der Baikalregion" vom 04. bis 06. September 1993 in Listwjanka / Baikalsee.

The Lake Baikal Region in the Twenty-first Century (1993): A Model of Sustainable Development or Continued Degradation?

The Ecotourism Society (1993): Ecotourism guidelines for nature tour operators. N. Bennington, USA.

The Russian Academy of Sciences (Siberian Branch), Irkutsk and Buryat Scientific Centres (1994): The Present and Future State of the Baikal Region. Possibilities for Sustainable Development, Part I-III. Ulan-Ude.

Sachs, I. (1980): Stratégies de l'écodéveloppement. Paris.

Scace, R.C., Grifone, E., Usher, R. (1992): Ecotourism in Canada. Hull, Quebec.

UNEP, UNESCO (1993): Managing Tourism in Natural Heritage Sites. Report of the International Workshop held in Dakar, Senegal, 22–26 November 1993. Paris.

The International Sector and Sustainable Development in the Lake Baikal Region

John Massey Stewart
IUCN—The World Conservation Union
20 Hillway
London N6 6QA
UK

What part does the international sector have to play in the sustainable development of the Lake Baikal region? Can foreign national and international organisations and programmes coupled with the high quality of Russian expertise help to solve the region's problems?

The answer is undoubtedly yes. However, it is difficult enough in the West to establish what Western organisations exist, both governmental and non-governmental, which can support particular projects and programmes and how to apply, so it must be infinitely harder to find such information in Russian itself. It therefore seems potentially useful to produce a short resumé of organisations, programmes and funding sources of potential and existing relevance to the Lake Baikal region.

UN Agencies

On a global scale are UN organisations such as *UNESCO, UNEP, UNDP*, etc. This workshop's co-chairman, Prof Martin Uppenbrink was, as it happens, until the beginning of 1994, head of *UNEP's Regional Office for Europe*. That Office has very recently approved a programme and funds for Central and Eastern Europe (which includes Russia beyond the Urals) on training, capacity building and meetings. It is also looking at many requests from the Russian Federation—for example on toxic chemicals and the building of information systems across the CIS, and is now preparing (in collaboration with the Russian Federation's Ministry of Environmental Protection and Natural Resources and Federal Forest Service) a project proposal on the Russian taiga, focusing on the maintenance of its carbon balance and the protection of its biological diversity. Reflecting its Russian commitment, UNEP's Regional Bulletin for Europe is published twice a year in English, French and Russian.

Less well known than UNEP is *UNDP* (the United Nations Development Programme) which in 1993 added seven more country offices in the former Soviet Union, bringing the total to 15. Its activities in the area have included economic development, transport-information systems, public-sector development and sustainable human development. Activities which have attracted substantial

NATO ASI Series, Partnership Sub-Series, 2. Environment – Vol. 6
Sustainable Development of the Lake Baikal Region
Edited by Valentin A. Koptyug and Martin Uppenbrink

funding have included privatisation, the development of small and medium-sized enterprises, and management development. Its new network of offices in the CIS provides advisory and programme-advisory services and its satellite-based Wide Area Network has been set up to link eleven CIS countries and strengthen communications for key professionals such as parliamentarians, scientists, etc.

A third relevant UN organisation, briefly, is *UNECE*, the United Nations Economic Commission for Europe, a pan-European forum for co-operation in the preservation of the environment and the promotion of sustainable development. In 1990 it decided to assist countries moving to a market economy and help integrate them into the European and world economy. The environment is listed as the first of its five priorities, and its policy at all times is to make sustainable development a guiding principle. Among its activities are workshops (largely funded by the countries in transition), advisory services and training.

International Funding Institutions

The two UN organisations UNEP and UNDP manage jointly with the World Bank the *GEF* (Global Environmental Facility) to address transnational global environmental issues. The GEF is a major contributor to the Baikal area with its $10.10m Lake Baikal Regional Programme, which sees the lake as 'fulfilling the criteria for the regional pilot natural resource management plan' and has selected the area as 'a regional model of integrated natural-resource management with the integration of social, economic, institutional, and sectoral issues'. The Japanese government has already funded a feasibility-study component for a regional ecotourism development plan. The World Bank itself has its own separate Baikal projects, including one of the three major elements in a $27.2m Biodiversity Conservation and National Resource Management project, part of the Bank's Environmental Management Project for the Russian Federation. (*Staff Appraisal Report*, p. 31). At the World Bank's request, the *Canadian Parks Service* recently identified the protected area system in Russia as being in immediate danger of disintegration and saw Baikal as an area of particular importance and under serious threat.

Europe has its own banking arm, the *EBRD* or European Bank for Reconstruction and Development, based in London, with 53 member countries. Its purpose is 'to foster the democratic and economic transition process and to promote private and entrepreneurial initiative' in the countries of Central and Eastern Europe, including the FSU, 'through loans, equity investments and technical co-operation'. One objective is to promote environmentally sound and sustainable development. Its environment investment strategy adopts both a regional and a project-based approach.

The EBRD undertakes environmental appraisal on all potential investment and technical co-operation projects both in the public and private sector, for instance a grant of 220,000 ECU for a feasibility study for improving Moscow's solid-waste disposal system and a loan of 18.99 million ECU to a private-sector joint-enterprise project to enhance crude-oil production in the Vasyugan area of Western Siberia. Its key priorities, however, include assistance to small and medium-based enterprises through locally based financial intermediaries. On the EBRD's international Advisory Council, incidentally, sits Professor V. A. Katasonov of Moscow's International Independent University of Ecology and Politology and Vice-president of the Baikal Academy.

Multilateral European Organisations

NATO's Scientific Programme has made possible this workshop as well as other workshops on key subjects. Turning to other Western European organisations, the European Union, with its Member States, accounts for two thirds of total world assistance to the FSU. The European Commission's *TACIS* Programme (Technical Assistance to the CIS), for instance, has been set up to help the Newly Independent States achieve the difficult transition to market economies and to strengthen their democratic societies. 'It provides support in the form of grant finance to foster the exchange of knowledge and expertise through partnerships, links and networks at all levels of society' and is the leading programme making grants of know-how to the FSU. By the end of 1994 TACIS had provided nearly 1,900 million ECU to launch more than 2,000 projects in the FSU. Russia is the largest recipient of its annual budget, 510 million ECU in 1994 (1.3 ECU = $1).

Last year (1994) the European Commission set up in Copenhagen the *European Environment Agency* (EAA) with an initial staff of seven. This new agency is intended to provide the Community and its Member States with objective and reliable information at the European level so that they can take the necessary measures to protect the environment and to assess the results of these measures. It regards as equally important its aim to fully inform the public about the state of the European environment. For the moment this may not be relevant to Baikal, but joint requests from more than one member state about a non-member state will carry weight, Russia may yet become a member of the Agency itself (though hardly the European Union) and special attention will be given to transfrontier, multinational and global phenomena—all certainly features of Baikal.

Turning now from Brussels and Copenhagen to Strasbourg: here sits the *Council of Europe* which has an Eastern European programme with a special environmental section which, for instance, organises training courses for foresters. The Council can also send experts if requested, including lawyers from its strong legal section, which helps new governments draw up environmental legislation. It is the only European organisation with horizontal co-operation between regions and local authorities, and receives requests, for example, from universities and local governments. One very powerful organ is its Congress of Local and Regional Authorities which meets two to three times a year and fosters co-operation across Europe among local and regional authorities in the field of the environment, planning, local finances, etc. Strong and well-developed systems of local government, for instance, in Germany and Spain, can help their weaker colleagues elsewhere. As Russia is soon to become a member of the Council of Europe, the local authorities of the Baikal region should consider collaboration in advance and attendance of Congress meetings as observers.

A recent addition to the multilateral European organisations is the *International Academy of the Environment* based in Geneva. Its mandate is in part 'to provide high-level decision-makers with the basic knowledge and management principles to enable them to take decisions that are consistent with sustainable development'. This independent foundation collaborates with the major international organisations on sustainable development and is funded primarily by the Swiss Confederation and other Swiss public sources with a present budget of 8 million Swiss francs (1.1 Swiss francs = $1).

The Government of the Netherlands is particularly committed to funding European nature conservation and in 1993 took the lead in establishing the *European Centre for Nature Conservation* in Tilburg. Its main sphere of interest is policy analysis and research into the social context of nature conservation and sustainable development. Through its network, national expertise becomes available in an international context. It seeks to contribute to the reconstruction of pan-European ecological networks through the exchange of expertise, by providing information services, and by initiating and co-ordinating international projects. In Russia it aims to develop closer links with both governmental and non-governmental bodies and is at present reviewing joint proposals towards landscape ecology projects in the Baikal basin in association with Baikal institutes, WCMC (the *World Conservation Monitoring Centre*) and the *MacArthur Project on Environmental and Cultural Conservation in Inner Asia*, both based in Cambridge, England.

International Scientific Collaboration

Besides these multilateral organisations are the international scientific programmes and organisations such as *START* (Global Change for Analysis, Research and Training), *IGBP* (International Geosphere-Biosphere Programme), and *MAB* (Man and the Biosphere) in all of which Siberia is already involved. Two international programmes, well known locally, focus specifically on Baikal and its surrounding area: *BICER* (the Baikal International Centre for Ecological Research), co-ordinated by the Limnological Institute in Irkutsk, and the *Russia-American Comprehensive Programme of Sustainable Land-use Policies for the Protection of the Lake Baikal Region* headed by Dr George Davis and Dr Sergei Shapkhaev. One of its component parts, largely funded by *USAID* (the United States Agency for International Development) and local administrations, is the Model Sustainable Development Planning for the Proposed Okinsky National

Park and National Anthropological Reserve. But USAID is working with local and national governments, NGO's and the Baikal Commission in the overall sustainable land-use planning project in the Baikal watershed.

Baikal has already benefited from *INTAS*, the International Association for the Promotion of Co-operation with Scientists from the Independent States of the Former Soviet Union. Founded by the European Union in 1993 as an international non-profit international association, its 12 member states and 5 other European countries committed by the end of 1994 over $60 million to more than 1000 collaborative West–East projects in all sciences. Among them was an international workshop in Irkutsk in 1993 on 'Baikal as a natural laboratory for global change', one of 400 applications which succeeded out of a total 4,000. It is worth noting that INTAS prefers multi-country applications and that a recent joint application from the Limnological Institute and only one European partner (a British university) failed.

IUCN

The world's largest and arguably most influential environmental organisation is *IUCN*—The World Conservation Union, which was founded in 1948. Based in Switzerland, it brings together states, government agencies and an extremely wide range of NGO's (non-governmental organisations) in a unique partnership of 800 members in some 125 countries. Through its six Commissions, it draws together over 6000 expert volunteers in project teams and action groups, focusing in particular on species and biodiversity conservation and the management of habitats and natural resources. IUCN's European Programme covers more than the name implies—all Russia indeed to the Bering Straits. In 1991 it produced a booklet *Lake Baikal: on the Brink?* which I had the honour to compile, and in the summer of 1994 its European Programme Co-ordinator, Dr Zbigniew Karpowicz, and I as IUCN consultant participated in the World Bank ecotourism project for the Baikal region.

IUCN works closely with another international organisation mentioned above, *WCMC* (the World Conservation Monitoring Centre), the world's leading data centre for information on species, protected areas and trade in wildlife and about to become a partner in a programme which incorporates Baikal. IUCN has helped many countries prepare National Conservation Strategies and is now producing a practical handbook for planning and implementing strategies for national sustainable development which lists both activities and lessons learned and is thus of great relevance to the Baikal region.

This handbook, *Strategies for Sustainable Development: A Handbook for Their Planning and Implementation,* is one of a series being prepared by IUCN covering various types of strategy and associated methods for sustainable development. This particular handbook distils lessons learnt from more than 60 national and provincial conservation strategies, environmental action and development plans, etc. in 50 countries since 1980. Funded by Sweden, Canada, UNDP and others, this major handbook is aimed at practitioners (governmental officials, citizens and community groups, educational institutes, businesses and international organisations) who are or could be involved in developing a multi-sectoral strategy on environment and development at the national or provincial (vide Baikal) level.

It describes how to use national multi-sectoral strategies in the integration of environmental, economic and social concerns in national development processes. It suggests ways of developing and implementing a National Sustainable Development Strategy (NSDS) either by building on an existing strategy or, if none exists, from scratch. Moreover, it provides ten top lessons and features for success. The 10th is: 'External agencies should be on tap, not on top', and, elaborating this, recommends that 'Recipient governments must be able to take the lead in co-ordinating assistance; low-level continuous support over a long period is almost always better than high-level support for a limited period' and 'donors should support the capacity-building process and not just the products of their strategy'. It also gives 10 lessons to donors, the first of which is: 'Co-ordinate: a recipient government must be supported in taking a leading role in co-ordinating external contributions to the strategy process'.

National Government Programmes

Looking now at bilateral programmes, almost certainly the largest is *USAID*. There are many other official US programmes, however, including the *US Fish and Wildlife* scientific exchange programmes, complemented by the efforts of the many non-governmental organisations and institutes solely focused on the FSU. The total list of all US organisations and institutions involved in the FSU environment indeed covers 13 pages. Among them are the active *ISAR* (founded as the Institute for Soviet American Relations), the *Centre for Citizens' Initiatives* with an office in Irkutsk and, of course, *Baikal Watch*.

To take one governmental example in Europe, the British Government's Department of the Environment has set up an *Environmental Know-how Fund* specifically to benefit Central and Eastern Europe which technically includes Siberia. Its aim is to provide technical assistance in certain key areas, e.g., environmental management where, in conjunction with the World Bank, over £950,000 has been provided to help Russia set up its National Pollution Abatement Fund. In due course this will trigger off World Bank loans to environmentally friendly processes and enterprises. With a relatively small budget of £2 million in 1994 and the same in 1995, many of its projects are deliberately in collaboration with other national or international donors.

The British Government has also set up the *Darwin Initiative* to assist biodiversity and conservation projects in Central and Eastern Europe (once more including all Russia) with British expertise and technical skills. A British *Charities Aid Foundation* also exists for specific collaborative projects. Both these can be of direct relevance and benefit to the Baikal region.

NGO's

Many non-governmental organisations have been set up in the UK to work with the FSU. Two of the most effective are the *BEARR Trust* (British Emergency Action for Russia and the Republics), primarily for humanitarian aims, and *HealthProm* (the Association for the Promotion of Healthcare in the Former Soviet Union), both relevant for the Baikal region because sustainable development and adequate health and welfare are both sides of the same coin.

I myself founded, together with London's *Conservation Foundation*, the non-governmental *London Initiative on the Russian Environment*. Despite small resources so far, we have high-level contacts and support and are working with international organisations such as UNEP and TACIS and with governmental and non-governmental organisations in both Britain and Russia. The primary aim is to co-ordinate and facilitate Western environmental collaboration with Russia, and we see the initiative as international. In June, 1994 we organised in Suzdal, together with the Socio-Ecological Union, what was probably the first West European–Russian NGO conference, attended by some 150 delegates from Europe, the USA and all over Russia. This was funded by TACIS to establish contacts and joint projects which are now being developed and we would be very glad to collaborate on small-scale schemes for the Baikal region.

One organisation present at Suzdal was the *Baikal Wave* from Irkutsk, one of the small but practical and hard-working Russian NGO's with deeply committed members, which it would be unwise to overlook in the sustainable development of this area. Such NGO's have an important part to play, often providing expertise and supplementing the work of state organisations, and they need support and encouragement. One British-based NGO, *Raleigh International*, has been working in the Baikal area for two summers with international teams of young volunteers—including Russians—on practical environmental projects including monitoring the Barguzin river, one of Baikal's northern tributaries.

Turning to the big and best known international NGO's: *WWF* (World Wildlife Fund) is developing a biodiversity programme in about 12 oblasts of the Russian Federation; WWF-US in collaboration with the *Canadian Parks Service* and Russia's Ministry for Environmental Protection and Nature Resources has established a framework to conserve Russia's biological diversity. Within this are potential three-year projects for Baikal totalling over $800,000, with the Barguzin and Baikalo-Lensky zapovedniks (state nature reserves) rated as the highest priority, supported by *The John D. and Catherine T. MacArthur Foundation.*

Other NGO's not yet active in the Baikal region but with Russian projects include *Greenpeace International*, for instance, which has been working on Russia's nuclear-radiation problems, and *Friends of the East (Japan),* which has begun a scientific study of Siberian forest resources to identify regions of worldwide biological and climatic significance not protected under the zapovednik or national-park systems. With an office in Vladivostok, the project will work closely with Russian scientists, government bodies and NGO's, and is made possible by the US *W. Alton Jones Foundation.*

There are also potentially useful pressure groups such as the *Taiga Rescue Network*, an international pressure group based in Scandinavia which publishes an informative newsletter, and *Survival International*, based in London, which helps the interests of small indigenous peoples. This latter organisation could perhaps help the sad plight of the Evenki people in the northern Baikal region. Sustainable development, after all, must be based on people and on their health. Also based in London is *MERLIN* (Medical Emergency Relief International) which, in collaboration with the UN's WHO, is helping to remedy the high incidence of tuberculosis in Siberia and in autumn 1994 worked in Irkutsk as well as Tuva and the Altai.

Foundations

Finally, as regards Western foundations: the best known Western private funding body in Russia must be the *Soros Foundation*, but there are many relevant foundations in the US including the two already mentioned, the *W. Alton Jones Foundation,* which sponsored an ecoforestry conference in Krasnoyarsk in 1994, and the *MacArthur Foundation* which has an excellent reputation in Moscow as solid and reliable and gives grants from $10,000 to $90,000, but is inevitably inundated with requests. MacArthur has a joint board with *IREX* (International Research and Exchanges Board), which sometimes applies to MacArthur for special projects beyond its programmes. IREX, a Washington-based NGO, part-funded by US government programmes like USAID, began organising scholarly exchanges with the Soviet Union many years ago and now has offices across the FSU. MacArthur is funding both the WWF/Canada Parks project on Russia's biological diversity and the Cambridge University MacArthur Project for Environmental and Cultural Conservation in Inner Asia (mentioned above) which includes the Baikal region in its remit: to enable policy makers to make better informed decisions on sustainable development in the area.

One new important funding agency is the *Eurasia Foundation*, established in 1993 with financial assistance from USAID. With offices in Moscow, Kiev, the Far East, etc., it supports technical assistance and training as well as educational and policy programmes in the New Independent States, covering many activities in economic and democratic reforms. It makes grants of a total $16 million each year to US organisations with partners in the NIS or directly to NIS organisations. One huge advantage is its policy of completing within one month preliminary reviews of proposals submitted.

As will be apparent, there are many Western governments, programmes and organisations of different kinds of actual or potential relevance to the Baikal region. West–East collaboration, however, is not without its frustrations on both sides, which it may be helpful to indicate in the hope of solving them. Russian frustrations include: lengthy meetings and raised expectations with no results; over-paid foreign consultants with little or no knowledge of Russia or the language; lack of information on which organisations can do what and how to apply; immensely long bureaucratic delays; strings attached to aid programmes aimed to benefit the donor country; the invidious position of being a proud country at the receiving end of aid programmes—Western sensitivity is essential, as is proper appreciation of levels of Russian expertise.

Western frustrations include: lack of clear prioritisation by the Russian side; lack of clearly structured, specific and properly budgeted proposals, with the result that Western agencies may have the funds but cannot responsibly commit them; resentment and ill will when Russian applicants do not tell prospective Western donors or partners that they are applying elsewhere as well—understandable, but lending to a loss of Western time and efforts and lack of trust.

The above is a summary survey of some of the leading national and international organisations and funding institutions which can be of potential help in the Baikal region. It is hoped that this will be of very practical use in the quest for the sustainable development of this unique and unrivalled part of the world's heritage.

[This paper has been updated in April, 1995 to incorporate developments since the NATO ARW in Ulan-Ude of November 1994.]

References

Gordina, Faina (project co-ordinator and general compiler) (1995) *Okhrana prirody: Mezhdunarodnye organisatsii, konventsii i programmy.* (Nature protection: International organisations, conventions and programmes). IUCN European Programme Moscow Office

IUCN, Strategies for Sustainable Development: A Handbook for their Planning and Implementation (1995) IUCN

Krever, Vladimir et al. (eds) (1994) *Conserving Russia's Biological Diversity: An Analytical Framework and Initial Investment Portfolio.* World Wildlife Fund-US

Massey Stewart, John (compiler) (1991) *Lake Baikal: on the Brink?* IUCN East European Programme. Environmental Research Series 2

Pendill, Jr, Dr C. Grant, et al. (compilers) (1990–1992 Edition) *The Directory of Environmental Groups in the Newly Independent States and Baltic Nations.* Kompass Resources International, Washington, DC

Staff Appraisal Report—Russian Federation Environmental Management Project. (October 19, 1994) Infrastructure, Energy and Environment Division, The World Bank

State of Environment of the Russian Federation, 1993. National Report. (1994) Ministry for Environmental Protection and Natural Resources of the Russian Federation

Combining Ecological and Social Studies in the Comparative Study of Pastoralism in Inner Asia

D. A. Sneath
University of Cambridge
Research co-ordinator
Newnham Road, Cambridge, C83 9EY
UK

This paper describes the research underway at the Cambridge University MacArthur ECCIA project, which involves both social and environmental research in the comparative study of the Inner Asian steppe and its peoples. It summarises the research strategy, and goes on to mention some of the points that are emerging from the project at this stage. The paper compares the numbers of livestock supported by the pastoral economies of Inner Asiabefore and after collectivisation, and contrasts the pastoral sectors of Buryat and Chita with the more mobile pastoral systems of Mongolia and Tuva, which appear more sustainable. It notes that the case-studies carried out by the project in Buryatia and Chita suggest that the conditions for the emergence of a sustainable market economy are largely lacking, and that so far attempts to introduce one have been unsuccessful. It questions the wisdom of continued top-down planning in an attempt to apply inappropriate western economic models, and suggests locally-developed plans for pragmatic agricultural development, learning from the experiences of neighbouring regions of Inner Asia.

1. Introduction

In his introduction Professor Uppenbrink reminded us of the excellent motto "think globally, act locally". Between the global and the local, however, there is an enormous gap. I suggest an intermediate step - "think globally, investigate regionally, then plan and act locally."

The Cambridge MacArthur ECCIA project has tried to "investigate regionally." We took Inner Asia as our frame of reference, so as to make comparisons between different attempts at development in Mongolia, Inner Mongolia, Buryatia, Chita, Tuva and Xinjiang. These regions have more in common with each other, environmentally and historically, than with western Europe or America. Framed by the same steppe and forest ecology, with a common continental climate, in the past the different parts of Inner Asia supported very similar rural economies. In this century, however, these pastoral areas have been developed in different ways by Russian, Mongolian and Chinese governments. These different experiences represent experiments, if you will, in developing the Inner Asian environment. In this context Buryatia and Chita are some of the most

NATO ASI Series, Partnership Sub-Series, 2. Environment – Vol. 6
Sustainable Development of the Lake Baikal Region
Edited by Valentin A. Koptyug and Martin Uppenbrink

technically developed areas, and already their agricultural sectors are facing problems with sustainability, as the paper by Korsunov and Salyaev described.[1]

The aim of this project was, therefore, to compare the environmental, cultural, and administrative conditions throughout Inner Asia, in an attempt to learn lessons for the sustainable development of the region. The pastoral populations of each sector of Inner Asia: Mongolia, Inner Mongolia, Buryatia, Chita, Tuva and Xinjiang are reforming their pastoral economic systems in line with their different national and local governments' policies. We considered that an investigation of these different experiments in managing pastoral society gives the best chance for evaluating the variety of reform options being discussed today, and their likely effects on the societies and environment of the Inner Asian steppe.

The project brought together local social and environmental scientists from these regions to design a research strategy that combined qualitative and quantative analysis. This research was carried out in collaboration with research institutes throughout Inner Asia, in Buryatia, for example, we worked with the Baikal Institute of Environmental Management. The resulting international research team consisted of two participants from Buryatia, Mongolia, Inner Mongolia and Xinjiang, and one from Tuva. In addition to a host of other collaborators and contributors there were four British research and administrative staff at the Cambridge office.

The team agreed on the need for in-depth studies of particular rural communities and we carried out detailed case studies of ten sites. Each of the researchers spent months living in the local communities concerned, collecting data through formal and informal interviews.

It was evident that we must study pastoral economies as part of social and cultural systems, rather than using a narrowly ecological or economic analysis. We agreed to pay special attention to the way in which indigenous explanatory models and cultural values generate pastoral practice. This should enable us to suggest possible alternative strategies for the future, taking into account the socio-cultural values of each region, and help answer the question "what can the people in different parts of the region learn from the experiences of the others?"

The sustainable development of the Baikal region, if it is to be achieved, will depend upon whatever new socioeconomic organisational forms arise as a result of the changes underway in Russia. In particular, a sustainable and productive agricultural sector must be a key element in the environmental planning for the future. The MacArthur ECCIA project has been carrying out research into the pastoral sector of Buryatia and Chita, and the actual social and economic

---

[1] In their paper "Opportunities for Sustainable Development of the Agricultural Complex in the Framework of the Lake Baikal Region," presented at the workshop.

conditions of the rural population. These local conditions are the reality that must form the basis of any workable plan for the future.

In this paper, because of the limitations of space, I intend only to describe the research strategy and methodology of the programme, and outline a few of the preliminary findings of the project. In particular, a brief comparative overview of changes in the pastoral sector in Buryatia, Mongolia and Inner Mongolia since collectivisation, and the results of rural reform on the case-study localities Buryatia and Chita, in comparison with developments in Mongolia.

Research Strategy

We concluded that along with gathering the relevant statistical material, it was necessary to conduct in-depth studies of chosen rural communities, looking at all factors relevant to sustainability.

The workshop produced a strategy for this comparative study composed of three elements:- (1) a common series of structured interviews for each case study site, directed primarily at rural district level, (2) projects for problem-focussed research of a comparative kind, and (3) Inner Asia-wide surveys of national statistics, maps, and climatological data. We have made some use of both Russian satellite photography and American remotely sensed data to trace general patterns of environmental change.

Case Study Design

Each participant carried out their case-study using a six hundred page document of questions which served as a guide to the interviews. This document was written after careful discussion of each point, and translated into all the relevant languages. The information from the ten case-studies has been entered into a database, where responses to questions we asked and all the other information collected is indexed to permit searches to be made on almost any term. This has allowed us to make informed comparisons between the different administrative areas studied, of everything from household budgets to perceptions of local environmental problems.

Comparative Problem-Focussed Research

The different topics for individual study were discussed and selected. Each participants studied a given issue within their own speciality, with the aid of the project database, and wrote a paper on this aspect of the situation for our final report. They have been gathering material for these individual studies since the beginning of the project, and most of them should have been able to visit other parts of Inner Asia to further their research.

Surveys

The aim has been to make possible three sorts of comparisons:

1. Geographical survey to show how the pastoral economy is adapted to climatic and other environmental conditions.

2. Administrative survey to show how different management systems affect pastoralism and the steppe environment.

3. Historical survey to show how pre-revolutionary pastoral economies differed from the modern ones.

Satellite images of most of the region have been collected, and a simple GIS of the study area has been generated, including ecological, economic, administrative and other data. Relevant statistics from each area, including human and animal populations, are being collected. The project has made some use of the Glasod database, which details soil degradation by cause, type, severity and rate; enabling us to examine patterns of soil erosion within the Inner Asia region and make some preliminary comparisons with the patterns of livestock density, although great care has to be taken when interpreting the database. It has also been possible to utilise the NOAA AVHRR-GAC Vegetation index for 1989 for this early stage of our work and gain some idea as to the crude biomass produced throughout the region.

Survey work of this kind has allowed us to identify areas that appear to face the greatest threat from pasture degradation, such as Buryatia, Chita and Inner Mongolia. We carried out livestock density mapping of the whole of the area and found that in general the greatest problems with grassland damage are found in areas with the highest density of livestock. This is, however, not by any means a perfect correlation, and the variations and exceptions are particularly interesting topics for further study.

The data collected from the case-study sites suggests that the pastoral sector of Buryatia and Chita are facing much more serious problems of the ecological sustainability than neighbouring Mongolian areas. Local officials estimated that over 70% of the pasture at both case-study sites had suffered degradation, whereas in the Mongolian and Tuvan case-studies the amount of degraded pasture was considered by interviewed officials to be negligible, considerably less than 10% in each case. This is confirmed by the analysis of satellite photographs that show that long the Russian - Mongolian border there appears to be a severe loss of topsoil on the Russian side, and almost none on the Mongolian side.

In Buryatia and Chita agriculture is generally more mechanised than in the other parts of Inner Asia. Livestock is relatively static and there is a relatively large amount of fodder grown, cut by machines, and moved to the animals. Mongolian and Tuvan pastoralism, on the other hand, involves a higher degree of livestock mobility, so as to make greater use of natural forage. One might expect such systems to be incapable of supporting the high numbers of livestock that the more mechanised Russian pastoral sector does, but in fact areas of Mongolia and Tuva support comparable densities of livestock but with apparently less damaging effects on the soil.

Preliminary Historical Comparative Work

A comparison of the histories of collectivisation of the pastoral economies of the different parts of Inner Asia shows some interesting parallels. If, for example, we quickly compare the changes in livestock numbers from pre-collective times to the present in Buryatia, Mongolia and Inner Mongolia, we can generate these crude graphs.

Buryatia

Buryatia was collectivised early, relative to other parts of Inner Asia, the process being virtually complete by 1932. The pastoral population has been largely settled in permanent housing, and there is a relatively high level of fodder and other inputs for livestock production. The graph of the total numbers of livestock shows a dramatic increase, and first glance it appears as if the pastoral sector is larger than in the pre-collective period. However, if livestock numbers are translated into standard stocking units (SSU's) a very different picture emerges.[2]

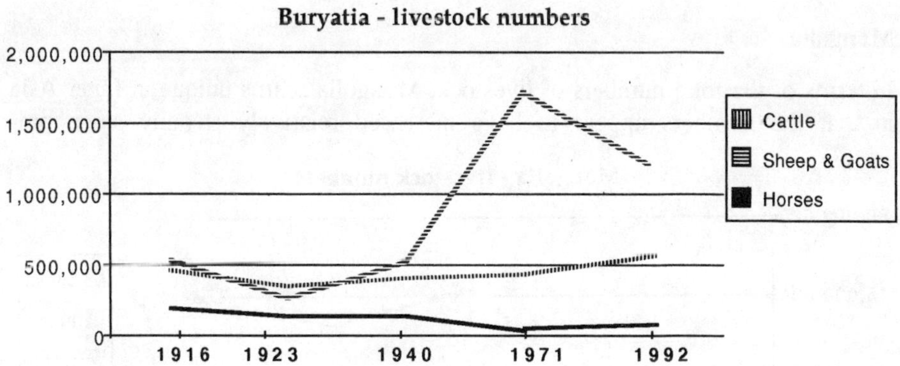

Graph 1

Source: p. 89 & p.62, Narodnoe Khozyaistvo Buryatskoi ASSR ,1963 & 1986.

---

[2] In choosing a standard for calculating stocking units, we adopted that used by local specialists in Mongolia for small animals. This gives the value of sheep as one, goats as 0.9, cattle as five, horses as six, and camels as seven.

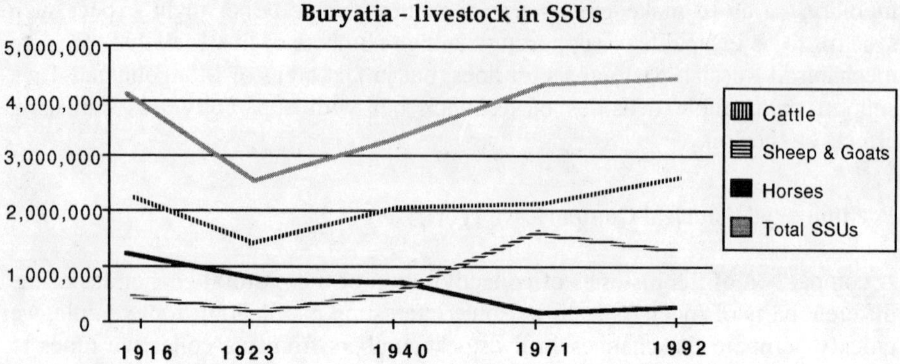

Graph 2

Source:p. 89 & p.62, Narodnoe Khozyaistvo Buryatskoi ASSR ,1963 & 1986.

These crude graphs suggest that the total numbers of animals (as measured in standard stocking units) is in fact similar today to the pre-revolutionary totals, but there has been a great change in the composition of the pastoral sector. The numbers of sheep and goats pastured in the region has risen dramatically, whilst the numbers of horses has declined steadily. The numbers of cattle (and indeed the overall numbers of domestic animals as measured in SSUs) has declined over the period of collectivisation, probably as a result of the disruption in production that this caused.

Mongolia

In terms of the total numbers of livestock, Mongolia seems unique in Inner Asia in that the numbers appear to have increased relatively steadily since pre-

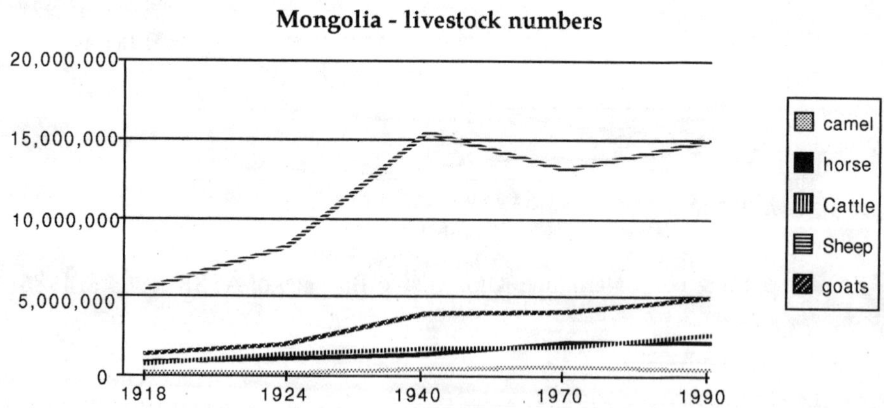

Graph 3

Source: Karamisheff & National Economy of the MPR for 70 years 1921-1991.

revolutionary times. At first glance Inner Mongolia seems to have had an even more marked increase in livestock numbers.

Inner Mongolia

If, however, one uses standard stocking units to evaluate the changes over time, again the picture is very different.[3] As in Buryatia, it seems that relatively high numbers of livestock were supported before collectivisation. Inner Mongolia again resembles Buryatia in that it has also seen an important increase in the amount of land cultivated for crops, (although even today the area amounts to less than 4.5% of the total area of the region).[4] As a result comparing the pre-revolutionary livestock totals with the current ones is difficult before we have finished our work comparing land-use then and now. The amount of pasture now

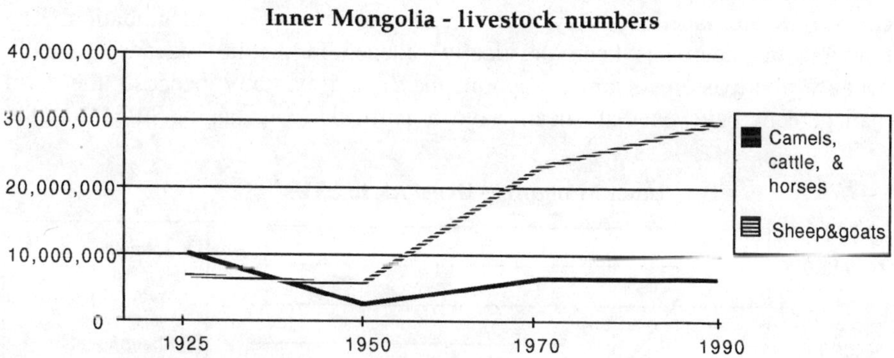

Graph 4

Source: Chang p. 84 &*Nei Menggu Guotu Ziyuan,*

in use in Inner Mongolia is less than early this century, and grazing pressure on the grassland is clearly likely to be greater as a result. However, it is interesting to note that the pre-revolutionary economy supported as many animals (as measured by standard stocking units) as modern Inner Mongolia - around 70 million SSUs.

---

[3]Not having separate figures for each species in pre-revolutionary Inner Mongolia we calculated each "big animal" as six SSUs, and each small animal as one.

[4] There are 53,000 square Km cultivated, and 880,000 square Km classified as grassland - 74.4% of the total area of IMAR, according to Inner Mongolia Association of Foreign Cultural Exchange publications. However, Yong Shipeng & Cui Haiting give the total area of grassland as 791,529 Km sq (only 75.58% of which is utilisable for grazing).

In Inner Mongolia today there is a very strong perception that the pastures are saturated, and that this overstocking is the cause of widespread grassland degradation.[5] Recent work by Inner Mongolian specialists suggests that as much as 35.6% of Inner Mongolian grassland has been degraded and attributes this to overgrazing.[6] It is interesting to note, then, that in terms of SSU's there has not been a dramatic increase over pre-revolutionary "stocking levels" at all.

In almost every case we see that the older pastoral economies of Inner Asia supported a large number of large animals, relative to sheep and goats. In the last fifty years or so, the numbers of these small herbivores have risen quickly in all the areas under study. If we are seeing widespread degradation of the pasturelands of Inner Asia (and in Inner Mongolia it seems that we are) then this change in the composition of herds may be of some importance. It might suggest, for instance, that a decrease in the sheep and goat population, and higher numbers of larger animals might be a more sustainable utilisation of pastureland.

This analysis is supported by the developments in Buryatia in recent years, where collectives are faced by very widespread problems of soil degradation. The numbers of sheep has been drastically reduced in the last decade, partially because the pastures cannot support them, and partially because the farm management believes that sheep have a particularly damaging effect on the

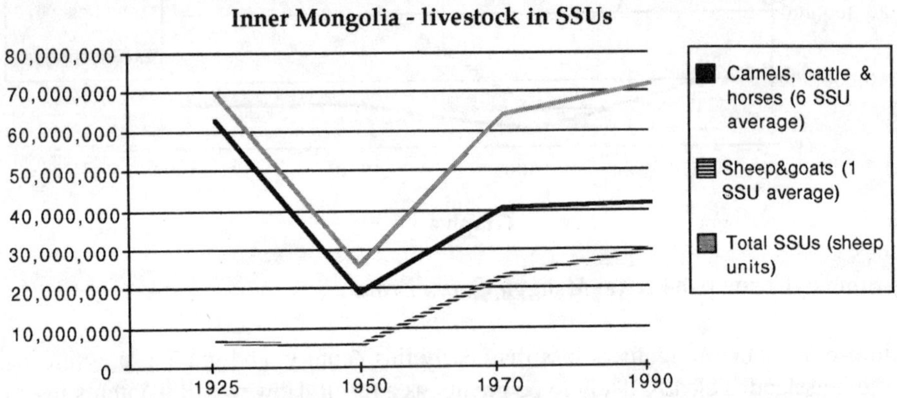

Graph 5

Source: Chang p. 84 &Nei Menggu Guotu Ziyuan,

[5]See Yong Shipeng & Cui Haiting IN Li Bo et al (eds) p. 94 They argue that of the total area of grassland (791,529 Km sq) only 75.58% is utilisable for grazing. They calculate the yearly forage yield as 91,286,657 tons. If stalks are also taken into account the total yearly forage is estimated as 107,666,000 tons. Based on this total yield they calculate a "carrying capacity" in stock of 70,663,000 units in sheep equivalent SSUs. The 1986 total figure was 69,373,600 SSUs.

[6] See Orie Loucks and Wu Jianguo, 1992, p. 71 of Grassland *and Grassland Sciences in Northern China.*

grassland. The largest collective farm in the region, Bayangol, for instance, has reduced its numbers of sheep by 30% over the last ten years.[7]

Overview of 20th Century Changes

In each case the pastoral regions under study have experienced a dramatic increase in the numbers of sheep and goats ("small animals"), and a stagnation, or in some cases (such as Inner Mongolia), a decline in the numbers of the "large" domestic animal species, principally horses, cattle and camels.

We have two processes that may be causally related. Firstly a general shift in the composition of the burden on the pastoral regions of Inner Asia, from large domestic species to sheep and goats. Secondly, a growing perception among both herdsmen and specialists that the grassland has experienced an increasing amount of degradation. In Inner Mongolia in particular, there appears to be serious environmental degradation in some regions, which is attributed to overgrazing.

One of the patterns to emerge is that in Buryatia, Tuva and Inner Mongolia the pastoral sector - in crude terms - supported something like the same numbers of animals (measured in standard stocking units) in pre-revolutionary times as they do today. This raises a number of important questions with regard to the perceived problem of overgrazing. If, for instance, Inner Mongolia, is facing an acute and historically unparalleled problem of environmental damage in many regions as a result of overgrazing, then we cannot assume that this is because it now supports an unprecidentedly high number of animals. This would suggest that it is the change in species of domestic animals, or some other aspect of herding practice was to blame. It may be that the greater numbers of sheep and goats generate different and more generally damaging effects on the grassland environment than an equivalent number (measured in stocking units) of large animals (horses, cattle and camels). We must be, however, extremely careful in reaching such a conclusion. Firstly, we cannot assume that there was little or no environmental damage caused by overgrazing in pre-collective times. What might be conceived of as the recent problem of degradation of pastures may be part of a much longer cycle. Grassland ecosystems are a product of the way they are grazed by different species, and what is perceived as "environmental damage" might be an example of environmental change in response to a different balance of domestic animals in the systems. Secondly, assessing the situation in terms of total numbers of animals (by means of an artificially constructed "standard stocking unit") may not be as relevant for grassland degradation as the understanding the local management of herding (e.g. herd sizes, over-reliance on certain pastures, and what use is made of local micro-resources such as water-sources, winter sheds etc.).

---

[7] MacArthur ECCIA preliminary report on Buryatia, summer 1993.

However, the information from Buryatia[8] traces a stronger link between the growth of sheep and goat numbers and loss of topsoil, as does some comparable material from other studies.[9]

Preliminary Findings on the Results of Rural Reform in Buryatia and Chita

The turmoil of the Russian economic reforms has created widespread economic disruption and the remains of the old kholkhoz collective farms and sovkhoz state farms remain the basic organisational structures for agriculture in many rural areas.

The collectives and state farms, and their successor institutions, continue to be the most important agricultural organisations. Since 1991 the kholkhoz collective farms and sovkhoz state farms have been able to re-register their status, so as to indicate when, and in what way, they will privatise. Many - over half at the beginning of 1994 - had not changed their status.[10] Even among these largely unreformed institutions, some households could voluntarily leave the farm with their allocated share of property, and start up private farms, either alone or jointly with others. The power and influence of the leadership of the collective farms has been extremely important in this process. In some cases the officials chose which households were to become private farmers, and in general those who have set-up independent family farms are usually those with the best relationships with local leaders, and those with exemplary work records. These households received special support, loans and provisions to try to make their small-scale enterprises viable. In many cases, such as the Argada case-study site, this has largely failed, and at the time of writing (October 1994) most of the private farmers have recently elected to reform the collective.

The future of this small private farming sector remains uncertain, and there are some indications that it has been unable to perform even as well as the ailing collectives. The official press reported that at the start of 1994 there were 2,770 new private farming enterprises, with 183,000 hectares of land. Although this is twice as many as existed a year earlier, it still represents a very small part of the agricultural sector. These enterprises had only 5% of entire agricultural area of the republic (although a slightly higher share - 7% - of ploughing land). They

---

[8] Tulokhonov, A.K. 1992, personal communication.

[9] Sandford, 1983, p.42

[10] In Buryatia, of the 206 former collective and state farms, 60 collective farms and 49 state farms (about 53%) retained their status, and made no formal change in their status. Six state farms became collective farms. The remaining 97 organisations changed their status. 60 adopted some form of collective or allocated (shared) ownership. 20 became limited companies, four became agricultural cooperatives, five became joint stock companies, eight became subsidiaries of other enterprises. Source -*Buryatia* Number 30; 17 Feb., 1994 (This paper is the daily Publication of the Supreme Soviet and the Soviet of ministers of the Buryat republic).

produced even less produce than one would expect from this small share of the arable land - only 4% of the grain, 1.3% of the potatos, and 2.3% of vegetables.[11]

Agricultural production has fallen over the last few years, and rural producers, private and collective, have increasingly turned to their own resources to supply themselves and their relatives with necessary provisions.

The Buryatian case-study was carried out by Bair Gomboev in and around the settlement of Argada in the Barguzin valley, (Kurumkanskiy raion). The Chita case-study was located in the Gigant collective in the Ononskiy raion , and was carried out by Balzhan Zhimbiev. Both researchers found that in this uncertain economic climate kin relations have become increasingly important, especially in rural districts. Not only does kinship provide a mutual aid network in the face of shortages, but it is the basis for the new property-owning corporations that have become possible. In both case-study sites the most important relationships are generally, those of kinship, both conceptually and practically. All the private farms are based on them, and in an atmosphere of mistrust and uncertainty, informants felt that they could trust their kin, and so go into business with them.

Buryatian rural residence forms reflect the relatively settled nature of the population, and the fact that most families live in static buildings means that residential groups are generally constant and composed of some part of the "conceptual family". In Buryatia a relatively large proportion of the rural population live in settlements, and in the Argada site, for example, only about 100 to 120 of the 500 households of the district can be considered "pastoral" households. Many of the features observed in the Buryat studies are, however, common to all the study sites. The nuclear family is the most common household form, often grouped into a larger residential family unit. The composition of this domestic group is generally constant, but the residential unit is in some important respects less significant than the extended family.[12]

In Argada the economic reforms led to the division of the collective in the early nineties. Twenty-seven small private farms were formed, each usually containing between three and five families, who were virtually always close kin. There were three larger private farms, one with 80 families and two with 30, and these too were largely composed of kin. Kin or neighbourhood networks do not generally organise much joint economic activity beyond the level of the extended family or private farm, although sometimes two or three small private farms would get together to carry out haymaking jointly.

---

[11] Source- *Buryatia* Number 30; 17 Feb., 1994 .

[12] The exact composition and permanence of the residential group depends, to a large degree, on the housing available, and the opportunities that the family has to use or gain property. If a family has a chance to gain a house then it will usually find a subsection to move into it. Several of the family farms formed in the early nineties included a large number of kin so as to have a sufficiently large allocation of formerly collective property to get a tractor or some other important capital asset.

The large farms made use of the wider kin network to gain machinery and, in effect, sections of the old collective, while small private farms were based upon the close kin of the conceptual family. In the face of economic restructuring, kinship provided a ready basis for an adaptive response in social organisation. As it was difficult for administrative and teaching staff who also received land and animals to get fodder and other agricultural requirements, they generally joined up with rural relatives so as to be able to make use of their assets.

Most families in the settlement have around 10 cows, for milk and supplementary meat. Some they keep in the autumn, summer, spring pastures around the settlement so that they can milk them regularly. The other cattle they usually send off to be herded by relatives in the more distant pastures. As a result of this tendency, locals say, the pasture around the settlement has deteriorated badly out to a radius of a kilometre or so, and people try to avoid keeping many animals on it.

In Gigant the collective did not divide all its assets between new private farms, but houses have become privately owned and a number of private family farms were created by officials. Although the collective remains, however, kin have become more important economically in recent years, as a result of the shortages and high prices of many goods which has increased the need for subsistence production and self-provisioning. The prospect of further privatisation has also meant that families have become important as potential property-owning corporations.

The pastoral households are often based around the elder couple of such an extended family who live at the otara (the allocated pastoral area at a distance from the settlement), and carry out herding. The married children of this couple often live in the settlement, although one couple may stay with the parental couple and form a stem family residential group. Family members often stay at each others homes, so that within the extended family each family residential group could be said to have a core membership, and peripheral membership that includes the members of the other residential groups. The (usually elder) section of this extended family stays most of the year around at the distant otara house, but may frequently visit the homes of the younger section in the settlement.

Most of the extended family usually comes together to carry out the heaviest pastoral tasks - haymaking, and shearing. They also share goods, such as fuel, even when this is a violation of regulations. Those in the otara area will herd livestock belonging to other sections of the family. Pre-school age children often stay in the otara with their grandparents, so there is also some sharing of childcare. The whole family usually comes together for New Year and some other celebrations. The pastoral section of the family provides meat and animal products, and the (generally younger) kin living in the settlement often have networks in the settlement and elsewhere through which they can sell animal produce from the parental herds so as to get a good price.

One of the reasons that the elder sections of the families tend to stay in the otara collective herding stations is that they wish to make sure they keep a claim to these buildings and animals in case of further privatisation.

Most people who live in the city have relatives who live in the rural areas. The kin networks between urban and rural households may be quite extended. A young city couple, for example, may have good relations with their parents, who in turn enjoy good relations with country relatives. This indirect relationship with rural producers allows them access to important goods such as meat.

Self-provisioning has had to increase as a result of the economic difficulties of recent years. This has lead to an increase in the amount of domestic work, which mainly falls to women, and children also generally do more work than they used to, because they often help with a wide range of self-provisioning activity, including herding.

Because of the increased reliance on individually produced and procured goods, and upon the personalised relations of networks for their distribution, the economy seems to have become more, not less, socially "embedded" in the opening years of the "age of the market". It does not resemble the impersonal transactional arena of the ideal-typical market economy.

A Comparative Look of the Reform Process in Mongolia

During the collective period the pastoral sector was, as in the Russian case, composed of collectives - negdels - and state farms - sangiin aj ahui . These were large economic units that owned most of the livestock and allocated labour, households were allowed a few private animals.[13] The recent economic reforms have been designed to privatise collective and state assets, and build a market-oriented economy. In 1992 the negdels were largely replaced by joint-stock companies, and a distribution of formerly collective assets was begun.

Throughout the country, Mongolia is carrying out reforms in rather different ways.[14] Some areas have dissolved the companies, in most they remain but having privatised a variable amount of the formally collective livestock. In other areas, however, although the collectives have become companies, and distributed much of the their livestock, they continue to own considerable numbers of animals. These are often herded by households under "lease" arrangements that are almost identical to the old "norm" by which the collective members were given production plans.

---

[13] 50 private animals were permitted in most of Mongolia, and 75 in Gov' regions.

[14] The method of privatisation was to issue *tasalbar* coupons worth a total of 10,000 tugrugs. The small and large coupons (representing 3,000 and 7,000 tg respectively) were used at different times to buy animals, and sometimes enclosures and other assets, from the company.

In areas that have divided formally collective assets a few horshoo (co-operative)[15] formations have arisen, it seems that these units tend to form around some section of the old negdel. For instance, the former negdel shop might become a co-operative, as might small factories or plants that the collective had set up, such as a sewing factory. The broker, who had become an important figure in most collectives, might also have formed a co-operative that is independent on paper, although most of his customary arrangements and contact will be the same. In the pastoral sector horshoo may form about some former brigade property, such as a tractor or some such. These tend to be very leader-focused, and as Caroline Humphrey's work in Dornod suggests, tend to form along the lines of kinship and friendship, rather than social groups based upon locality and the joint use of certain resources.[16] If the experience of Inner Mongolia is anything to go by, these assets will tend to become the property of one individual (who may buy-out other joint-owners).

These formations may not be without precedent. In the fifties "primary co-operatives" were formed in Inner Mongolia. On the exterior they were a collaborative grouping of equals. In fact, an examination of the internal structure of these early co-operatives suggests that they were more like clusters of poor clients about a rich patron. This may be highly comparable to the Mongolian horshoo because in an environment of scarce resources, and highly personalised economic arrangements, a certain individual's standing in the local government and community may give them largely exclusive access to important resources - such as transportation. However, in many places, - just how many we will not know until we get all our research material from Mongolia - the companies have not split-off sections to become such horshoo . Indeed, it would be premature to assume that this decentralisation will continue, as in many regions where it has occurred, the breakup of the collectives seems to have been unpopular.

Historically, it can be seen that powerful herd-owning institutions tend to increase their holdings, through a variety of mechanisms, including the extension of credit. It may be that the future of Mongolia will resemble the past in some ways. Experience from Inner Mongolia demonstrates that the pastoral rich have quickly got richer, and Shahrani's study of the Kirghiz of the Afghan pamirs suggests that a marked differentiation of herd wealth followed migration into an area of grassland. My own work in Inner Mongolia suggests that in conditions of private ownership of domestic livestock, a variety of factors tend toward concentration of herdwealth in the hands of a few. In a competitive arena, such large owners may

---

[15] It should be noted that this term can also be used for marketing organisations based upon old state structures.

[16] This picture is confirmed to some degree by the work of Potkanski and Szynkiewicz, who found that in Erdene sum, Dornogov', the horshoo was not based upon the residents of one valley (half of which had refused to join the new co-op) and that it was very much centred upon the founder, who commanded formally collective resources - especially tractors. pp. 62 - 66.

have a number of advantages of this sort, giving rise to a tendency for them to increase their holdings over time, relative to smaller herdowners.

It may be, then, that companies remain as the economic wing of local government in the sum. They may retain a large number of animals, leased out to various herding families, especially the poor who have few animals of their own. The profits from these animals, will certainly not be sufficient to maintain all the managerial resources and rural services that most Mongolians would welcome, so it may be that by acting as the principle conduit for the marketing of pastoral produce, and by charging for the hay and other services they provide the herdsman, the company could support itself, and the limited welfare function that it would tend to provide by acting as a patron for poorer families. In fact, at the moment, this appears to be exactly what is happening in most sums in Mongolia, with local government struggling to keep health, education and social services going with a shrinking resource-base. Because there are few if any alternative operable marketing organisations, the company can influence prices to a very large degree, as individual herdsmen have difficulty moving their produce far enough to sell to a neighbouring company. Brokers, on the other hand, could generate competition to the company, except that in my experience they tend to enjoy very close relationship with the company and local government bosses.

The experience of Buryatia, Chita and Tuva suggests that the collective and state farms remain even more important there, despite recent decrees on land reform. Given the central role of collectives and former collectives, the importance of personal and kin networks, and the renewed emphasis on subsistence and self-provisioning production, the case-studies suggest that local rural communities have few of the necessary conditions to operate a modern market economy, as this term is understood in the west.

Policy Recommendations – a Cautionary Note from Mongolia

Like many rural inhabitants, the experience of the last few years must make us question the assumptions that have influenced the reform of the agricultural sector of Buryatia and Mongolia, and consider the danger that rather than the prosperity that it promised, the economic revolution that formerly Soviet regions are undergoing runs the risk of creating impoverished nations.

The experience of Mongolia provides an example of some of the difficulties of introducing reforms designed to bring about a market economy. In Mongolia, the "Age of the Market" has meant a loss of almost all the benefits of a Soviet-style society - limited though these may appear to westerners. The safety of cities and rural areas, created by the high level of social control, appears much reduced to Mongolians, who feel besieged by the crime-wave of the last few years. From the pastoral herders in their tents, to taxi-drivers in the city, there is a widespread perception that young people are growing up lawless and without respect for figures of authority. Policemen complained to me that people, young and old, no longer fear or respect them, and that bribery is increasingly common. There has been a slow collapse of effective state social security. Today people can, all too easily, become destitute in a way that was unthinkable ten years ago.

Education has also suffered. In the area where I did fieldwork the local schools are teaching fewer and fewer children each year. The numbers have been dropping steadily since the boarding facilities closed for lack of money. Some teachers have lost their jobs in the reforms of the last few years, and the disillusioned remainder try to manage on pay that inflation has made so low as to be barely viable. There are few jobs available for the educated, and at the moment only the rich can afford to support their children through university.

Cheap fuel oil was another advantage of the old regime, although not one for which the old Mongolian government could claim much credit. The Soviet Union had sold oil very cheaply to its satellite states, and for Mongolia this meant that mechanical transport, a key factor in such a huge and thinly populated country, was affordable and widely available to local government, national services and productive enterprises alike. With the introduction of something approaching the international market price of oil, Mongolia found it could no longer afford to use the quantities it did in the past. As a result there is a severe shortage, high prices, and petrol rationing. The knock-on effects of this has caused endless difficulties in agriculture and industry, as well as in local government and private spheres.

The comprehensive health and vetinary services were also modest, but important, achievements of the old regime. The standard of health care may have varied widely, but the doctors and vets generally got to where they were needed, and treatment affordable, if rather basic and unsophisticated. Today medicine is becoming more and more expensive, and the problems of transport means that it is available to fewer and fewer people. Doctors and vets continue to loose their jobs, and young medical students no longer travel to Moscow for training.

It is ironic, then, that at a time when the Mongolian government is savagely cutting back public spending on services such as education, and concentrating on introducing new legislation to make the markets as free as possible, that Western economists are beginning to realise that this approach is flawed. As Robert Chote notes in his article "Free markets are slowly being superseded by investment in education and training as the fashionable cure-all for economic ills. In part this is because the old theory of economic growth is being ousted by new theories which argue that education spending significantly boosts technical progress." [17]

One of the difficulties with making recommendations concerning the pastoral economies of Inner Asia is that they are not easily separable from other aspects of rural society. Any analytical framework must recognise the importance of networks in Inner Asian rural life. Relations between kin and friends are widely

[17] *The Independent on Sunday*, 26 September 1993, (Business Section p.12) Robert Chote

extended, partly as a response to shortages. When one considers that a pastoral household may give away up to a third of its entire animal offtake to family and friends, it becomes clear that the informal movement of goods in networks is a key area of study.

Assessing the "efficiency", therefore, of any given pastoral economy is very difficult without some very serious thought as to what we consider to be efficient. To take two extreme examples:- If the primary aim of a Mongolian rural reform programme was to maximise re-investment in the pastoral economy, then reform options will tend to be those concerned with concentrating sufficient surpluses in the hands of individuals or institutions who have an interest in re-investment. If, on the other hand, the aim is to keep all rural Mongolians above the poverty line then reform options will probably revolve around retaining a pattern of ownership that will keep all pastoral households herding some animals either through some redistributive mechanism, or through some large central holdings that can lease animals to the poor. One possibility might be that the companies lease animals to poor households as matter of priority i.e. act as socially concerned, paternalistic "super-patron" to support the poorest. In this option the collectives should not be assessed purely in terms of western-style economic efficiency, and thus compared directly to the productivity of private herdsmen - or expected to compete with other institutions who do not have these obligations. In the second scenario local obligations to the poor are of great importance, in the first they are likely to be an obstacle.

Mongolian specialists are well aware of these problems. They are developing a broader, more socially-conscious approach. "It is our belief that privatisation and the redistribution of the property of the collectives should not have begun with livestock. The first objective should be to determine the appropriate form of socioeconomic organisation."[18]

The "appropriate form of socioeconomic organisation" is a normative question that cannot be decided by outsiders, but must be answered by the people of Buryatia, Chita and Mongolia themselves. Specialists can and should, however, attempt to improve our analysis of the situation, so as to be able to make a constructive critique of the concepts that western and western-influenced policy-makers use in their attempts to analyse and develop post-Soviet societies.

Conclusion

The case-studies suggests that attempts to rapidly introduce agricultural reforms based upon western economic experience have driven many rural households to increase subsistence production. Measures designed to increase the economic

---

[18] p.4 Bazargür, D., Shiirevadja, C., Chinbat, B.

importance of the market have actually depressed it by forcing rural people to be more self-sufficient, and reducing the circulation of agricultural and industrial commodities. The Baikal region, indeed the whole of Inner Asia, has a long history of top-down central planning, in the past this has created what might be called a "reality gap" - the gulf between what was supposed to be happening and what actually was.[19]

There may be a danger of recreating such a "reality gap" if unrealistic plans are drawn-up for a transition to a sustainable market economy. The history of Inner Asian development to date demonstrates that the local administration is often very skilled at fulfilling the requirements of planners, at the expense of real development in local communities. If the priorities of planning do not reflect the interests or capabilities of local rural people it is unlikely that "sustainable development" will benefit local communities.

While on the one hand environmental studies suggest that in Buryatia and Chita current pastoral land-use is not sustainable, social studies suggest that it would be a mistake to subject the sector to further radical change based upon social economic and legal models developed for other parts of the world. This suggests to me that any steps towards sustainable development should be based upon the particular circumstances of the Baikal region, and directed towards solving concrete problems of local rural communities. In making these small practical steps it may be informative to look at the experience of neighbouring regions of Inner Asia that also have a history of collectivisation, to see what sort of development might be both feasible and sustainable.

References

Bazargür, D. , Shiirevadja, C., Chinbat, B., 1993. "Territorial Organisation of Mongolian Pastoral livestock Husbandry in the transition to a Market Economy". Working paper, January 1993. Institute of Geography and Geocryology & Deptartment of Geography, Mongolian State University.

Buryat ASSR Statistical Department Narodnoe Khozyaistvo Buryatskoi ASSR : statisticheskii sbornik. (The Peoples Economy of the Buryat ASSR: statistical report) Ulan-Ude, 1963.

Buryat ASSR Statistical Department Narodnoe Khozyaistvo Buryatskoi ASSR : v odinnadtsatoi pyatiletke (1981 - 1985) (The Peoples Economy of the Buryat ASSR: the eleventh five-year plan, 1981 - 1985) Ulan-Ude 1986.

---

[19] See Humphrey (1983) pp. 196 - 200

Chang, Y.T., 1933. The Economic Development and Prospects of Inner Mongolia. The Commercial Press, Shanghai.

Chote, R. The Independent on Sunday, 26 September 1993, (Business Section p.12) London

Gomboev, B. 1993, "Database of results from fieldwork in Argada", Unpublished working materials, MacArthur ECCIA, Cambridge, UK.

Humphrey, C., 1983 Karl Marx Collective: Economy, society and religion in a siberian collective farm, Cambridge, 1983.

Humphrey, C., 1993. "Report on fieldwork in Dornod, summer 1993" Unpublished working paper, MacArthur ECCIA Cambridge UK.

Loucks, O. and Jianguo, W., National Research Council, 1992. Grasslands and Grassland Sciences in Northern China.. National Academy Press, Washington, D.C..

Inner Mongolian Territorial Resources Compilation Committee Nei Menggu Guotu Ziyuan, (Inner Mongolian Territorial Resources Survey), Hohhot, 1987.

Inner Mongolia Association of Foreign Cultural Exchange, 1987. China's Inner Mongolia. Inner Mongolia People's Publishing House, Hohhot, China.

Karamisheff, W., 1925 Mongolia and Western China; Social and Economic Study, Tientsin, 1925.

MPR Central Statistical Board under the Council of Ministers of the MPR, 1974. National Economy MPR 1973. State Publishing House, Ulaanbaatar, Mongolia.

MPR State Statistical Office, 1991. National Economy of the MPR for 70 years 1921-1991. State Publishing House, Ulaanbaatar, Mongolia.

Potkanski, T. and Szynkiewicz, S. 1993 "The Social Context of Liberalisation of the Mongolian Pastoral Economy: Report of Anthropological Fieldwork" PALD Research report No.4 Working paper produced by IDS, Sussex, UK.

Shahrani, M. 1979.The Kirghiz and Wakhi of Afghanistan: Adaption to Closed Frontiers, Seattle and London, 1979.

Tulokhonov, A.K. 1992, "The Historico-geographical aspects of the connection of agriculture of the Baikal region with the environment" personal communication.

Yong Shipeng & Cui Haiting "The Explanation of the Vegetation Map (1:500,000) of Inner Mongolia Autonomous Region" IN Li Bo et al (eds) (Editorial Committee of Series resources Maps of IMAR) Explanation of Series Resources Maps of Inner Mongolia Autonomous Region Science

Zhimbiev, B. 1993, "Database of results from fieldwork in Gigant", Unpublished working materials, MacArthur ECCIA, Cambridge, UK.

Preservation of the Folklore Monuments and Endangered Languages as a Major Factor of World Culture

A.B. Soktoyev
Institute of Philology of SB RAS, Novosibirsk, Russia

The paper gives information about the preparation of the 60-volume bilingual series "Folklore Monuments of the Peoples of Siberia and the Far East", a new type of folklore publication, carried out by the Institute of Philology of SB RAS with folklore specialists of RAS and humanitarian institutes of the peoples and ethnic groups of Siberia.

We are all well aware of the difficult cultural situation of peoples in Siberia and the Far East, especially during the last decades. Everyone can notice a considerable decline in the role and practical use of different spheres of the spiritual life of these peoples: language and folklore extinction, destruction of national traditions and centuries-old customs.

For the ethnic communities of this huge Siberian and Far Eastern region, all this would mean the ruin of everything that in the old era constituted their national uniqueness and originality.

The mere thought that the spiritual heritage of the Siberian and Far Eastern peoples could disappear without a trace sounds incredible and shocking; for world culture it would be an irreplaceable loss and an unjustified impoverishment. Could humankind's conscience resign itself to the fact that everything which had been created and developed not only in centuries but millennia ceased to exist?

The problems concern cultures which have long histories and could be traced back to ancient times. That has been proved by the successes of Siberian archaeology. Excavations and findings of the members of the Russian Academy of Sciences, A.P. Okladnikov, A.P. Derevyanko, V.I. Molodin and others made known to the world the rock paintings (petroglyphs), gold jewellery, beautiful wood carving and metal-treatment samples, armaments, articles of peaceful labour, wooden vessels, wonderfully carved animal figures, carpets, thick felt using extremely complicated ornamental patterns and composition. Especially impressive are the treasures from the Pasyryk burial mound in the Altai Mountains where monuments of civilisations from the 7th to 8th centuries B.C. have been found.

The most noteworthy thing in the archaeological findings is that rock paintings and other works of art created 3–4 millennia ago vividly and convincingly illustrate nothing less than folklore topics and motifs. The folklore of the peoples of Siberia and the Far East can very likely have the most ancient history. This

NATO ASI Series, Partnership Sub-Series, 2. Environment – Vol. 6
Sustainable Development of the Lake Baikal Region
Edited by Valentin A. Koptyug and Martin Uppenbrink

allows us with good reason to consider illiterate peoples' art to be the most fertile power in the whole system of their native cultural values.

Invaluable and great is folklore's significance in preserving the national language which turned out to be most vulnerable and unprotected in the period of assimilation processes. Folklore served as a protection for the language, granted it a "shelter", saved it from disintegration and erosion, and, by its measured polishing, thorough choice of the best potentialities of the living national languages, from complete disappearance.

Linguists who try to understand the structure and inner regularities in the development of these languages have to make use of the folklore texts, especially now that many languages of ethnic minorities in Siberia and the Far East have suffered considerable losses and started functioning only in the narrow sphere at home and family circles, and even then only among some individuals of the older generation. Folklore for the linguists is a historical document in which all the major language manifestations are fixed, a source of the most valuable data about that language development.

In getting a little ahead of my report, I would like to mention that inclusion of the folklore masterpieces in a 60-volume series "Folklore Monuments of the Peoples of Siberia and the Far East" will promote not only the preservation of the Siberian and the Far East regions' languages but will also provide an opportunity to conduct comparative and typological investigations of different language systems. The linguists of the Institute of Philology of the Siberian Branch of the Russian Academy of Sciences are involved in solving all these problems.

Scientific preparation for publication of the folklore texts of some peoples of the Far North (the Dolgans, the Yukagirs, the Evens) and the Far East (the Tofalars, the Udege, the Negidalians) has already stimulated the researchers' work on the systemisation of the national written languages and compilation of the dictionaries.

However, the danger of traceless disappearance threatens folklore itself: once strong schools of local narrators, storytellers, performers of folk songs become extinct, the last outstanding representatives of the national folklore performing art are dying out. Now the voices of the remarkable Altai kaichi S.I.Savdin, the Nganasanian narrator T.A.Kostyorkin, the Tuvinian storyteller M.Ochuur, and many others can no longer be heard.

Nikolai Germogenovich Trofimov, one of the most talented performers of the Evenk folklore, watched with a heavy heart the loss of interest in his native people's oral art. He used to be a common reindeer breeder who had all his life worked in the most remote pastures. At the end of 1970's, feeling that little time was left, that after his death the splendid tales his own people loved so much would soon be forgotten and in spite of the fact that he was semiliterate and ill, he began as best he could to write down, by the light of the oil lamp after a hard day's work, those Evenk heroic epic works which he still remembered. He was convinced that all these works ought to live as long as the nation that created them exists. When his illness reached a state of crisis, Trofimov was taken by helicopter to the nearest regional hospital. And there the life of the storyteller came to an end. In his little knapsack people found a thick notebook with the text "Sodani-bogatyr the Brave". This masterpiece Evenk epic was of such high poetic quality that it aroused great interest with Anna Nikolayevna Myreeva, the Evenk folklore expert. Later, in 1990, after all necessary textual changes, this work was included in the first volume of the academic bilingual series "Folklore Monuments of the Peoples of Siberia and the Far East".

The folklore experts in Siberia were deeply worried by the fate of the national folklore. So the first actions towards saving the national folklore heritage started, first spontaneously, from the base, and then at one of the scholarly conferences devoted to the folklore problems of the peoples of Siberia and the Far East (Yakutsk, June, 1977) the Buryat folklore specialists' proposal was accepted to organise the publication of the best folklore monuments of the peoples of Siberia and the Far East in a unified scholarly edition.

On March 31, 1981, the publication of the series "Folklore Monuments of the Peoples of Siberia and the Far East" was approved by the Presidium of the Siberian branch of the U.S.S.R. Academy of Sciences, and in October, 1981 by the Editorial and Publishing Council of the Presidium of the U.S.S.R. Academy of Sciences in Moscow.

By now the following eight volumes of the series "Folklore Monuments of the Peoples of Siberia and the Far East" have already been published: The Evenk Heroic Epic; The Buryat Heroic Epic "Alamji Mergen"; Russian Epic Poetry of Siberia and the Far East; Russian Tales of Siberia and the Far East: Fairy Tales and Animal Tales; Russian Tales of Siberia and the Far East: Legendary and Everyday Life; The Yakut Heroic Epic "Kyys Debiliye"; Tuvinian Folk Tales; Buryat Fairy Tales. Three more volumes: Yakut Myths and Legends, Tales and Songs of the Dolgans, and The Yakut Heroic Epic "Er Sogotokh the Brave" will be published in 1995.

Unprecedented in scale, this large scientific and publishing undertaking is being realised under the auspices of the Siberian Branch of the Russian Academy of Sciences. The leading research and editing institution which carries out the publication of the series is the Institute of Philology of SB RAS. The Department of Literature and Language of the Russian Academy of Sciences, the Scientific Council on Folklore of the Russian Academy of Sciences and the Editorial Board of the series are responsible for general scholarly and methodological guidance.

14 research institutes of the Russian Academy of Sciences, as well as the Gorno-Altai, Tuva and Khakass Institutes of Language and Literature are taking part in preparation of the series for publication. Among the editors-in-chief and the

authors' teams of the series are the leading folklore experts, who have gained broad recognition in Russia and abroad through their works.

It goes without saying that it is not possible to include in the 60-volume series everything from the immense folklore heritage of Siberia and the Far East. This series is not a collection of folklore works, but a publication of the monuments, the most remarkable artistic objects of national folklore, i.e., the works which are significant both for history and modern times. This requires a thorough selection for the publication of the folklore classics with a parallel translation of the original national text into Russian (except the Russian folklore texts).

Unlike other typical books popularising the works of oral art of different peoples, the series "Folklore Monuments of the Peoples of Siberia and the Far East" represents a special type of scientific publication. The researchers taking part in preparation of the series used the experience of collecting, studying, editing and publishing folklore works by outstanding scientists who investigated the culture of the peoples of Siberia and the Far East (P.A. Rovinsky, A.A. Savelyev, A.A. Makarenko, G.N. Potanin, S.I. Gulayev, M.K. Azadovsky, I.A. Khudyakov, V.I. Yohelson, E.K. Pekarsky, L.Ya. Shternberg, V.V. Radlov, V.G. Tan- Bogoraz, V.K. Arsenyev, M.N. Khangalov, T.Z. Zhamtsarano, N.F. Katanov, G.U. Ergis, S.S. Surazakov and many others).

At the same time, the publication of the series "Folklore Monuments of the Peoples of Siberia and the Far East" demanded that its organisers work out new scholarly and methodological principles reflecting present achievements in the field of selecting, presenting, interpreting the folklore data.

The preparation of each volume in this bilingual series is, first of all, a huge amount of textual work. The major requirement for the publication of the texts is to strictly preserve the folkloric authenticity and to adhere to the linguistic and dialectal precision and peculiarity of the work.

The most crucial and difficult moment is the translation into Russian of the poetic gems of the text, often expressed in highly original and sometimes almost inexpressible national idioms, in phraseological and figurative expressions. All this requires from the teams of authors a deep penetration into the national text and high philological culture.

III. Legislative and Administrative Support of Sustainable Development

International Conventions — Tools for Sustainability

Harold K. Eidsvik
Senior Programme Specialist
UNESCO World Heritage Centre
7 Place de Fontenoy
75007 Paris
France

Introduction

Lake Baikal, with its 31,500 sq. km surface and 560,000 sq. km catchment basin, its 1,800 endemic species and its high biological diversity, has been the focus of hundreds of meetings and thousands of scientific papers. Its future has been discussed on many occasions. Central to these discussions has been what type and how much use it can sustain without prejudicing its unique qualities.

In essence there appear to be two Baikals, one "the Pearl of Russia", a naturalist's paradise, and the second an "industrial sink" with air- and water-borne pollution gradually destroying unique species and natural systems of "outstanding universal value". Has a solution been found to the political and pollution issues? Is legislation in place for future management? Is the "Baikal Commission" working? These are key questions.

Questions of sustainability have been at the forefront of industrial forest policies for generations. It can be said that they have been more ignored than practised. Sustainable development as it is currently used is more a reflection of "The World Conservation Strategy", IUCN, 1980, the "Brundtland Commission", "Caring for the Earth", IUCN, 1993, and the Rio or UNCED meeting of 1992. From my perception, the "Davis Report"[1] correctly questions the concept of sustainable development and substitutes the more appropriate term, "ecologically sustainable economic development".

However, perhaps we should go one step further and add "ethics" to the term. This would be in keeping with reflections arising from IUCN's General Assembly held in Buenos Aires in February, 1994. At that time, many delegates expressed their concern about "Sustainable Development" and the focus many countries tend to place on the development aspect of their programmes. Perhaps more important for decision-makers here is the concept of "ethically, environmentally and economically

[1] "Davis Report": The Lake Baikal Region in the Twenty-First Century: A Model of Sustainable Development or Continued Degradation? Center for Citizen Initiatives-USA, Centre for Socio-Ecological Issues of the Baikal Region, Davis Associates, Russian Academy of Sciences, March 1993.

NATO ASI Series, Partnership Sub-Series, 2. Environment – Vol. 6
Sustainable Development of the Lake Baikal Region
Edited by Valentin A. Koptyug and Martin Uppenbrink

sustainable use," of Lake Baikal and its catchment. By ethically I mean with full consideration of all impacts of development on the people who have traditionally lived in the area.

If a sustainable level of use can be determined for the catchment basin and the lake, what mechanisms exist to ensure adherence to standards and how can the inevitable deleterious side effects which arise from industrial and urban pollutants be controlled? This will be an extremely critical issue in evaluating the potential World Heritage status.

As the "Davis" report suggests, the management of natural resources and control of their impact rest principally with people at the local, regional and national levels. This is the major role of the "Baikal Commission". This is a reflection of sovereignty. Without the full commitment of the responsible governments, the international community has its hands tied. However, as catchment basins often cross local, regional, and even national jurisdictions, there exists a large body of knowledge related to treaties and agreements which can assist in management.

The importance of agreements between states can be underlined by noting that there are more than 2,000 such agreements covering 214 catchment basins shared by two or more countries. (World Resources Institute, 1994–95, p.182).

In 1970, there were 59 international environmental conventions, and today there are more than 170. Guidelines for the protection of Baikal's wetlands and migratory birds are at the heart of some of these conventions. The value of these international conventions has become more and more evident as global communications improve and scientific knowledge expands.

However, international conventions will be of little help in ensuring the sustainability of Lake Baikal if primary sources of pollution or uncontrolled resource or industrial development are not under control. Thus, the effective management of human settlement, industry and mining, agriculture, logging and fisheries must remain the primary consideration in ensuring the sustainability of the region.

International Conventions

There is no place on earth which remains free of human-generated pollutants, be they CFC's, other green-house gases, chemicals or nuclear radiation. There are global conventions relating to all of these factors. These conventions can prove useful in building government and public commitments to conservation and sustainable use.

The UN Framework Convention on Climate Change (CCC)

The CCC was signed by 166 countries during the UNCED Conference in Rio de Janeiro. There are now more than 63 ratifications, and an interim secretariat has been established in Geneva. A conference of the parties is expected in 1995.

The CCC concerns man-made emissions of green-house gases such as CO_2, SO_2, NO_2 and the potential impacts of climate change on people as well as on natural resources and sea levels. While some of these issues may seem distant from concerns at Lake Baikal, they do however underline the universal concern for our physical environment as well as the need for legal mechanisms for effective cooperation. If effectively implemented, this Convention could have a major impact on air quality in the Baikal basin.

The Convention on Biological Diversity (CBD)

This was also a product of the Rio meeting. The CBD was signed by 157 governments, and after ratification by 30 countries it went into effect on December 29, 1993. An interim secretariat is located in Geneva. The first conference of the parties is planned for later this year.

The CBD focuses on the identification and monitoring of processes and activities which are likely to have a significant impact on the conservation of biological resources. In this respect, the convention deals with protected areas as well as ecosystems within or outside of those areas.

Other elements of the convention include public education and awareness as well as the obligatory preparation of environmental impact statements. It may thus play an important role in influencing effective government action on issues of biological diversity, protection and use.

Financially, both the CCC and the CBD are linked to the Global Environment Facility. Thus, they differ from other conventions in having an established financial mechanism for their implementation. This may be of considerable assistance for sustainable use programmes in the Baikal area.

The Convention on Wetlands of International Importance Especially as Water Fowl Habitat (RAMSAR)

This is one of the earlier conservation conventions, dating from 1971. There are 80 contracting parties which have designated more than 421 sites covering more than 30 million hectares. The signatories deposit is at UNESCO Headquarters, whereas the Secretariat is located at IUCN, in Gland, Switzerland.

Under the Convention, states agree to make “wise use” of their wetlands and to designate at least one wetland of international importance under criteria established by the Convention. In 1990, a “Wetland Conservation Fund” was established to assist developing countries.

Parties to the Convention meet every three years and will meet next in Budapest in October, 1994. The listing of some of Baikal’s wetlands would be a major contribution to the implementation of this Convention. Furthermore, it would provide an excellent mechanism for the exchange of technical information on migratory species.

The Convention on International Trade in Endangered Species of Wild Flora and Fauna (CITES), (Washington, 1973)

This Convention, which is perhaps the best known of the international conservation conventions, has been in force since 1975. It has 122 States Parties which meet every two years. The next meeting will be in November, 1994 with an anticipated attendance of 2,000–3,000 people. Its secretariat is located in Geneva.

The primary function of the Convention is to combat illegal trade and overexploitation of species. This is done by establishing lists of endangered species for which international commercial trade is either prohibited or regulated. Another important function is the establishment of management and scientific authorities to grant and review export or import permits.

Decisions related to species listings are often controversial, involving significant differences of opinion between governments and NGO's as well as between governments. Thus, the Convention garners considerable public and political attention.

The Convention Concerning the Protection of the World's Cultural and Natural Heritage (Paris, 1972)

The World Heritage Convention provides for the listing of cultural and natural sites of "outstanding universal value". In addition to World Heritage listing, there is provision for a World Heritage in Danger listing, as well, a World Heritage Fund which has been established to provide for technical and emergency assistance, as well as the training of specialists. I will address this topic further as a following presentation.

I have touched on several conventions which would be applicable to Lake Baikal: Ramsar, CITES, Biodiversity. There are, of course, others which have not been dealt with, such as The "Hague Convention" relating to cultural property. I have also not dwelt on the "Man and the Biosphere" programme, as I believe it is familiar to most of you.

Potential Benefits of International Conventions

International conventions, when effectively implemented, can provide considerable support to politicians and managers, they:

– Provide mechanisms for international cooperation

– Enhance sharing of scientific and technical information

– Provide legal instruments for international action

– Enhance protection of natural and cultural heritage values

– Strengthen access to national and international funds

– Facilitate control of international trade in species

– Increase public and political awareness of critical issues

– Stimulate increased public involvement and tourism

– Encourage the establishment of scientific networks

– Enhance site/area or regional management.

Conclusion

I have touched on several conventions which would be applicable to Lake Baikal: Ramsar, CITES, Biodiversity. There are of course others which have not been dealt with such as The "Hague Convention" related to cultural property. I have also not dwelt on the Biosphere Reserve Programme, as I believe it is familiar to most of you.

As mentioned earlier, the key recommendation of the 1993 "Davis" Report is the nomination of Lake Baikal to the World Heritage List. This has been broadly endorsed by scientists, politicians and the 1990 UNESCO mission. The boundaries proposed in the report with a core and buffer zone appear to be in keeping with previous discussions, as does the creation of the Baikal Commission.

Participation of the Local Population in Area Protection

Helmut Grimm
National Park Office
Schleswig-Holstein Wadden Sea
Schlossgarten 1
D-25837 Toenning
Germany

The Wadden Sea National Park in Schleswig-Holstein is the largest national park in Central Europe. It is also a Man and Biosphere Reserve and Ramsar site. It covers an area of about 2,850 square kilometers on the the west coast of Schleswig-Holstein, North Germany. It is directly subordinate to the Minister for Nature and Environment of Schleswig-Holstein and not to the Federal Government. The National Park Administration, which is situated in the region, is responsible for all types of nature-conservation measures, issuance of permits, public-relations work and environmental education.

In 1985 the parliament of the State of Schleswig-Holstein passed the law for the National Park of the Wadden Sea. Because the draft of this law was not discussed sufficiently with the local population and because they were not sufficiently involved in the process of establishing the National Park, they feel that the National Park was a decision made by "outsiders" against their will. Although the National Park has been generally accepted, increasing opposition to the National Park and its aims is the result. As a consequence, we are now attempting to include the participation of the local population in our work.

Experience has shown that it is necessary to make a distinction between two main groups of people. On the one hand is the local population whose families have often lived for generations in the region. They were in the past a part of the "natural" system and in some cases still use it as a resource. They work in the region and have often, at least in the past, had to struggle against natural forces in order to survive.

On the other hand there are the tourists, who visit the region one or more times for a more or less short period of time each year. They use nature for recreation, relaxation, adventure, education, cultural activities etc.

These two groups are affected in different ways by the introduction of new conservation measures. For the local population they often mean new restrictions and changes in living habits and sometimes lead to restrictions on resource usage and loss of jobs.

NATO ASI Series, Partnership Sub-Series, 2. Environment – Vol. 6
Sustainable Development of the Lake Baikal Region
Edited by Valentin A. Koptyug and Martin Uppenbrink

Although for the tourists nature conservation measures almost always increase the attractiveness of the area for recreation and other activities, they can also lead to some restrictions on use of the area.

Since the local population lives permanently in the area, this group must be integrated into the decision-making process. Nature-conservation measures will only be successful if local people accept them and identify with their aims. Therefore their interests must be taken into account. In particular, consideration should be taken of:

– The quality of life of the resident population

– Commercial structure

– Support of traditional uses and typical culture

– Land ownership should remain in the region; however, investments should be regulated, for example to avoid the construction of buildings which are not typical for the area

– The measures should either produce or preserve jobs within the region or raise the quality of work

– Preservation of the natural character of the landscape

– Limits on touristic activities have to be defined

– Tabu zones have to be established.

These are principles of soft tourism as set forth in the Concept for the Development of Tourism in Schleswig-Holstein. However, we do not use the name soft tourism any longer, but rather the term nature-conformable tourism.

Cooperation with the local population, to ensure that their interests are given due consideration when introducing new conservation measures, is carried out in two ways in Schleswig-Holstein:

1. Boards of Trustees exist, one for each of the two counties which border on the National Park. They consist of 24 representatives of most of the interest groups, district councils, communities, scientists and NGO's within the National Park. They are chaired by the District Administration Officer and not by the National Park administration. Questions of principle and long-term planning must be decided by the National Park administration in agreement with these advisory boards. If agreement is not reached at the board meetings, the State Minister of the Environment makes the final decision. In all cases in the past, an agreement was reached at the meetings of the Boards of Trustees.

2. In addition to these advisory boards, we have installed ten working groups to deal with special subjects and problems. The groups are made up of local inhabitants, representatives of interest groups, communities and experts. They do not make decisions on management measures but advise the administration and Boards of Trustees. Such groups exist for all types of usage in the National Park and the adjacent areas, for example hunting, fisheries, agriculture, tourism, shipping, hiking, seal management and for special problems, for example damage caused by geese to agricultural crops in the farmlands adjacent to the National Park or the resiting of car parks outside the National Park.

This requires the administration to come up with good arguments for the implementation of nature-conservation measures.

This method has been successful in many cases, e.g., a ban on waterfowl hunting and cockle fisheries, a reduction of sheep grazing on saltmarshes and the regulation of sightseeing boat trips to seal banks.

The working groups are also of crucial importance for the development of soft tourism. And tourism can be an important commercial factor for the region: in the Schleswig-Holstein National Park region, the number of tourists has increased in the last 8 years from 700,000 to 1.8 million. The number of overnight stays increased from 9.8 to 16 million in the same period. As of now (1993), overnight stays are estimated at about 19 million. In addition, we registered about 11 million excursions into the National Park region in 1992.

We have made an estimate of total income through tourism. Whereas in former times agriculture was the main source of income of the local population in the National Park area, it is now tourism. Tourism accounts for about 20 % of the total income of the region. By comparison, agriculture only accounts for about 5 % and fisheries, 1 %.

A more thorough analysis shows that about 93 % of total revenue can be attributed to tourists who stay for more than one day in the area. On average, people stay for about 10 days in the region.

Two more figures are of interest:

About 57 % of the income derived from tourism is earned by private persons renting flats or rooms and 17 % by commercial hotels with more than 9 beds. The renting of holiday accomodations is thus dominated by the private sector with small units. This seems to me important when developing “soft tourism”. A good infrastructure, for example a bathroom, is necessary.

In a socio-economic investigation of the National Park region, we made a comparison of money spent for nature conservation and that derived from tourism. The result is striking: the benefits are double the expenditure! In addition, about 50 % of the jobs in the region depend on tourism.

These figures show that tourism is the main commercial factor in the Schleswig-Holstein Wadden Sea region. It is also the only industry showing growth. But one important point must be emphasized. It is not enough to put the label ELO-Tourism on the programs, as many tourist organizations now do. The quality of the program must be carefully scrutinized.

We asked tourists about their reasons for visiting our region. Nearly all of them visit the national park area for recreational purposes in the "intact Wadden Sea environment". This type of "national-park tourism" or "nature tourism" will become more popular in the future and will, I am afraid, create more problems: on the one hand tourism will allow for the commercial development of the region. But on the other hand hordes of tourists disturb nature and may destroy valuable biotopes and the typical landscape, thus destroying the very values that were a decisive factor in their choice of a holiday region. This is a reciprocal process of great importance!

It is necessary to prevent negative developments such as those, for example, which have taken place in the Mediterranean. These main problems can only be solved with the help of the local population. They must learn that an intact nature and landscape constitute the foundation of a commercial system based on tourism. Tourism has to be developed slowly and softly and to such an extent that it brings long-lasting commercial advantage and not just short-term gains.

The quality of life and the identity of the region should not suffer. A social analysis is just as important as an environmental-impact study.

The evaluation of limits on the number of tourists is of importance, but unfortunately no objective method or measure exists. A strategy of zonation is indicated. The local population must be consulted when defining the number of tourists the region can support. Besides zones of zero use, less sensitive zones for touristic use are necessary. Another problem is the funneling of tourists into less sensitive areas. Limits on the number of tourists, e.g., 50 per day, seem to be a suitable measure in many cases; also, trails can help.

We have had good results on a small island of less than two square kilometers. The problem was that thousands of tourists did not keep to the only road on the island, but walked all over the saltmeadows. We employed a local farmer, sponsored by a large bank, to act as a warden on the island. She welcomes the visitors at the pontoon bridge, guides them over the island and tells them about her life on this island. This has had two positive effects: the tourists now keep together and stay on the road, and the employment of a local person also convinced the other 25 inhabitants of the island that the National Park also has positive side effects. They now recognize that the National Park also produces jobs. Moreover, the only restaurant on the island makes more profits.

Cooperation with the local residents requires a well coordinated public-relations and environmental-education plan. This must be addressed first to the local population, to generate an awareness of the value of nature, a responsibility for the region and an acceptance of the aims of conservation. The target groups are mainly users, the media, administrators, political decision makers, tourism managers and tourist guides. Public- relations work should, however, also be directed at tourists, with the aim of guiding them into less sensitive areas.

To achieve our aims, we organize educational courses, give lectures, issue publications. We publish a newspaper, the *Nationalpark Report,* especially for the local population. Once a year it is delivered into nearly every household.

Our information centers are the nuclei for our environmental education work. They are also established in cooperation with communities, NGO's or interest groups.

Acceptance, landscape, local population, Man and Biosphere Reserve, nature, nature- comformable tourism, nature conservation measures, participation, soft tourism.

The paper deals with the experience of the Schleswig-Holstein National Park, a lesson in the complexities of managing land use in a protected area. Local resistance to conservation measures in the past showed how important it was to seek active participation and support of the residents.

Considerations Regarding Nominating Lake Baikal as a World Heritage Site

Bernd von Droste, Harold K. Eidsvik
Director/Senior Programme Specialist
UNESCO World Heritage Centre
7 Place de Fontenoy
75007 PARIS
France

We have already heard much about the wonders and the problems of Lake Baikal and its environs. There is little doubt that it qualifies as a World Heritage site. That will not be a subject of great debate.

Listing as a World Heritage site is recognition by the States Parties that an area is of immense local and national importance as well as being of "outstanding universal value". Areas of "outstanding universal value" are placed on the World Heritage List by the World Heritage Committee. These are the 21 elected representatives of the 139 states which are parties to the Convention. In this paper, I propose to present some of the considerations related to such a listing and to explore some of the problems which could arise.

Nominations of World Heritage sites must be made by the State Party to the Convention, that is, in this case, the Russian Federation. They cannot be made by individuals or other organizations. According to the Operational Guidelines, states must file a "tentative" or preliminary list of proposals. This is, in particular, required for cultural sites. For Lake Baikal a tentative list according to the format has been submitted.

Russia already has seven cultural sites on the World Heritage List. These are the historic centre of St. Petersburg and related groups of monuments, the Kremlin and Red Square, the historic monuments of Novgorod, the Solovetsky Islands, Kizhi Pogost, the White Monuments of Vladimir and Suzdal and, finally, architectural monuments in Sergiev Posad. To date no natural areas have been nominated.

The criteria for the inclusion of natural and cultural properties in the World Heritage List are attached as an annex to this paper. Briefly, there are four:

(i) sites must be outstanding examples representing major stages of the earth's history, including the record of life, e.g., the Galapagos (Ecuador), Kakadu (Australia), the Everglades (United States);

NATO ASI Series, Partnership Sub-Series, 2. Environment – Vol. 6
Sustainable Development of the Lake Baikal Region
Edited by Valentin A. Koptyug and Martin Uppenbrink

(ii) sites must be outstanding examples representing on-going ecological and biological processes in the evolutionary sense, e.g., the Grand Canyon (United States), Lake Malawi (Malawi), the Galapagos (Ecuador);

(iii) sites must contain exceptional natural phenomena or areas of exceptional natural beauty, e.g., Canadian Rocky Mountains (Canada); Mount Kilimanjaro (Tanzania); the Great Barrier Reef (Australia);

(iv) sites must contain the most important and significant natural habitats for in-situ conservation of biological diversity, e.g., Manu (Peru); Kahuzi-Biega (Zaire); Tasmanian Wilderness (Australia).

There are six criteria for cultural sites which include cultural landscapes. This concept may be applicable to Baikal; however this requires further examination.

In addition to meeting at least one of the above criteria, sites must fulfil several criteria relating to integrity or authenticity.

Revisions to the Operational Guidelines of the Convention place more emphasis on the criteria related to biological diversity than previously. This, of course, will not per se effect a nomination of Lake Baikal, which obviously meets these criteria. Cultural criteria were also revised to promote greater recognition of cultures which have evolved over time as a result of their physical environment. The long period of cultural development and human settlement surrounding Lake Baikal may warrant further consideration vis-à-vis the cultural-landscape criteria of the World Heritage Convention.

A nomination, once made, is reviewed by the World Heritage Centre for basic compliance with the Convention. It is then submitted to IUCN (The World Conservation Union) and/or to ICOMOS (International Council on Monuments and Sites) for evaluation against the established criteria for cultural or natural properties.

The evaluation process is rigorous, involving literature reviews, on-site evaluations, peer- group reviews and the deliberation of the World Heritage Bureau as well as the Committee.

Under the Convention, the State Party is obliged to "protect", through legislative and other means, the nominated area. Such legislation should spell out the authorities and responsibilities for the management of the major zones of the proposed World Heritage site. Adequate legislation is one of the critical elements for effective management. It must form a central element of any nomination. The Baikal region is extremely complex socially, economically and ecologically; therefore it would be extremely helpful to spell out in some detail exactly what is meant by protection of the site, so that future misunderstandings may be avoided. Certainly the high water quality of the lake must be assured and the natural qualities existent in wildlife and habitats would need to be protected. Furthermore, the protection of the core zone is essential, and this will no doubt require some careful analysis as to how it relates to urban, industrial and mining areas. Central to all of these decisions is the social, cultural and economic future of the people living in the region.

The technical advisors to the Convention (IUCN, ICOMOS), report their findings to the Bureau of the World Heritage Committee, which generally meets each July and submits its recommendations to the World Heritage Committee, which meets in December of each year. The 21-member Committee, which is elected by the States Parties, is the decision-making body of the Convention.

Professional, technical and financial assistance have been made available to some of these sites on the request of the government. In addition, special promotional events have been carried out to raise public awareness of the importance of the sites. These are the most common benefits arising from World Heritage designation.

As many of you know, in May of 1990 a five-person mission reviewed the World Heritage potential of Lake Baikal and concluded that it met the standard of "outstanding universal value" essential for World Heritage listing. The mission also raised several questions related to the lack of an effective legal and administrative framework. It called for a revised research structure, improvements in planning and zonation as well as strengthening of legislation, planning and administration of the catchment area.

There have been dramatic changes in the government structure since the 1990 mission. In addition, there have been significant changes in the administrative and research structure. There have also been changes in the Operational Guidelines of the World Heritage Convention and its administration.

This conference is an important step in providing more information on the current situation regarding Lake Baikal, its administrative and legal structures as well as other activities that have taken place since 1990.

Perhaps the most difficult issues with respect to the nomination are its boundaries. Can they be as proposed by the May, 1990 UNESCO mission and include the entire watershed? If so, how do we resolve urban and industrial issues such as the mining areas and the paper mill? Can we exclude them, or would that create a Swiss cheese? That is to say, a site with many holes in it. Can we accept some areas as anomalies which will be eliminated in the future. Does a solution lie in a combined nomination with separate sectors for the natural and cultural heritage? Could the latter be recognized as a living, evolving landscape? These are some of the many questions that will need to be resolved during the nomination and evaluation process.

Listing of a site obliges the State Party to file a management plan and to ensure the effective management of the area in such a manner that it will retain the characteristics for which it was nominated.

Conclusion

Lake Baikal undoubtedly deserves a place on the World Heritage List. Considerable effort has already been made to ensure that this will occur.

The nomination of Lake Baikal for World Heritage listing will signify a major commitment to the people living in the region, as well as to the international community, that its resources will be managed in an ethically, environmentally and economically sustainable fashion for this and future generations. At UNESCO's World Heritage Centre we anticipate and welcome the Baikal nomination.

Sustainable Development Within Area Protection and Management of National Parks

Werner d'Oleire-Oltmanns
National Park Berchtesgaden
Doktorberg 6
D-83471 Berchtesgaden
Germany

The paper deals with the possibilities of integrating sustainable development and nature protection within a regional planning process. The results of the MAB6 project at Berchtesgaden and the guidelines for protection, management and development of biosphere reserves in Germany are used as basic information. The general approach is adapted to the Lake Baikal Region.

The subject shows the wide range between the legal status of different protected areas and the desire for sustainable development. Thus, it provides a first framework for discussing the protection approach in use in a region. According to the accepted definition of the IUCN, national parks do not permit commercial activity within their borders. Therefore, sustainable development is not possible in national parks. In discussing a larger region such as the Lake Baikal Region, national parks can be one kind of protected area on a gradient of intensity of protection.

Before beginning the discussion in more detail, we have to be clear about the viewpoint from which we will be looking at the Lake Baikal Region. That of a nature conservationist is different from that of a logger or a developer, and a hunter also sees an area differently. It even happens that the same person reacts differently to a landscape, depending on which function is in question, the official or private one. Therefore, we need a method for looking at a landscape more or less objectively. The program Man and the Biosphere (MAB), initiated by UNESCO, endeavored to develop approaches toward this problem. Messerli and Messerli published in 1979 a model for a regional ecological-economic system. Their viewpoint initiated a long discussion. In the Berchtesgaden MAB6 project (Kerner et al. 1991) this approach was adopted and further developed. It actually did change the way of looking at ecosystems. For a natural scientist, a landscape no longer consisted only of biotic and abiotic factors, but also land use such as forestry, agriculture, hunting and tourism had to be included. So the ecosystem has been subdivided into the subsystem of the natural abiotic and biotic resources and the land-use subsystem. In addition, a third subsystem is part of the approach, that of sociodemography and economy. In contrast to the approach of the International Biological Program (IBP), man was included here as part of the ecology. Because all ecological systems are steady-state, open systems, external factors both from the biotic/abiotic complex as well as from the socio-economic complex influence such a regional system. For Central Europe such

NATO ASI Series, Partnership Sub-Series, 2. Environment – Vol. 6
Sustainable Development of the Lake Baikal Region
Edited by Valentin A. Koptyug and Martin Uppenbrink

parameters can be the decisions of the European Community regarding agriculture as well as air pollution. This dissection of a region gives the chance to integrate different research results into a complex view of the region. Because the approach is open, different results and different viewpoints can be integrated to produce an understanding for an ecologic/economic system. It also provides the chance to describe the actual status of a region versus a "natural" or ideal one. The descripton of the actual status is a summing up of all events occurring in a region. After the dissection of a region in the described way, one does know what sort of ecologic/economic system the region under concern is. Now the question of sustainable development and the role of protection can be objectively discussed. Coming from the side of protection, we find very different levels of nature protection. A stringent approach is the national park as defined by the IUCN, something like a zapovednik, where man is excluded rather than integrated. In a certain contrast is the idea of a biosphere reserve initiated by UNESCO, where man is part of the concept. A biosphere reserve is a reaction to the human use of a landscape, where the result is a very differentiated structure along a gradient of use intensity. Therefore biosphere reserves have a zonation concept. In the core zone, strict nature conservation is practiced. In the buffer zone the management of the landscape is predominant. Here the actual resources will be used sustainably.

In the transition zone, sustainable development is the goal. Before deciding which concept is to be used, it is necessary to note that any reserve needs a surrounding. One may also say "the surrounding defines the interior". From the point of view of zoology, no reserve can be really large enough to support a population of large vertebrates. Therefore, any core zone is a nucleus which needs buffers to create the sustainability of a region. This thinking is supported by the mosaic-cycle concept for ecosystems (Remmert 1991). Its thesis, in brief, is that all succession stages of a system are rotating in time and space. A complete reserve would need to incorporate all of them. In practically no reserve is this possible. For this reason, we need management in most reserves, and in the buffer zone management does stabilize certain sucession stages we want protected. Therefore strictly protected areas, where succession is free, will always be a small part of the region where sustainable development is to take place. This does not diminish their importance.

Let us discuss these theoretical concepts for the Lake Baikal Region under consideration here. They indicate that if we engage in planning in this region, different areas of intensity – meaning the human use – have to be defined. Step by step, using the German approach without explaining it here in detail, one can proceed using a hierarchical approach. With the aid of regional planning, the overall intention of the region can be fixed and regional entities defined.

In order to plan any detail, a system of goals applying to any subdivision must have been previously defined. This is of importance because there is always a difference between wish and reality. To reduce this difference, the system of desired goals is unalterable. Otherwise the different viewpoints of different users pointed out in the beginning will dominate the discussion without any chance of solution. The result must be a zonation plan for the whole region. One solution could be the biosphere concept, which defines zones of different usage. If we define these zones today, we have to ensure that succession within these zones can in time change their scope. Then we have to redefine our region. This indicates that our planning today does not fix the actual situation forever, but defines it as of today and a development towards tomorrow.

If we are to succeed in the future, we will have to redefine our zonation as mentioned above.

What zonation criteria do we have? In Germany we have established guidlines for the protection, management and the development of biosphere reserves (Arbeitsgruppe der Biosphärenreservate in Deutschland, in press). For Germany the core zone will always be small. This is because the intensive use of landscape in Central Europe has only left small areas in a natural or near-natural state. Therefore, only small areas remain where self- regulation or free succession of nature are dominant. In referring back to the mosaic-cycle concept, this means that within these areas succession can take place in a manner which is unusual in Europe, because most areas are under management. Within this zone species are being allowed to disappear, and the ecosystem may possibly change its structure completely without any human interference. Much larger are the buffer zones in Germany which are defined as cultural landscapes. Mostly they are areas which have been used extensively. Especially in the German hills, for example the Rhoen biosphere reserve, special kinds of vegetation on limestone have enlarged their area due to usage. They are defined as very important from the view of biodiversty. Sheep browsing of local sheep breeds is the management approach. The animals are used with all their products on the local market. In poor areas here, sustainable development is being tried. In view of the overall European development of agriculture, these regions would have no chance of surviving economically because of their low productivity. Here economy and nature conservation merge to achieve sustainable development. Another example can be found in the Spreewald. This former flood area of the Spree river was cultivated about two hundred and fifty years ago. Man changed the whole landscape and settled there. The very specific structures of the different ecosystems today also result in an important biodiversity which is greater than in the natural system. Here, too, sustainable use is the management approach for nature conservation. The largest areas in the German biosphere concept are the transition zones. Here sustainable development is the goal to be reached. These areas are model regions for this idea. Beside the marketing of local products, there will be a process of planning with the incorporation of local organizations. This is intended to encourage regions of small productivity to use their resources for production, marketing the local products and building up a certain form of "soft" or eco-tourism. To ensure the quality of the products, methods of marketing are being discussed to ensure quality standards and a solid price system. This is only possible if the tourists understand the specific special characteristics of a region. For this interpretation of the local ecologic/economic system is an obligation of tourism marketing. Thus, these regions can support a potential of jobs for the local human population.

The dissection of a region into different zones as described above can bolster the local economy and thus preserve jobs and in addition generate new income, e.g., from tourism. On the other hand, this means management of the cultural landscape with all its natural and agricultural genetic resources and the preservation of specific man-made landscapes with all their animal and plant species. In certain areas nature is allotted areas for free succession.

The cluster will be different in the Lake Baikal Region than in Germany, but the method of use could result in making the region highly attractive. The idea of nature under protection and the sustainable development of the other regions under discussion could become very attractive. A small remark should be made that this plan also incorporates the local historical buildings as well as the typical construction style of houses or any other human activity and special characteristics. Within this framework a form of tourism could be developed which is adapted to all these local idiosyncrasies. For the Lake Baikal Region this probably would be a very widely divergent tourism in complete contrast to the mass tourism of the Mediterranean region in Europe. If the numbers are lower, the quality of the tourists and their budget must be higher; therefore the offer must include a quality standard to satisfy them.

Discussing the approach and possible results entails cross-checking some thoughts from a European viewpoint with the local ideas for the development of the region. Only then could a concept for sustainable development incorporating nature conservation and use of the natural resources be successfully developed. This could provide the opportunity of using an approach which allows a plan to be developed from a viewpoint which tries to accept different ideas from the beginning. So a more integrated planning could be the result. Another necessity is the identification of the local community with the region and its accepting the task of defining certain basic parameters for further development. Otherwise the sustainable approach will not be successful. Sustainable development is a concept, an idea which always needs a very local specification. Along a gradient from natural to urban industrial ecosystems, nature conservation was given its part on the one side of the gradient. The natural to seminatural part, but the forest and agricultural segments, too, are areas where nature conservation is important for sustainable development. This importance ranges from biodiversity to genetic resources. Sustainable development is a man-made idea. Therefore it does not alter any scientific viewpoints, but it provides the chance of incorporating ecological knowledge into our ethical values.

References

Ständige Arbeitsgruppe der Biosphärenreservate in Deutschland (in press) Leitlinien für Schutz, Pflege und Entwicklung der Biosphärenreservate in Deutschland.

Kerner H. F., Spandau L., Koeppel J. G. (1991) Methoden zur angewandten Ökosystemforschung entwickelt im MAB-Projekt 6 "Ökosystemforschung Berchtesgaden" 1981–1991. Deutsches Nationalkomitee MAB-Mitteilungen 35: 1–176.

Messerli B., Messerli P. (1979) Wirtschaftliche Entwicklung und ökologische Belastbarkeit im Berggebiet (MAB Schweiz). Deutsches Nationalkomitee MAB-Mitteilungen 4: 2–22.

Remmert H. (ed.) (1991) The mosaic-cycle concept of ecosystems. Springer, Berlin/Heidelberg/New York.

Political and Legal Base for Nature Protection in a Federalist System

Winfried Krahl
Institute of Environmental Protection
of the State of Baden-Württemberg
Griesbachstrasse 3
D-76185 Karlsruhe
Germany

The Federal Republic of Germany in its present structure is comprised of 16 federal states, the "Länder". These Länder participate in the legislative process through the "Bundesrat".

Concerning nature-conservation legislation the federal government adopts framework laws; the Länder have to pass their own laws, valid for each individual state. Consequently, the Länder are responsible for administration and financing.

A committee of experts (LANA) is responsible for coordination and uniformity between the Länder.

The instruments of nature conservation and protection are as follows: protection of special areas, landscape planning and infringement regulations.

The structure and organisation of the nature-conservation administrative body differs between the Länder. This report presents Baden-Württemberg as an example.

Living together in a political system as we do, we need a legal basis for all our actions. Due to the different political systems all over the world, we also have different legal bases. As you know, I am from Germany, and my republic has a federalist system. And it is now my task to explain how nature conservation is regulated in such a federal system. I know that we have many different federalist systems in the world, and perhaps you will find that another country has a better one. All I will try to do is to give you one example, and I choose the example of the Federal Republic of Germany.

The Federal Republic of Germany has existed in its present structure of 16 federal states— the so-called Länder—for four years. One of these "Länder" is Baden-Württemberg, where I live and work. It is situated in the Southwest of the Federal Republic.

The Constitution of the Federal Republic regulates the interconnection between the federal government and the so-called "Länder". According to the Constitution, the individual "Länder" are autonomous on principle. Their autonomy is only limited, if at all, for the sake of the unity and continued existence of the whole country. This applies, for example, to the fields of criminal law, foreign and defence policy, but

NATO ASI Series, Partnership Sub-Series, 2. Environment – Vol. 6
Sustainable Development of the Lake Baikal Region
Edited by Valentin A. Koptyug and Martin Uppenbrink

also to the important fields of economy and transport policy. In all the other fields the "Länder" are autonomous. Over and above that they have important tasks and a special clout in regard to legislation and administration of the whole Republic. I would like to use the example of legislation to give you a few rough details.

The individual branches of the federal government are:

1. The President of the Federal Republic. He is the representative of our country;

2. The government, consisting of the Chancellor, the ministers and their respective ministries;

3. The Parliament, we call it the "Bundestag", where we find the elected representatives of the people;

4. The "Bundesrat", which consists of representatives of the governments of the "Länder".

Thus, the "Länder" have their own chamber in the federal government, which plays an important role in the legislative process:

1. The Länder are in a position to initiate laws via the Bundesrat. These are then transmitted to Parliament via the federal government.

2. The Bundesrat, the representatives of the governments of the Länder, has to agree to most of the laws adopted by Parliament. If the Bundesrat does not agree to the laws as adopted, i.e., if it rejects them, there will be an attempt to pass the law in a conciliation process. If there is still no agreement, Parliament with its majority may reject the opposition of Bundesrat concerning important legislative material and then pass the law, and the President can sign the law.

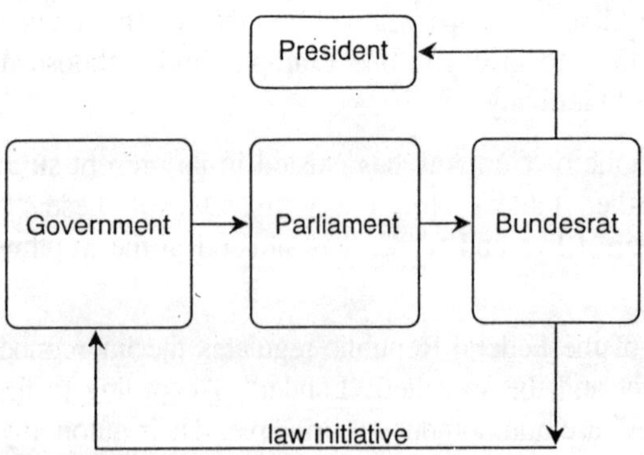

Fig.1 Law intiative via the Bundesrat

3. In many fields the federal government has only the power to pass framework laws. The Länder adopt the actually binding laws on the basis of them. All such laws require the agreement of the Bundesrat on principle. Should the Länder not agree with what the elected representatives in Parliament have decided, they can reject the law in the Bundesrat, and the law does not pass.

In most fields concerning environmental protection, the federal government is only in a position to adopt framework laws, which provide a certain legal framework in which the Länder can act on the basis of their legislative sovereignty. Under these framework laws the Länder can pass more detailed rules and laws. This applies, for example, to the areas of soil, water, air (immission protection), waste and also to nature conservation.

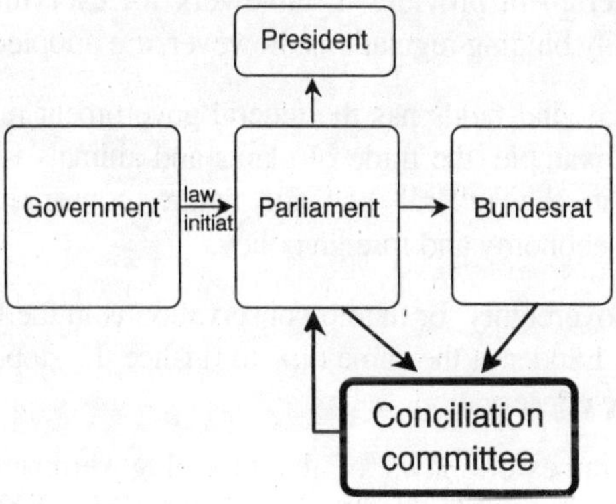

Fig.2: The conciliation process

Other important socio-political fields are organised along the same lines, for instance culture (schools, universities, museums, etc.), the police force, agriculture and forestry. In all these important fields, the federal government only has the competence to initiate framework laws.

In our republic we also have uniform regulations on a national level where this is necessary. In the field of environmental protection the nuclear laws are an example for this. Here we need the strong hand of the central government. And therefore the federal government can intervene directly at the level of the Länder. Not only does it have legislative competency, but also control of the administration, it has the say. I would like to take the example of legislation on nuclear issues to explain this in more detail.

When talking about nuclear waste, control over the Länder is necessary because of the dangers involved; hence a uniform regulation on a national level is required. This implies that permission for transportation, storage and disposal of nuclear material is granted by the federal government. In recent time this has created tensions and led to conflicts between certain Länder and the federal government. Several Länder are ruled by governments that have a different opinion with respect to atomic energy than the federal government. For this reason these Länder tried to oppose to the federal government's granting such permits. However, the federal government was always able to effect implementation by citing its "federal obligation".

This has been a brief introduction into the distribution of competencies between federal government and Länder in the Federal Republic of Germany. It is especially important for the field of nature conservation, because the federal government only has the competency to adopt framework laws, whereas the individual Länder pass the real binding laws for nature conservation. What does this mean?

1. The federal government provides a framework for everything that needs to be regulated. The legally binding regulations, however, are adopted by the Länder.

2. Only in individual vital fields has the federal government reserved the right to a final decision. For example, the trade of plants and animals is regulated uniformly and conclusively in the federal law for nature conservation because of its significance for the economy and foreign policy.

3. Administrative sovereignty for nature conservation is in the hands of the Länder. But this obliges the Länder at the same time to finance the steps that are to be taken in the field of nature conservation.

It is true that the framework laws of the federal government roughly outline a uniformity. Nevertheless the binding legal regulations can differ widely in various fields because of the sovereignty of the Länder. An example for this is the case of the "infringement regulation". This provides that whenever there is an infringement on nature—e.g., when building a new industrial enterprise or roads and all damages thus resulting to the balance of nature—it must be compensated for at a different place, for example by removing a road or setting up new wetlands to replace lost wetlands. These compensation measures cost money. Depending on how you fix the compensation measures, they can have different financial consequences. For certain industrial operations this can even lead to location disadvantages so that they try to move to other Länder. At the moment this is happening in the new, formerly East German Länder, which—in order to create new jobs—are partially refraining from compensations. This is the reason why some operations are shifting their production sites from the old to the new Länder. In order to curtail this development as far as possible, there is an attempt to harmonise crucial provisions of the laws of the individual Länder and the resulting administrative regulations in a so-called "Länder cooperation".

This is done by two panels (Fig.:3):

1. The conference of the ministers for the environment, where the ministers for the environment of the Länder and of the federal government meet twice a year and

2. The “LANA”, which is a working group of the Länder for nature conservation (similar working groups also exist for the fields of water, waste, immission protection, etc.).

In the “LANA” the experts from the ministries of the Länder work together. They deal with topical questions and discuss compromises for regulations and the need for regulations. They prepare recommendations for the conference of the environment ministers.

Depending on the importance of individual projects, the ministers for the environment can use the representations of their respective state in Bonn to launch an initiative of the Bundesrat, which means that they can make a suggestion for a law via the Bundesrat.

The agreements of the Länder amongst themselves, especially in the field of nature conservation, are important for a further reason: due to the fact that the enforcement of environmental protection is up to the sovereign Länder, it is possible that its effects are worse as compared to more centrally organised political fields. This applies, for example, to the fields of economy, transportation and housing construction. If the

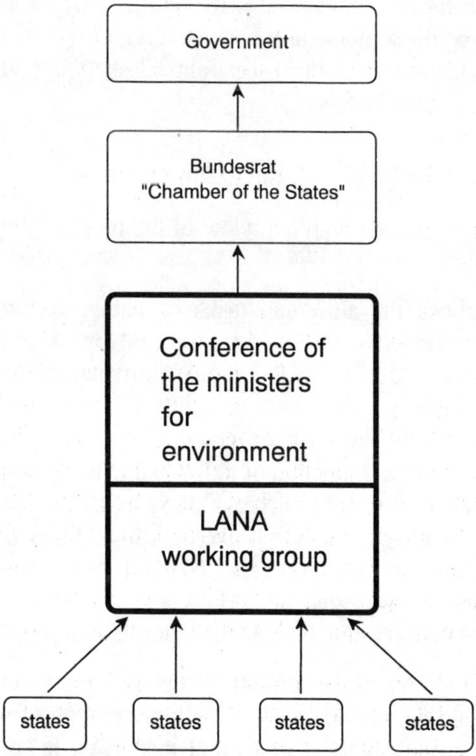

Fig 3. The „Länder“ cooperation

federal government would like to achieve something in these fields against the interests of nature conservation, it could profit from the different opinions in the Länder. It is therefore necessary to try to reach an agreement among as many Länder as possible, so that they work into the same direction and nature conservation can present itself as a strong socio-political power.

Obviously, the functioning of this federal system is not readily comprehensible to outsiders. But I would like to add from my own experience that it is a proven system that works efficiently despite its complexity. If you favour the federalist model, it should be mentioned that from the point of view of nature conservation its needs can best be met on a regional, local level, close to the citizens, which entails regulating and implementing them directly in the region. This widely increases the level of acceptance in the population. Furthermore, the central government, owing to its competition with the community federal states in such a federal system, cannot just set up guidelines that might be a disadvantage for nature conservation, because the federal states always participate in the legislative process. A more centrally governed system is in a greater position to disregard regional interests because it is far removed from the citizens. And it is my opinion that in much bigger countries the regional competencies in the field of nature conservation should show even more positive effects than in a country as small as Germany. This is why Baden-Württemberg is in favour of a Europe of the regions, in which each region will have its own tasks and thus will have considerable weight in the political shaping and administration of a united Europe.

Legal Regulations in the Field of Nature Conservation

Below I would like to give a rough overview of the most important areas of the law on nature conservation.

§ 1 of the law defines the aims and tasks of nature conservation. It says that undeveloped and developed land is to be protected by nature conservation as the basis of the life of mankind in a way that the performance of the natural balance, the usefulness of natural goods (soil, water, air, climate, fauna and flora) as well as the beauty of the landscape are sustainably secured. Furthermore, the habitats of wildlife are to be maintained and the extinction of individual animal and plant species is to be efficiently counteracted. In order to achieve these aims, the citizens are required to behave in a way that damage to nature is avoided. In addition, the law states that the authorities for nature conservation are charged with enforcing these rules. Furthermore the law provides that authorities and planning bodies other than the authorities for nature conservation include the latter in their projects.

What does this mean? The administration for nature conservation is responsible for 100% of the area, and its foremost task is the conservation of natural landscapes as well as ensuring a sustainable use of natural resources. It possesses mainly three instruments to do so efficiently: first, the designation of protected areas, second, landscape planning and third, the infringement regulation.

Concerning protected areas, the law on nature conservation distinguishes two main categories: first, total reserves, that is national parks and nature reserves, and second, rural parks and protected landscapes of outstanding natural beauty that are mainly intended for conserving the landscape and precautions for recovery. The law requires an ordinance for each protected area to be designated, detailing the subject to be protected, the necessary measures against illegal acts and measures for their protection and maintenance. Furthermore, it provides which administrative level would be responsible for the adoption of the ordinances and therefore also for their future control. For a national park, for example, the ministry for the environment or the government of a Land might be responsible in some of the Länder, for a nature reserve it would be the authorities at the intermediate level—called "Regierungspräsidium" in Germany—and for a protected landscape area, the lowest administrative level—in Germany the "Landratsamt".

What is important to these natural reserves? For each area an individual ordinance is to be adopted in an individual process. There must be hearings with the persons affected, landowners, municipalities, etc. The legal process usually requires much time and effort. It is therefore not surprising that in Baden-Württemberg, for example, only 1.5% of the area of the land is designated as nature reserve, although a much larger part of it could and should be designated with respect to its natural richness, namely 3%. Very often the areas are very small. They are spread over the land like pimples on the skin. Natural networks are rare. Even though the number and area of natural reserves has been doubled during the last ten years with considerable administrative effort, the aim of qualified nature conservation of at least 10% of the Land's area could not be reached.

For this reason, there are attempts to embark upon a new course to apply the aims of nature conservation to large areas: in our cultivated areas there is mainly extensive agricultural use. In order to be able to preserve these valuable landscapes, we need a new protection category. We call it the biosphere park, in imitation of the biosphere reserves of UNESCO. According to a plan by the administration for nature conservation, twenty of these valuable, almost unspoilt, extensively used cultivated areas in Baden-Württemberg could be turned into such biosphere parks. There are, however, considerable problems still to be overcome concerning acceptance by politicians and the population, which last—but not least—is at a very high population density of about 300 inhabitants per hectare.

The second important instrument to achieve the aims of nature conservation is landscape planning. It is an independent field of planning, apart from regional planning and urban-development planning. Landscape planning is intended as a basis for land-development planning. It is intended to show which areas of a landscape should be protected because of their natural richness and where potential building developments could take place and where areas could be ecologically improved and restored through landscape conservation and manipulation. From this point of view landscape planning is an important instrument to determine the development of landscapes.

The third instrument is the infringement regulation. What is that? Here is an example: if you want to build a house outside of a populated area, you infringe on the balance of nature. For nature and the landscape this is usually not very good. And therefore we need an instrument that enables the administration to prohibit the building. And this instrument is the infringement regulation. It says that infringements may be inadmissible from the start, namely if they contradict important interests of nature conservation, for example if they destroy a protected area. If this is not the case, an infringement on nature may be permissible, if there is an overriding public interest in the building measure. This could be the case when setting up overhead electrical cables to provide electricity to a town. In order to keep the infringement as low as possible, the perpetrator is obliged to compensate for the consequences as far as possible with as little harm to the landscape as possible. If this is not possible, he is required to implement compensation measures at a different place. Example: if a person destroys a biotope through building measures, he is required to create a new biotope in a different place. If this is not possible, he has to pay a corresponding compensation fee, the amount of which will be fixed by the authority for nature conservation.

Structures of Authorities for Nature Conservation

Finally, a word on the structures of authorities for nature conservation and their tasks. Due to the fact that each Land applies different rules, we have no uniform administration for nature conservation in Germany. This is why I would like to present here the structure of the administration for nature conservation in Baden-Württemberg. We find a dual administration on various levels.

On the one hand we have the authorities for nature conservation with the Ministry for the Environment on top; on the level below we find the "Regierungspräsidium", which is the authority for nature conservation at the intermediate level; and the third authority for nature conservation, on the lowest level, is the "Landratsamt". On the other hand, we have expert bodies organised along the same lines with the "Institute for Environmental Protection" on top, where I come from. At the intermediate level we have a district office for nature conservation and on the lowest level governmental agents for nature conservation working as volunteers. In the authorities we find legal experts and administrative staff, whereas in the expert bodies we see biologists, landscape conservationists, land planners—in brief, experts. The task of these experts consists in giving expert advice to the authorities for nature conservation in all questions concerning the protection of nature. The most important point is that these experts are not bound by directives, so that they cannot be given "political directives". Just imagine that for certain interests the administration does not want to accept the expert advice. There would be a danger of damage to nature. In order to prevent this as far as possible, the expert bodies have the right to turn to the next

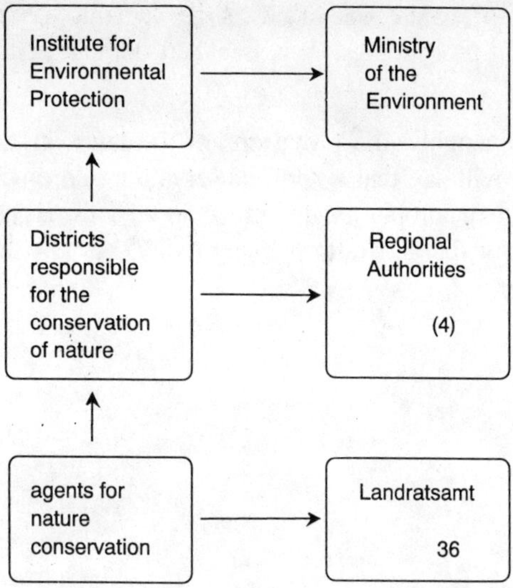

Fig.4: The authorities for nature conservation in Baden-Württemberg

higher authority and ask it to make a decision. This so-called "devolution right" has proven to be very efficient in practise when talking about nature conservation. In controversial issues the side of the authorities was thus forced to either seek a compromise or to take the important interests of the experts on nature conservation into consideration. It is a human characteristic not to like being forced to make decisions by third parties. From the experience in Baden-Württemberg, I can only recommend this dual organisational structure, together with the "devolution right".

There are, however, also a large number of private associations for nature conservation actively at work in this area. They have been successful in fighting for new rights in recent years. In most procedures they have their say today and they are often asked to give their opinions. In some Länder they have even acquired the right to go to court if their objections have not been taken into consideration. However, the "association's right to legal action" is still highly controversial among the Länder, and Baden-Württemberg has not instated it thus far.

One final remark should be added. In our governmental system nature as such has no legally enforceable rights. This is why nature is very often on the losing side. If you want to strengthen the issue of nature, there are, in my opinion, only two possible means: either you set up a strong, independent administration for nature conservation, which defends the interests of nature in all instances, or you make it possible for anybody to turn to the courts for legal action in favour of the conservation of nature. This could be "the association's right to legal action", so that associations for nature conservation can go to court, or the possibility that each citizen may go to court to fight for the cause of nature inasmuch as he feels it has not been taken into consideration in legal proceedings. This is called popular action and is possible in France, for example.

This has been, very roughly, the provisions of the laws on nature conservation in Germany. Finally, I will say that nature conservation and environmental protection must be awarded their proper status in society. This is part of precautionary measures, not only for the generations living today, but also for the generations to come.

Legislating Sustainable Development: People, Politics and Process

G. Gordon Davis
Senior Vice President and Counsel,
Ecologically Sustainable Development, Inc.,
Elizabethtown, New York, U.S.A.

This conference is dedicated to the proposition that Lake Baikal is precious to Russians, and indeed, to the world, and that its future should be governed by principles of sustainable development. Lurking, though, in every presentation at this conference, implicit in the very premise of the conference, is a question not only relevant to the future of Lake Baikal, but to the future of all environmentally sensitive regions of the world where people coexist with nature and where resource conflicts are therefore inevitable. The question, of course, is simply "How in the world do we achieve sustainable development? How can we make it happen?" And how can we do it quickly, for, my friends, we have emergency conditions. Rome is burning.

Is the idea of sustainable development going to turn out like Christianity, a great system that's never been put into practice, or like democracy, the worst system around, except for all the others? Or is it going to be stillborn because we lack the practical skill to breathe life into it?

Others have identified the major threats to the natural systems at work here, and have shared their visions of sustainable development in relation to this extraordinary region. I want to address the question of process. I want to provoke you into thinking critically about process: precisely how do we achieve sustainable development? How in the world can we make it happen? For whatever the design, despite any consensus we may reach on definitions or objectives, there must be an effective process governing every land use in the Baikal basin and the airshed to the west, if the Baikal region is to have sustainable development. Can we depend solely upon legislation? I think not, at least not legislation at the Russian Federation level of government.

Built into any definition of sustainable development are three assumptions. If we think of sustainable development as the cumulative outcome of all human activities affecting land, the first assumption would be that each such activity must have a human benefit: it improves the quality of human life.

The second assumption is that no such activity, whether by reason of its essential character, its size, or its specific location, should be allowed to overstress or degrade the local ecosystem.

NATO ASI Series, Partnership Sub-Series, 2. Environment – Vol. 6
Sustainable Development of the Lake Baikal Region
Edited by Valentin A. Koptyug and Martin Uppenbrink

The third, and essential, assumption is that someone is affirmatively evaluating all land-use activities, allowing those which fulfill the first two assumptions, and prohibiting those which do not.

In other words, there must be a mechanism to identify and prevent inappropriate land uses. Such a mechanism is not alone sufficient to achieve sustainable development, but it is a necessary component of a comprehensive policy to achieve that goal.

Why does the definition of sustainable development assume the need for an umpire to test land-use activities for compliance with applicable standards? Why, in short, must we have a coercive process for prohibiting noncomplying land uses? Because without such a process, short-sighted and selfish land use and resource decisions driven by special interests will inevitably defeat sustainability and degrade the environment. It happened under Communism, it is inevitable under Capitalism. It has already happened here, and without such a process it will continue to happen here, regardless of the character of the mixed economic system now emerging.

If we are to balance economic and ecological needs, at least until such distant future time as a strong land ethic voluntarily constrains the special interests, we need to impose upon government the task of channelling and focussing land use and development into acceptable paths.

Now, there are many ways in which we might harness the institutions of government to mediate resource conflicts in support of the sustainable development goal. Why can't we simply grasp gratefully at any such mechanism offered up by the parliament?

The answer to that question, I think, is that not just any government mechanism is likely to work, and the stakes—a trashed Lake Baikal—are too high to take foolish chances. We should expect from our politicians in the parliament a carefully calibrated, hand-tailored, well thought-out, workable legal structure within which Baikal regional resource-conflict mediation should occur. Clearly a strong, well-drafted Baikal Law would be the preferable solution. But if we cannot achieve that result, then we must work with the relevant oblast and Buryat Republic parliaments or otherwise fashion an effective substitute. And in the process of developing such a structure, we must reject half measures, and fearlessly expose and neutralize careerism and self-interested negativism. Baikal is a world resource. Its value compels us to make a superhuman effort, and its future depends upon it.

So, in the words of a famous Russian, shto dialets (what is to be done)? Let me try to answer that question, even though I am an American.

For my answer I look to the Comprehensive Program of Land-use Policies for the Russian Portion of the Lake Baikal Region. Now I know that the Comprehensive

Program has been described as a Trojan Horse full of Americans. In fact, it was prepared by Russians with American assistance, and it resonates sufficiently among Russians to have been adopted by two out of three regional governments here.

Shto dialets? Enact a federal Baikal Law which forms a Baikal Commission with juridical authority. Structure its governing board so as to achieve meaningful representation by Irkutsk, Chita and the Buryat Republic as well as meaningful representation from outside the immediate area. It should have full authority to coördinate the regulation of all land uses within the Russian portion of the Lake Baikal basin and within the airshed to the west of the Lake. More specifically, give the Baikal Commission the legal authority and responsibility to implement and enforce a comprehensive land-use program based upon a sustainable-development map which zones the region according to the land's carrying capacity, takes into account the need for biodiversity and makes adequate provision for the economic development needs of the region. Require that the map be refined in accordance with sound carrying-capacity principles at an appropriate scale to function as an effective decision-making tool. Incidentally, this last step sounds very much like the landscape-zoning concept being suggested in Irkutsk by the Germans and which has been described here, but that process, in my opinion, cannot stand alone.

Have the Baikal Commission recommend to the government and parliament such norms and standards as are necessary to the sustainable development of the region, and see that they are adopted with the force of law. These norms and standards should be uniform throughout the region, and should be uniformly higher than those applicable to the rest of Russia, reflecting the special character and sensitivity of Baikal. And the Commission should see that they are applied uniformly throughout the region.

How should the norms and standards be applied? To existing land uses by means of an ecological passport system. To new land uses by means of a new permit system Both systems should be administered by the regional Committees on Ecology, subject to the overall supervision and coördination of the Baikal Commission. Under the ecological passport system, violators of norms would be subjected to progressive fines and if not brought into compliance, they would be closed down. Under the new permit system, any new land-use activity within a given zone would fall into one of three categories. It would either be a "preferred use", in which case it would not be regulated, or a "conditional use", in which case it would require a permit—to be granted by the Committee on Ecology if found after review of an ecological expertise to meet all applicable norms and standards—or a prohibited use, in which case it would not be allowed in that zone. Such an approach makes provision for development within approved areas and, depending on the nature and size of the proposal, allows for economic activity while safeguarding ecological concerns. The question of sustainability is addressed efficiently, and within the context of a region-wide system of accountability.

In cases where the ecological expertise raises questions about the environmental impact of a proposal, have the Committee on Ecology hold a formal fact-finding hearing where testimony is taken from government officials and scientists and medical professionals and neighbors. Keep a careful record of this proceeding, so that a court can determine on appeal whether or not the Committee accurately applied the pertinent norms and standards in deciding whether to grant or deny the permit.

Of paramount importance: create an effective role for citizens and non-governmental organizations in the land-use decision-making process, including access to information, the right to comment on all proposals, the right to participate fully in all fact-finding hearings, and the right to appeal adverse decisions to the Commission and to a court of law. Give the Commission the authority to halt for up to six months any activity for which a permit has been issued contrary to the uniform norms and standards until such permit has been reconsidered by the Committee on Ecology that issued it, and where appropriate, by a court of law.

Why should this new Baikal Commission be created? Couldn't such a regional program be implemented by the Ministry of Ecology in Moscow? Or by the existing Baikal Commission created by Presidential decree? Or by the regional Committees on Ecology acting alone?

In my opinion, such a program could not be effectively implemented by the Ministry of Ecology in Moscow for some of the same reasons that the problems of the Baikalsk Kombinat have never been solved by the Ministry of Industry or indeed any other ministry in Moscow. It is too remote from the action. Moreover, it is too easily influenced by national considerations, and its finances are subject to too many conflicting demands and pressures to insure consistent long-term financial support for such a program.

What of the existing Baikal Commission created in December, 1992 by Presidential decree? Well, in the two years of its existence it has yet to make any meaningful contribution, it lacks funding, has no juridical authority and little representation from the region. It has the remoteness problem that afflicts the Ministry of Ecology. Its effectiveness would depend upon ratification of its existence and empowerment through adoption of a Baikal Law by the Russian Parliament. But a dozen such laws have been considered in Moscow, each weaker than its predecessor, and the process of getting an effective Baikal Law may now have been totally thwarted by special interests. Indeed, though the process I have described (of passports, permits, regional zoning, a strong Baikal Commission, etc.) should be specifically enacted in a Baikal Law, it seems most unlikely that an acceptable Baikal Law will be passed. The prospects now seem quite dim to me for any federal legislation which would create an effective program for regulating land uses in the Baikal basin. Frankly, the draft Baikal Law dated July, 1994 and submitted to this Conference, is a very weak law and its passage would likely do more harm than good.

How about the regional Committees on Ecology? Could they administer such a program? At present, these Committees lack the legal authority. If this problem were remedied, there is the question whether the regional Committees could control land uses proposed by or under the auspices of a ministry of the Russian Federation. There is also the problem of uniform enforcement of norms and standards. What incentive, for instance, would the Irkutsk Committee have to impose stringent air pollution standards if the only beneficiaries were down the air stream and outside of the oblast? And what incentive would the Chita Committee have to impose water-pollution standards on land uses that dump their effluent into rivers which empty into the Lake outside the Chita Oblast?

On the other hand, the regional Committees on Ecology are located within the Baikal region, they have personnel familiar with the area and they are accustomed to administering regulatory programs. With appropriate regional legislation establishing effective ecological passport and permit systems, and with appropriate regional policy cordination, they may represent the best present hope for an effective regulatory structure.

What would be the source of policy guidance and coördination? Why, the two Baikal Departments established, respectively, by Chita and the Buryat Republic. These two entities are hard at work implementing the Comprehensive Program. Their governments are squarely behind their efforts. If the region is Rome on fire, then all the other players are fiddlers. These two Baikal Departments are the bucket brigade, and in the absence of a Federation level effort, and a viable Baikal Commission, the initiative to fashion an effective regulatory structure at Baikal should be built around them by means of regional legislation.

In a legal workshop just held in Irkutsk, an impressive group of Russian and American lawyers and policymakers debated for three days alternative strategies for developing the land use regulatory structure so necessary for the Baikal region. It was clearly acknowledged at that workshop that these Baikal Departments, in Chita and Buryatia, have seized the initiative in Baikal sustainable development. The Irkutsk Oblast, having failed to create such a department, may be losing its opportunity to play a significant regional role.

It was the consensus of that workshop that the best approach to sustainable development at Baikal would be Russian Federation legislation: a strong Baikal Law which creates and empowers a Baikal Commission to enforce an ecological passport system and a permit system. The workshop participants considered this unlikely, however. And the serious problem of legal implementation and enforcement, which we can all agree is a persistent problem in Russia, would still have to be addressed. The workshop concluded that an effective regulatory system at Baikal could be established by means of regional legislation which empowers the regional Committees on Ecology to manage and enforce a land-use regulatory program. In other words, an effective ad hoc regulatory system with a coordinated approach to regional sustainable development can be based on regional legislation. But let me emphasize that legislation without implementation and enforcement is a joke, a farce, a cruel delusion.

Shtoh dialets? If we cannot have a strong Baikal Commission created by federal legislation, let's pass the necessary regional legislation to fashion an effective regulatory structure in Chita, Buryatia and Irkutsk. Let's use the laws we have. Let's harness public opinion, challenge the courts, make non-negotiable demands that the administrators in the region obey and enforce existing environmental laws. Let's sue polluters and the public officials who condone pollution. Let's work for change by using all the legal tools at our disposal.

As one considers the present Baikal situation, one's thoughts return to the phenomenon which gave momentum to Russian politics in the late 1960's and thereafter, particularly concerning Baikal issues: citizen environmental activism. Russian lawyers are now building on that base. They are beginning to take advantage of recent legislation allowing citizens and NGO's to enforce environmental laws in court. If citizens and NGO's, aided by able environmental lawyers, work to make it happen, the existing environmental laws can be invoked and the courts enlisted as partners in achieving sustainable development in the region.

In summary, reliance on legislation alone would be a profound mistake. A multifaceted approach emphasizing legal implementation and enforcement of sustainable development principles is necessary. Citizens must coerce politicians to enact appropriate legislation; NGO's must coerce judges to decide cases against polluters; we must all coerce the media to keep these issues alive; each of us must speak out with courage and with conviction. We have the opportunity to make Baikal an international model of sustainable development. We have a lot to lose if we do not.

The Federal Law on Lake Baikal as a Legislative Basis for Sustainable Development of the Lake Baikal Region

I.I. Maksimova, M.A. Grachev, A.N. Suturin[1]
Limnological Institute of SB RAS, Irkutsk, Russia

The Law on Lake Baikal should become the basis of the concept and legal foundation of the sustainable development of the Lake Baikal Region. The draft law has been worked out on the principles of the UN Convention on World Heritage Sites. Its framework character implies that direct normative acts be passed at federal and regional levels.

Sustainable development is based on the thesis that a certain measure of social welfare not diminishing with time (from generation to generation) can be achieved on condition of co-ordination of long-term and immediate interests of the world, country and region.

In the present situation of Russia, the concept of sustainable development can be implemented through a rational regional policy comprising co-ordinated actions on the part of governmental organs of the Russian Federation and its intermediate governmental entitities, and local administrative organs should be aimed at maintaining and developing Russia as an integrated state, at solving problems of regional development and the guaranteeing of human rights to each person.

One of the main tasks of the government's regional policy is ensuring a unity of the economic and social sphere, with conservation and augmentation of economic, social, natural, infrastructural and cultural potential of each region in the interests of its population. The goal of regional governmental policy must be sustainable development as set forth in Agenda 21.

The Lake Baikal Region is an ideal world model of a territory of sustainable development for testing the regional policy of Russia to make it meet all the requirements of the model (fig. 1).

---

[1] Note: the collective of authors also includes : A.V. Belov, V.V. Vorobyev, G.I. Glazii, A.V. Martynov, A.S. Kozlov, Irkutsk Scientific Center of SB RAS; Yu.G. Kurin, I.I. Shirobokov, Administration of the Irkutsk region; I.P. Glasyrina, Administration of the Chita region; V.E. Vikulov, A.K. Tulokhonov, Buryat Scientific Center of SB RAS; V.N. Vasilyev, M.I. Semenov, Government of the Republic of Buryatia; S.A. Bogoliubov, M.M. Brinchuk, K.K. Gofman, V.N. Leksin, R.I. Perelet, Russian Academy of Sciences, Russia; K. Berendess, Ministry for Environmental Protection; H. Schmauder, Federal Agency for Nature Conservation, R. Waal, University of Freiburg; L.Gundling, IHCN, Germany.

NATO ASI Series, Partnership Sub-Series, 2. Environment – Vol. 6
Sustainable Development of the Lake Baikal Region
Edited by Valentin A. Koptyug and Martin Uppenbrink

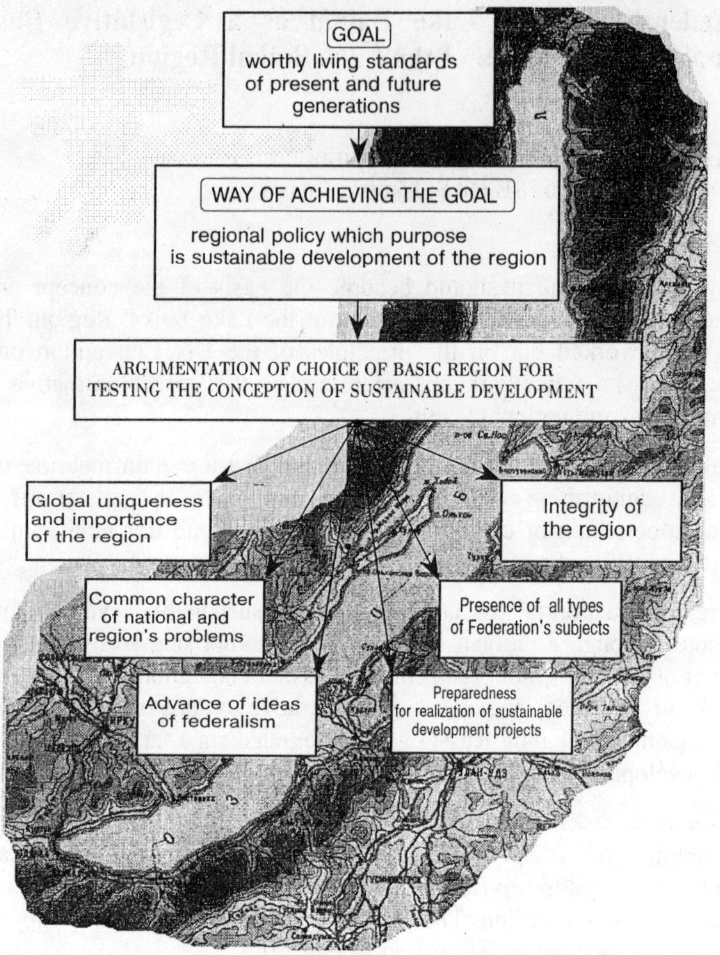

Fig. 1. **Argumentation of choice of the Lake Baikal Region as a world's model territory of sustainable development.**

The global significance of the region stems both from Lake Baikal and its unique water and forest resources, the largest national deposits of gold and copper, of natural gas and non-ferrous metals, power resources, production of aluminium, cellulose, etc.

In the region, all types of administrative entities of the Russian Federation—republic, regions and ethnic districts—are represented. The integrity of the Region is determined by an objective link and core: Lake Baikal.

Over 70 decrees of the Central Committee of the Communist Party, of the Council of Ministers of the USSR and RSFSR concerned problems of Lake

Baikal alone. They were signed by Stalin, Kosygin, Brezhnev, Gorbachev, Yeltsyn, Khasbulatov and Chernomyrdin. Much attention has been devoted to Baikal at all times.

Practically all problems of the country (property and investments, reprofiling and closure of enterprises for economic and ecological reasons, plant breeding, fodder production and cattle breeding, heterogeneous ethnic composition, serious ecological problems) are characteristic of the region.

At the same time, it is precisely the Lake Baikal Region that is the most prepared for the implementation of the concept of sustainable development: the Baikal Commission has been set up and the Governmental Programme of Lake Baikal Protection has been approved by the Government of the Russian Federation. The various governmental entitities of the Russian Federation are experienced in the joint solution of problems, and here a number of fundamental regional normative acts have been elaborated and are currently in effect; a considerable potential of academic, applied and university science is being employed; many foreign scientists are participating in studies on Lake Baikal and finding solutions to its problems. The Lake Baikal Region is known all over the world and attracts politicians and businessmen.

A necessary condition of implementation of the sustainable-development concept is legislative support, i.e., formation of a complex of international, federal and regional normative acts permitting the interests of the world community and of Russia, its regions, and people to be co-ordinated.

The first step to the formation of a legal basis of development of the Lake Baikal Region is enaction of the Federal Law on Lake Baikal. The Federal Law on Lake Baikal takes into account the world's experience and is a part of the current legislative system in Russia. At the same time, it must become a legislative basis for elaboration of a system of normative acts at different levels: from local government, the intermediate entitities of the Russian Federation, to federal and international levels (fig. 2).

The draft of the Federal Law on Lake Baikal was prepared by more than 100 people in co-ordination with the intermediate levels of the Federation and international expertise and was submitted by the Parliament as a legislative initiative. It was approved by the Parliamentary committees for regional policy and ecology and is ready for the first plenary reading.

The present draft law is based on the principles of the UNO Convention on World Heritage Sites and Conception of Sustainable Development. It has a framework nature which determines the necessity of accepting normative acts of direct effect at federal and regional levels.

The draft law distinguishes in the Lake Baikal Region the central zone, buffer zone and zone of atmospheric impact (fig. 3). In the draft law, a legal regime in

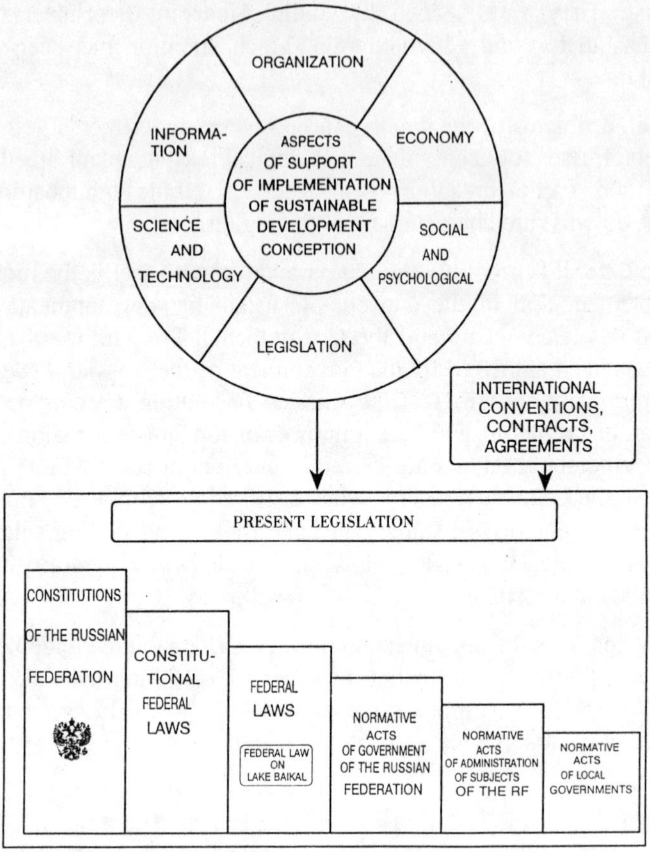

Fig. 2. Support of implementation of sustainable development conception in the Lake Baikal Region.

the economic zones is established which determines the limitations and rules of economic activities. The principal elements of the implementation mechanism of the law are: the co-ordinating body, the Lake Baikal Commission, special budgetary and extrabudgetary financing, ecosystem monitoring, landscape and comprehensive planning, ecological expertise and certification of enterprises, a redevelopment system, reprofiling and liquidation of ecologically hazardous enterprises and a system of social guarantees for the region's population (fig. 4).

Official proposals have been received from 53 representative and executive bodies of the Federation, scientific and social organisations concerning the draft law, 48 of which are constructive. Most criticisms have been taken into account in the draft law submitted to the editors.

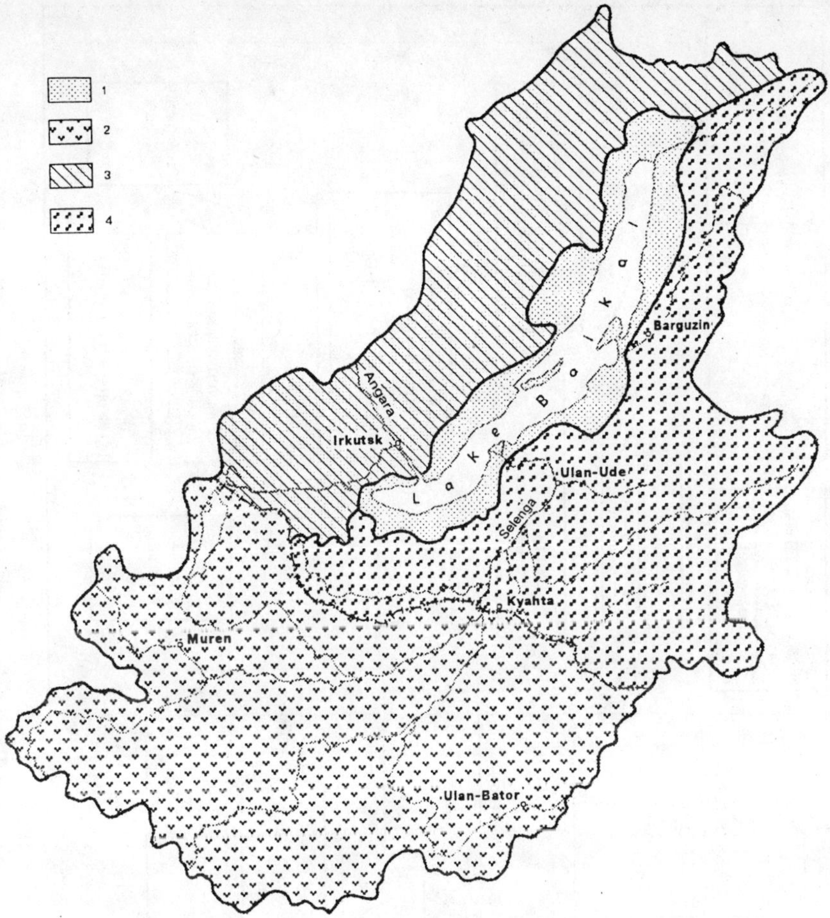

Fig.3. Zones of the Baikal Region.
1. Central Zone
2a. Buffer Zone (Russian territory)
2b. Buffer Zone (Mongolian territory)
3. Zone of Atmospheric Impact

The team of authors is grateful to the Members of Parliament and the Council of the Federation for the initiative of submitting the draft as a legislation initiative; to the Presidium of the Siberian Branch of the Russian Academy of Sciences and to the regional administrations of the Lake Baikal Region for their active stance that has made it possible to overcome some negative moments in the Parliament; to the head of the committee for regional policy for consistently defending the comprehensive character of the law; to the heads of the committees for ecology for successfully co-ordinating their positions, and finally, to the authorities of the Ministry of Environmental Protection and Natural Resources of the RF for the

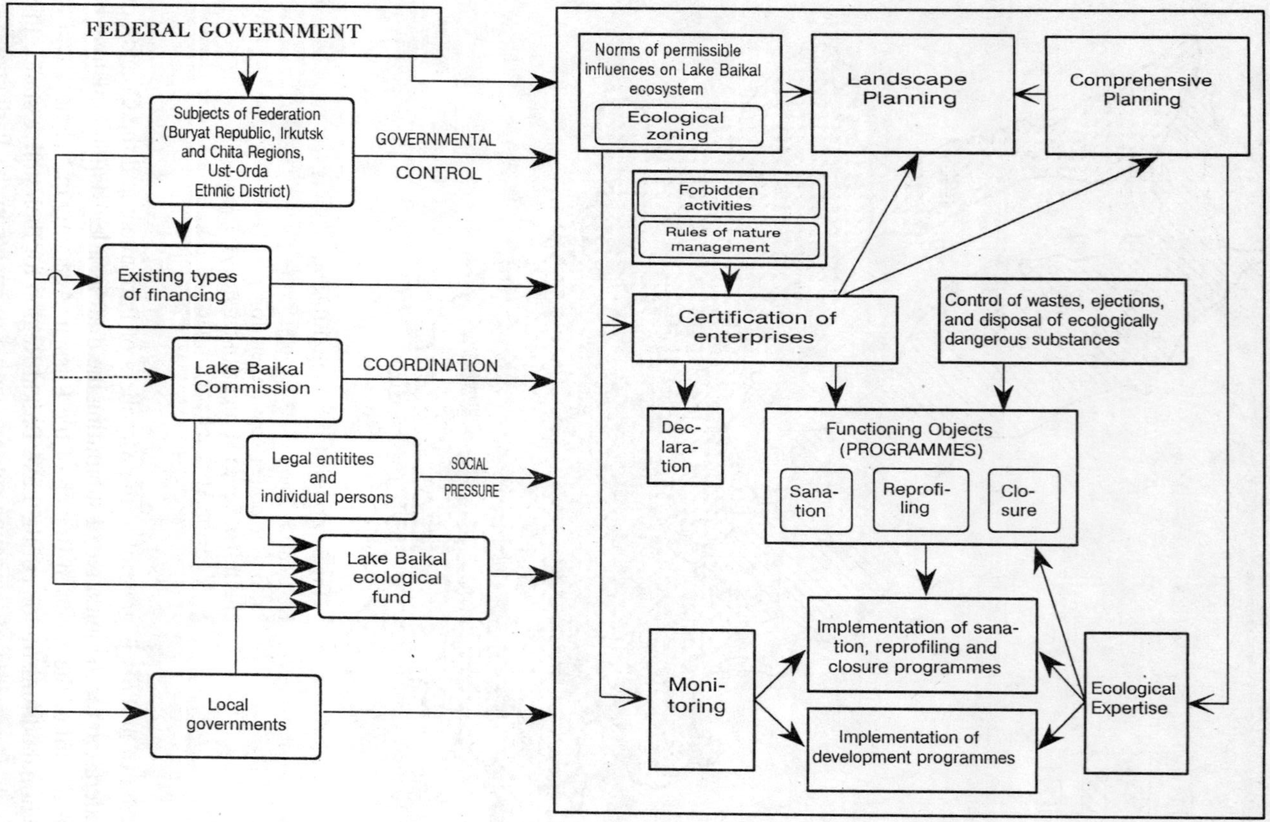

Fig 4. Stucture and mechanism of action of the draft of the Federal Law on Lake Baikal.

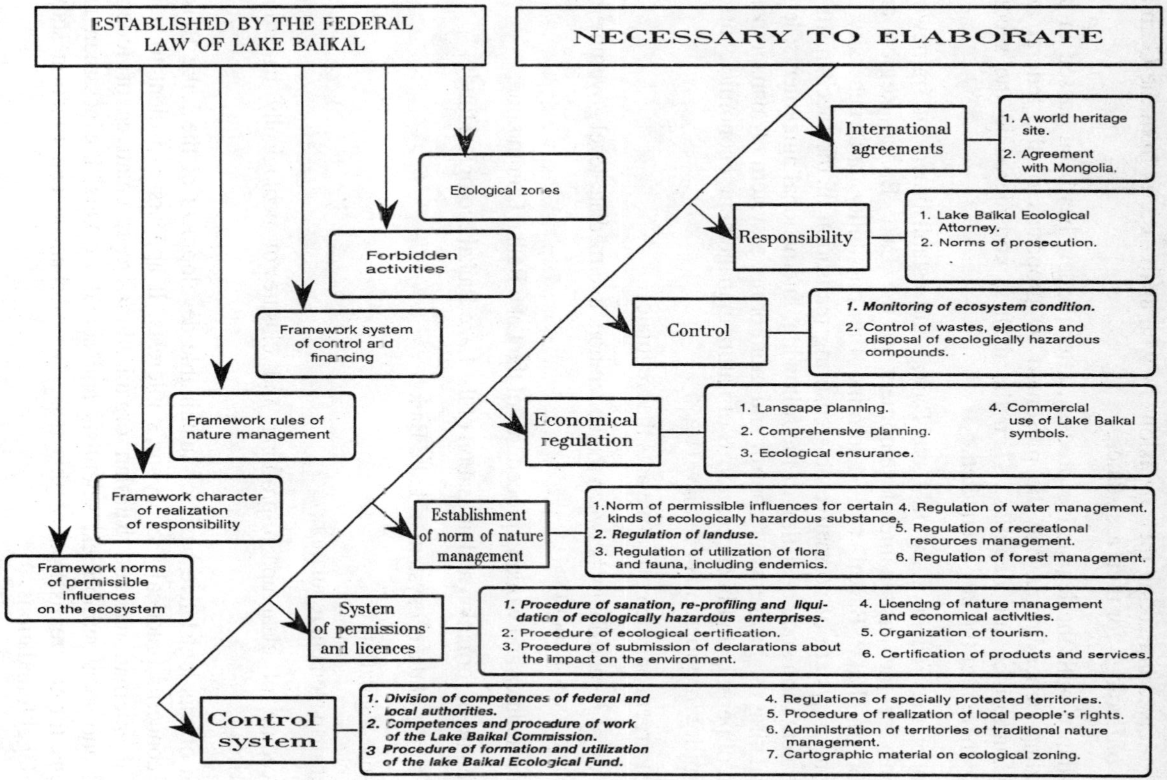

Fig. 5. A tentative system of normative acts, for the formation of the legislative basis of introducing the sustainable development conception in Lake Baikal Region.

reason exercised in making their decisions as well as to experts from the German Ministry of Ecology for their assistance.

In accordance with the legislative process, after the law has been passed in the first reading, amendments may be presented by proposals of the federal and regional authorities and deputies. Then the draft is considered in the second reading, whereafter the law is enacted.

Enaction of the Federal Law on Lake Baikal and establishment of framework conditions for the sustainable development of the Lake Baikal Region and protection of Lake Baikal will determine the possibility and necessity of elaborating a system of normative acts ensuring realisation of the mechanism of its implementation and consisting of the following: international agreements; legal acts regulating the problems of control and monitoring; a system of management; a system of permits and licences; norms of nature management; economic control; responsibility for violation of the law (fig. 5).

The most urgent are the normative acts that regulate:

– division of responsibility between the Federation, intermediate-level governmental bodies and local authorities;

– responsibilities and working procedures of the Lake Baikal Commission;

– procedure of formation and utilisation of the Lake Baikal ecological fund;

– monitoring of the ecosystem of Lake Baikal;

– land-use regulation;

– water-management regulation;

– procedure of clean-up, re-profiling and closure of ecologically hazardous enterprises.

Creation of this legislative basis of sustainable development of the Lake Baikal Region concerns practically everyone's interests. It appears expedient to unite efforts of international and Russian organisations, representatives of federal, regional and local authorities, scientists and social workers, businessmen and politicians for the realisation of this task. No mechanism can function stably without legal foundations.

It seems expedient to set up an international advisory council on the Lake Baikal Region as a world model territory of sustainable development.

Principles and Conditions for Implementation of the Sustainable-development Programmes of the Lake Baikal Region

A.K.Tulokhonov
Baikal Institute of Rational Nature Management of SB RAS
Ulan-Ude, Russia

The main principles of sustainable development in the Lake Baikal Region are revealed in the report. The dominant factors of sustainable development are the following:

– elaboration of an adaptive nature-management system in accordance with the natural conditions of the region;

– environmental education;

– ecological culture;

– calculation of the economic parameters of sustainable development while considering the interests of both economy and ecology.

The solution of the problem of sustainable development in a specific region is connected, first of all, with three main questions:

1. What do we have?

2. What do we want?

3. How can we reach our goal?

1. We have a unique object, Lake Baikal, which by virtue of all its characteristics and parameters is worthy of being included in the list of World Natural and Cultural Heritage sites. As a result of long-term studies by members of academic, educational and applied-research institutions, much information about the nature, economy and culture of this region has been accumulated, some of which is contained in the Atlas of Lake Baikal (Moscow, Roskartografiya, 1993), and in numerous monographs and articles.

It is also necessary to note that the results of these studies are given in very important governmental documents, including:

– a comprehensive territorial environmental-protection plan in the Lake Baikal basin;

– a general scheme of development of productive forces;

– norms of permissible influences on the ecosystem of Lake Baikal;

NATO ASI Series, Partnership Sub-Series, 2. Environment – Vol. 6
Sustainable Development of the Lake Baikal Region
Edited by Valentin A. Koptyug and Martin Uppenbrink

– a comprehensive programme of nature protection and natural-resources management of the Lake Baikal basin;

– the first Russian regional ecological law discussed by the Parliament (the State Duma), "The Law of Lake Baikal".

So it may be stated that the basic purport of the programme of sustainable development of the Lake Baikal Region fully corresponds to the main requirements of its realisation.

2. The ultimate goal of the sustainable-development programme is a decent level of life and culture for the people and of environmental protection. However, a practical implementation of this programme is extremely difficult, although the theory of the question has been elucidated in many publications. In the most general form, the problem of sustainable development can be represented as follows (Fig.1).

The scheme is based on the ideology of interaction between the natural and social components of development. The goal of the sustainable-development programme's realisation strategy is the solution of one three-fold task: cultural development, ensuring a decent standard of living, and conservation of human habitation patterns. Direct indicators of this process may be public health,

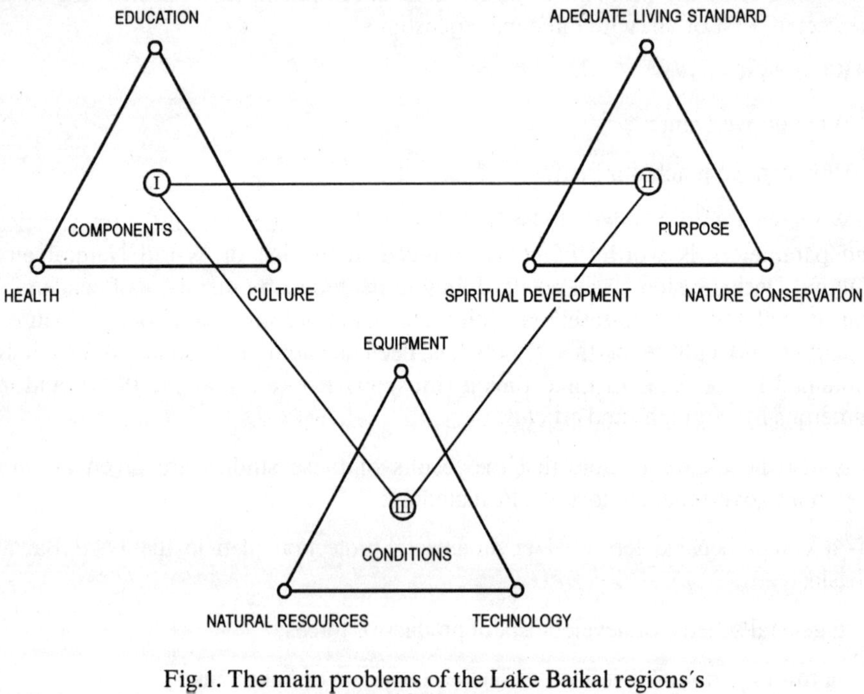

Fig.1. The main problems of the Lake Baikal regions´s sustainable development

education and culture. Achievement of these goals is realised on condition that appropriate natural resources are available and modern equipment and advanced technologies are applied.

Having assumed a priori such a formula of sustainable development, it is necessary to determine the priorities of our actions and to answer the third question: "How can we reach our goal?"

3. As we know, the world is polarised with respect to economic development, and this is reflected in the motives and principles of behaviour that guide population groups at different social levels. As a rule, in societies with low living standards, hard priorities based on extensive utilisation of natural resources prevail (Table 1), and only at higher levels of evolutionary development do the elements of environmental concern appear in the public consciousness, such as the necessity of limiting resource consumption or global changes in the geosphere. Unfortunately, calculations show that the "Third World" countries cannot achieve the economic strength of the advanced countries solely by means of the extensive use of natural resources.

Table 1 Social priorities determined by the living standards of society

	Low Living Standards	**High Living Standards**
Primary Priorities	Food, housing, consumer goods, ethnic problems	Disarmament, environmental protection, healthy way of life, culture and language
Secondary Priorities	Environmental protection, disarmament, healthy way of life, education and medicine, culture, language	
Ways Of Solution	Political (declarative), administrative ecological problems are solved at the expense of social ones	Legal, economical
Mentality	First satisfy my own needs, then attend to environmental protection	Solution of environmental problems is impossible in an isolated region of the planet

That is why it is so important to determine the ways to increase the living standards of the major part of the planet's population, and also to identify the global stages of such evolution as the main problem of sustainable development. First of all, it is necessary to bring, within the shortest possible time, the major part of society to the starting level of real understanding of the regional and global ecological problem. There may be different ways to the solution of this problem: through education, through religion, through introduction of sophisticated technologies, or after natural or anthropogenic catastrophes have occurred. In any case, the final result will be the same, and it is a new awareness: "It is impossible to go on living this way."

Applying all these conditions to the Lake Baikal Region, one may say that here the level of eco-economic organisations is still higher than in other regions, because the main goal of sustainable development is conservation of man and his habitat around Lake Baikal. If this is so, we must build here the most efficient waste-treatment facilities, keep to a minimum the use of mineral fertilisers and pesticides in agriculture, not develop the mining industry, limit industrial felling, and so on, i.e., incur considerable additional expenses. At present, the Republic of Buryatia, which occupies the major part of the Lake Baikal basin territory, completely depends economically on centralised sources of financing. And this dependence is increasing as the economic crisis heightens. In this way, the present state of the economy and a solution of environmental problems are far removed from the concept of sustainable development. What is the way out of this dilemma?

First of all, one has to recognise the ethnic and natural specificity of the Lake Baikal Region, situated as it is in the centre of the Asian continent, with its specific landscapes and methods of utilising its natural resources. The Trans-Baikal area is a territory that is almost equally distant from moisture-bearing air masses from the West and East, with a severe continental climate. That is why precipitation here does not exceed 300 mm. However, in years when cyclones from the East break through, precipitation sharply increases in the form of showers causing catastrophic inundations. Winters are cold and have little snow. Light sandy soils susceptible to erosion are widespread here. Agriculture is developed in the southern, mostly steppe, areas. All these factors also determine the specificity of development of the agricultural complex of the Lake Baikal Region, which until the 30's, i.e., the beginning of collectivisation, sharply differed from that of all the other regions of Siberia.

Under conditions of complete dependence of agriculture on natural climatic factors and absence of governmental support, agricultural workers always had to be oriented only to the types of production which yield sufficient produce at minimal expense.

In the Trans-Baikal area, people have been engaged in livestock raising, and nomadic or migratory forms such as horse and cattle breeding always prevailed.

The development of sheep breeding that negatively affected the productivity of pastures has been restrained. In plant breeding, fodder cultures dominated, primarily rye and oats, as the most undemanding cultures.

However, subsequent events associated with development of virgin lands, construction of light-industry enterprises, have completely changed the structure of the economy and made it similar to the central Russian model.

First of all, coarse-wooled local sheep disappeared: they were replaced by fine-fleeced breeds which could not feed on natural pastures all year round. Creation of gigantic light industry required an increase of sheep stock. As a result, the stock of horses was reduced, and camels disappeared. The pastures and haymaking lands sharply diminished, but ploughlands increased. The social conditions of the rural area changed. As a result of the liquidation of small villages, in Buryatiaalone, 1220 small settlements disappeared between 1939 and 1959.

As a consequence of changes in the structure of agriculture, erosion processes of the main land resources accelerated, especially near large cattle farms. Fodder became deficient, and fodder stores shifted to Mongolia, the Altai and Amur regions. Epizootic diseases increased.

Today, with governmental support reduced to a minimum, agriculture has to be re-oriented to its own resources on an entirely new basis. Achievement of harmony is possible only under conditions of adaptive nature management, when its final result is obtained under minimum influence on nature. As past experience demonstrates, this entails purposive anthropogenic activity in the natural systems and changing the natural potential to satisfy demands of the society without disturbing the general stability of the geosystem.

The ultimate goal of such ecological policy must be the development of a new ecological awareness in which the interests of society and its individuals are closely correlated with nature, and the criterion for this is the physical and moral health of the people. However, society can reach such a level only at a certain stage of economic development, when man realises that any of his wishes can decrease the quality of life because of the destruction of the "home" in which he lives. Special importance in the elaboration of the strategy of sustainable development of the Lake Baikal Region is paid to methodological aspects. From our point of view its scientific verification implies the solution of the following tasks:

a) monitoring of the ecological and economic situation in the region;

b) landscape planning, improvement of land use, and organisation of a system of specially protected natural objects and territories;

c) search and development of new resource-saving and low-waste technologies and equipment for processing natural resources;

d) development of economic and legislative mechanisms for nature-management control;

e) solution of social aspects of sustainable development;

f) elaboration of a model of sustainable development.

All these factors represent an informational basis for performing concrete eco-economical calculations which have to be realised both at the level of the Lake Baikal Region and of its parts, such as the Gusinozersk, Kamensk, Ulan-Ude industrial centres. According to the preliminary calculations carried out by a large team of researchers, there are three models of possible future development of the Lake Baikal Region.

The first model is based on the revival of the former economic structure by means of additional investments. Such a strategy is favoured by the inertia of economic development, existence of a broad spectrum of natural resources, availability of manpower, etc. However, in this case the growth of total production is accompanied by negative influences on the environment. Besides, the depletion of basic funds and the low level of infrastructure do not allow the world economic and ecological standards to be reached.

As a comparative economic model, the development of a recreational complex in the region may be considered, an industry that directly uses unique natural and cultural landscapes of the territory. However, this model of development presents excessive ecological and high-quality demands on the infrastructure of recreational complexes. It is clearly understood that severe climatic conditions, remoteness of the region and elevated vulnerability of natural landscapes rule out the development of the European model of mass tourism. Besides, the low level of material and technical facilities requires large investments with long-term amortisation.

All this precludes considering the development of the recreation industry as a basic branch in the economy of the region. From our point of view, it is more reasonable to regard it as an auxiliary branch, with respect to its real contribution, and to raise its marketing level. Of special interest for this territory are particular forms of educational, ecological and sport tourism which do not require large capital investments and do not have any considerable impact on the environment.

The requirements of sustainable development are most completely satisfied by the third model. It is based on the deep restructuring of the regional economy in favour of industries with thorough processing of raw materials and using modern sophisticated technologies. The practical realisation of such a development model is complicated by the necessity of incurring major material expenses in the diversification of production, as well as supplementary expenses for the training of personnel, and other objective factors. At the same time, the comparative analysis of the three development schemes indicates that the sustainable development of the Lake Baikal Region is possible only if the third, perspective model is implemented. The first two variants may be regarded as partial ones, or as stages in the solution of the main task.

The strategy of the sustainable development of the Lake Baikal Region may ultimately be reduced to the following:

1. Sustainable development in the Lake Baikal Region implies, to equal degrees, conservation of the unique nature of Lake Baikal and development of the economy and culture of the population.

2. The solution of ecological problems comprises not only the purification of sewage and gas emissions, but, first of all, solution of social problems and raising the living standards to a level which permits initial ecological problems to be understood.
Under modern conditions, it is impossible to solve ecological problems completely until personal priorities (housing, food, consumer goods) are substituted by social priorities (ecology, disarmament, policy of non-violence).

3. The solution of concrete ecological problems includes, first of all, working out a general concept of the sustainable development of the region, which consists of at least six interrelated factors: monitoring, economy, technology, culture, ecological education and health care.
The first stage of the realisation and formulation of all ecological programmes will be the estimation of the existing situation both in the natural environment and society (monitoring).

4. Optimisation of nature management and solution of present technical and technological problems means, especially in agriculture, returning to a new level of traditional forms of adaptive economy (meat-livestock raising, buckwheat, rye, fodder grasses).

5. The key factor in solving the ecological problems is to set up, on the example of the Lake Baikal Region, a model territory with social-programmes priorities and a high level of economic and legislative norms in nature management, and, also, the adoption of the Law of Lake Baikal.

6. A necessary condition for the real improvement of the ecological situation in the Lake Baikal Region is close international co-operation and including the Lake Baikal Region in the List of World Heritage Sites.

IV. Statement

Ulan-Ude (Russian Federation)
September 11-17, 1994

STATEMENT

Ulan-Ude (Russian Federation)
September 11-17, 1994

Introduction

The UN Conference on Environment and Development held in June 1992 in Rio de Janeiro on the level of Heads of State pointed out that global changes in the environment and economic disparity between and within countries reached at the threshold of the 21st century the point at which the mankind, with regard to the explosion-like growth of the world population, became aware of a catastrophic threat for our civilization. The previous path of the development cannot be used any further and the mankind must accept a new paradigm for the future development of our civilization. The new path of the development presumes achieving a necessary balance between socioeconomic development and the conservation of nature. As a base for this paradigm the UN Conference proposed the concept of sustainable development.

The Rio Conference called on all Governments to elaborate and adopt national strategies for sustainable development. For large countries it is also very desirable to elaborate new strategies for the development of particularly important territories, bearing in mind that they should be further on made compatible with the national strategy.

The Lake Baikal Region

The NATO Advanced Research Workshop held in Ulan-Ude on September 11-17, 1994, was devoted to the presentation of papers both Russian and from other countries on problems and perspectives of sustainable development of the Lake Baikal Region which is of great value not only for the Russian Federation, but also for all mankind as a unique natural heritage. Baikal is the most ancient lake of the planet, about 30 million years old. It is the deepest with maximum depth of 1637 m, and the largest in volumes - it contains 20% of the world reserve of fresh water. Lake Baikal is inhabited by 2000 species of living organisms, 2/3 of which are not found elsewhere. Baikal's sediments, up to 8 km deep, hold important information on the paleo- climate of Central Asia encompassing many millions of years.

The beauty of landscapes surrounding the lake and its basin is well known throughout the world.

The people of the Lake Baikal Region have very interesting historical and cultural traditions regarding nature protection.

The NATO Advanced Research Workshop

The world-wide importance of Lake Baikal makes it advisable to discuss the region's existing problems and perspectives of sustainable development with the participation of representatives of the international scientific community. In the NATO Advanced Research Workshop under consideration took part representatives from governmental and scientific administrations, scientists of the Russian Federation and its constituent territories (the Buryat Republic, Irkutsk and Chita regions) and from Germany; scientists from the Czech Republic, Mongolian Republic, United States, United Kingdom and Japan; representatives of UNESCO, IUCN, the European Commission in Russia and the NATO's Scientific Committee.

The scientific communities of the Irkutsk and Buryat Scientific Centres of the Siberian Branch of the Russian Academy of Sciences presented to the workshop for consideration and discussion a draft of the report "The Present and Future State of the Baikal Region: Possibilities for Sustainable Development" (Part I - Natural Sector, Part II - Economical Sector; Part III - Perspectives for Sustainable Development) based on the results of previous studies and discussions. More detailed information and proposals relating to environmental monitoring and the rational use of the main natural resources of the Lake Baikal Region, its protected ter- ritories, the restructure and modernization of its industrial sector, sustainable land use, potential legislative and administrative support of sustainable development, demographic and cultural aspects, etc. were presented in many papers. It is very important that a large number of papers were devoted to other countries' experience and to general aspects and requirements of sustainable development.

Baikal - A Unique World Natural Heritage Site

The participants of the Workshop agreed after a comprehensive discussions that the Lake Baikal basin has a world-wide importance as a unique natural heritage site and it is desirable to consider this region as a world's model territory for sustainable development, taking into account the following additional factors:

- the region is characterized by a long history of struggle against anthropogenic degradation of nature;
- the existence on the federal level of the coordinating Joint Governmental Commission on Problems of the Lake Baikal basin and of the special federal programme devoted to environmental protection in the region;
- the region has a developed scientific network for the solution of problems that will be encountered on the route to sustainable development;
- there is in the region quite successful experience and conditions for broad international collaboration.

The main difficulty on the way to sustainable development for this region is connected with the current overall economic crisis in the Russian Federation, lack of the appropriate legislative basis and insufficient understanding by population the inescapability of moving the region, country and whole civilization towards sustainable development.

Recommendations

In order to support the Lake Baikal Region's progress towards sustainable development, the participants of the NATO Advanced Research Workshop recommend the following measures and actions:

1. Regulations

It is necessary to adopt, as soon as possible, federal and regional legislative acts, rules and regulations within the Russian Federation and within the Mongolian Republic, determining legal, economic and administrative principles oriented to an acceptable balance between socioeconomic development and the conservation of nature, that are inseparable parts of sustainability.

Workshop's participants agreed that the presented draft of the Federal Law "On Lake Baikal" is an acceptable pioneering action in this direction.

2. A World Natural Heritage Site

The acknowledgment by UNESCO of Lake Baikal and its surroundings as a World Natural Heritage Site are extremely helpful for the creation of a legal base for the balanced development of the region.

3. Environmental Strategy for Industry

Taking into account the current economic situation in the Russian Federation and Lake Baikal Region, it seems reasonable to restore regional economics in the framework of the existed industrial complex ("basic scenario") with the accent on the conversion or modernization of industrial plants, applying Best Available Technology and Best Environmental Practice (as agreed by the Russian Federation in the Helsinki Convention, 1992). At the same time it is necessary to elaborate an integrated programme for the partial restructuring of the region's economic sector. Due to the long range transport of air and water pollutants, an ambitious approach should be taken to embrace all harmful and hazardous sources of emissions. The advanced process of a technological and ecological inventory of the industrial sector should be completed, and the list of industrial plants to be converted or modernized should be approved. The legal status of the inventory documents should be confirmed.

It is especially important, in particular, to define as soon as possible the future fate of the Baikal Pulp and Paper Plant on the shore of Lake Baikal. It seems, in particular, that a proposal on the conjugated re-profiling of the Baikal Plant and modernization of the Selenginsk Pulp and Cardboard Plant should be attentively expertised by the authorities of the Buryat Republic and the Irkutsk region. In case the agreement on this proposal is reached, very stringent environmental requirement to both plants are necessary.

4. Energy

The energy-supplying sector is very important for economic development, and at the same time it is characterized in many cases by a profound impact on the environment. One of the region's problems consists in the territorial disproportions of power resources. For example, the Pre- Baikal has an excess and Trans-Baikal a deficit. Therefore it is necessary to develop a federal programme of regional power production and transport of electric and other energy resources. Special attention should be paid to the sustainable production and use of energy, such as the energy saving and the use of renewable resources.

The second problem concerns the technological modernization of coal-fired power stations in order to decrease their negative impact on the environment. It is worthwhile to use in the greatest possible extent the natural gas resources of the Irkutsk region. The corresponding measures should be also included in the above-mentioned federal programme.

5. Best Available Technologies and Best Environment Practice

It seems desirable to ask UNEP and UNIDO for support to organize in the region a special workshop devoted to possibilities of access to the international Best Available Technologies and Best Environmental Practice, particularly in the areas of fuel and power production, machine-building and in the agricultural sector.

6. Agriculture

Structural changes in the sector of agriculture can be started earlier than in industry. They should include in particular the restoration of the existing structure of land use (e.g. transformation of heavily eroded lands into meadows and pastures and the reduction of the flocks of sheep with the simultaneous increase of herds of horned cattle and horses). This should be based on a study of the natural capacity of the project region and combined with a fairly simple set of ecological guidelines (the reduction of fertilizers and pesticides use through the implementa- tion of appropriate traditional or modern methods of agricultural engineering oriented to soil fertility conservation and plant nourishment optimization).

Farmers should be trained to use the farmland under sustainable conditions. The regional marketing of the products could be another objective of the project. The authorities of the Republic Buryatia as well as of the Chita region support "A Comprehensive Programme of Land Use Policies for the Russian Portion of the Lake Baikal Region", developed by a group of Russian and American specialists. At present it seems desirable to pay appropriate attention to the analysis and discussion of the efficiency of model projects implemented in the framework of this programme.

7. Nature Conservation and Landscape Planning

The planning and establishment of an appropriate institutional framework for its implementation can be considered as an important basis for the development of the Lake Baikal region. An overall regional strategy for the most important land use types (agriculture, forestry, industry, settlement, tourism) is necessary. Above all, the number and size of protected areas should be enlarged. Each area needs detailed zoning for multiple use. This should be done together with the local population (open planning), and would be developed under the overall goal of sustain- able development.

Due to the large area of the Lake Baikal Region it seems adequate to test in some areas the whole process from planning to implementation as a pilot project. This should be done for each land use type (see above).

8. Forestry

The sustainable development of forestry can be performed only on the basis of the eco-systemic approach. In forestry the transition from clearcutting to a sustainable forest management system should be required. Sustainable forestry means that all ecological functions are fulfilled besides the use of timber. For this purpose it is necessary that:

- when planning the scales of cutting, the ecological role of forests, their condition, and economical and technological accessibility be taken into account;
- new environmentally safe technologies of lumbering be used;
- small-size and economically sound plants, based on wasteless technology, be created;
- the bulk of using the leaf-bearing forest timber be enlarged up to the present proportion in the forest fund;
- the bulk of forest care cutting be enlarged as their resources covered? a considerable part of the regional market needs.

Serious attention should be paid to the amelioration methods for the prevention of the erosion process and soil deflation, and to making forest parks and shelf forests on the erosion lands.

9. Eco-tourism

Eco-tourism may be considered as an important element of socioeconomic development and an instrument of nature protection. But tourism in the region should be organized on a new base of nature protecting principles, and within an appropriate institutional framework providing for sufficient control of tourism activities. Based on the results of market analysis for eco-tourism and an assessment of the tourist potential in the region, it may be initially worthwhile to select 3-5 appropriate points where a radically new type of tourist settlements could be located, and to announce an international competition for the best project of autonomous eco-settlement for the Lake Baikal Region based on ethno-cultural traditions of the region, and on modern achievements in the spheres of advanced materials, renewable energy sources, recycling and disposal of waste, etc.

10. Monitoring and Control

The monitoring of nature in the Lake Baikal Region should be expanded further. Special attention should be paid to the extension of the control system network, the means and methods of space monitoring, the coupling monitoring systems of the Subjects of the Federation with formation of the united network and to the creation of a modern integrated geo-information system (GIS).

Some measures should be taken for the prevention of anthropogenic hazards (for example, industrial and railways accidents nearby the lake) and for the reduction of negative consequences of natural catastrophes, bearing in mind the high level of seismicity of the region.

11. Indigenous People

The conservation of the cultural and environmental traditions of the indigenous people of the region is a very important element of sustainable development. It is necessary to support all efforts that can be made in this sphere. As suitable examples, it is possible to mention the publications of "The Atlas of Tibetan Medicine" and "The Folklore Heritage of the People of Siberia and Far East".

12. Education

The NATO Advanced Research Workshop welcomes the regional initiatives on environmental education in the Lake Baikal Region, especially in the Buryat Republic, to raise the level of awareness, acceptance and support for environmental protection measures, including cultural traditions, and their development. In project areas this supports the success of all steps too. This education work should be done for all groups of the society, especially to our children.

13. Human Health

Human health is an important criterion of sustainable development. Therefore it is necessary to support, in every possible way, scientific investigations and practical actions oriented towards the improvement of human health in the region.

Participants of the NATO Advanced Research Workshop believe that it will give an additional impulse to broadening international collaboration in the field of science, administration and management in the Lake Baikal Region. Some priority areas for such collaboration were mentioned in the supplement to the Statement of the NATO

Advanced Research Workshop on Science Policy held by the Siberian Branch of the Russian Academy of Sciences in November 1993 in Novosibirsk (Russia), and in the course of the INTAS/SB RAS International Workshop "Baikal as a Natural Laboratory for Global Change" held in May 1994 in Irkutsk. In addition, it is important to underline the highly desirable collaboration in the areas of environmental economics and the elaboration of integrated indicators of sustainable development.

The range of cooperation should extend from the exchange of experts and organization of technical meetings to the development of demonstration projects in order to enforce the implementation process.

The pathway towards sustainable development in the basin of Lake Baikal presumes very close collaboration of the Russian Federation and Mongolian Republic in the areas of environmental, cultural and economic problems. It is necessary to broaden this collaboration with regards to the recommendations of the Agenda-21.

The participants of the NATO Advanced Research Workshop are hopeful that all programmes, plans and measures relating to the region's progress towards sustainable development will be discussed with the widest possible participation of society.

РЕШЕНИЕ

Улан-Удэ (Российская Федерация) 11—17 сентября 1994г.

Введение

Конференция ООН по окружающей среде и развитию, состоявшаяся в июне 1992 г. в Рио-де-Жанейро на уровне Глав государств или правительств, констатировала, что на пороге XXI века глобальные изменения в окружающей среде и слишком большое экономическое неравенство между различными странами и внутри стран достигли уровня, который с учетом взрывообразного роста численности населения мира заставил человечество осознать, что цивилизации на планете грозит катастрофа. Предыдущий путь развития не может быть продолжен далее и человечество должно принять новую парадигму будущего развития цивилизации. Новый путь развития предполагает достижение необходимого баланса между социально-экономическим прогрессом и сохранением природы. В качестве основы этой парадигмы Конференция ООН предложила концепцию устойчивого развития.

Конференция в Рио призвала все правительства разработать и утвердить национальные стратегии устойчивого развития. Для больших стран весьма целесообразно также разработать новые стратегии развития особо важных территорий, имея в виду, что они должны быть в дальнейшем согласованы с общенациональной стратегией.

Регион озера Байкал

На рабочем совещании, состоявшемся 11-17 сентября 1994 г. в Улан-Удэ, были рассмотрены доклады российских и зарубежных специалистов по проблемам и перспективам устойчивого развития региона озера Байкал, который представляет собой огромную ценность как уникальное природное достояние не только для Российской Федерации, но и для всего человечества. Байкал является наиболее древним озером планеты, его возраст — примерно 30 миллионов лет. Это самое глубоководное озеро (максимальная глубина 1637 м) и одно из самых больших. Оно содержит 20% мировых запасов пресной воды. В Байкале насчитывается около 2000 видов живых организмов, 2/3 которых нигде более на планете не встречаются. Донные отложения Байкала достигают мощности в 8 км и содержат информацию об изменениях климата Центральной Азии за многие миллионы лет.

Красота ландшафтов, окружающих Байкал, и в его бассейне хорошо известна во всем мире.

Население Байкальского региона имеет интереснейшие исторические и культурные традиции, связанные с защитой природы.

Международное рабочее совещание

Мировая значимость озера Байкал обусловливает целесообразность обсуждения существующих проблем и перспектив устойчивого развития региона с участием представителей международного научного сообщества. В работе совещания приняли участие представители правительственных кругов и науки Российской Федерации и ее субъектов (Республики Бурятия, Иркутской и Читинской областей) и Германии; ученые Республики Чехия, Республики Монголия, Соединенных Штатов, Великобритании и Японии, представители ЮНЕСКО, Международного союза защиты природы, Европейской комиссии в России и Научного комитета НАТО, а также ряда неправительственных организаций.

Научные коллективы Иркутского и Бурятского научных центров Сибирского отделения РАН представили на совещание для рассмотрения и обсуждения проект документа "Настоящее и будущее Байкальского региона — возможности устойчивого развития" (часть I — природный комплекс; часть II — экономический комплекс; часть III — перспективы устойчивого развития), подготовленный на основе результатов многолетних предшествующих исследований и обсуждений. В докладах участников была представлена обширная информация и предложения, касающиеся мониторинга окружающей среды и рационального использования основных природных ресурсов Байкальского региона, его особо охраняемых территорий, структурных преобразований и модернизации промышленного сектора, обеспечения устойчивого землепользования, законодательной и административной поддержки устойчивого развития, демографических и культурных аспектов и т.д. Ряд докладов был посвящен опыту других стран, а также общим вопросам и требованиям устойчивого развития.

Байкал — уникальный участок мирового природного наследия

Участники рабочего совещания после обстоятельного обсуждения всех вопросов согласились, что озеро Байкал имеет общечеловеческую ценность как уникальное природное явление и его регион, принимая во внимание ряд дополнительных факторов, целесообразно рассматривать в качестве модельной территории устойчивого развития мировой значимости. Среди благоприятных дополнительных обстоятельств можно упомянуть следующие:

- регион характеризуется большим опытом борьбы против антропогенного разрушения природы;
- наличие на федеральном уровне координирующей межведомственной правительственной комиссии по проблемам озера Байкал и специальной федеральной программы по защите природы бассейна озера Байкал;
- регион имеет серьезный научный потенциал для решения проблем, возникающих на пути к устойчивому развитию;
- в регионе накоплен достаточно серьезный опыт и созданы условия для расширения международного сотрудничества.

Основные трудности на пути региона к устойчивому развитию связаны с нынешним общим экономическим кризисом в Российской Федерации, отсутствием должной законодательно-правовой базы и недостаточным осознанием населением безальтернативности пути устойчивого развития для региона, страны и всей цивилизации.

Рекомендации

Чтобы поддержать движение Байкальского региона в направлении устойчивого развития, участники совещания рекомендуют следующие меры и действия:

1. Законодательное регулирование

Как можно скорее принять федеральные и региональные законы и другие нормативные акты в Российской Федерации и в Республике Монголия, определяющие правовое, экономическое и административное регулирование и ориентированные на достижение приемлемого баланса социально-экономического развития и сохранения природы, т.е. баланса двух неотделимых частей устойчивого развития.

Участники совещания согласились, что проект Закона РФ "Об озере Байкал" представляет собой приемлемую основу в качестве первого шага в этом направлении.

2. Участок Мирового природного наследия

Для формирования правовой базы сбалансированного развития региона было бы важным признание ЮНЕСКО озера Байкал и его окружения в качестве Участка Мирового Природного Наследия.

3. Экологическая стратегия в промышленности

Принимая во внимание кризисную экономическую ситуацию в Российской Федерации и в Байкальском регионе, необходимо прежде всего восстановить экономический потенциал региона на основе существовавшего народнохозяйственного комплекса ("базовый сценарий"), но с обязательным перепрофилированием и модернизацией промышленных предприятий с использованием наилучших мировых ресурсосберегающих и природощадящих технологий (в соответствии с Хельсинкской конвенцией 1992 года, подписанной Российской Федерацией). Одновременно необходимо разработать комплексную программу структурных изменений в экономическом комплексе региона для следующего этапа.

Кроме того, поскольку существует дальний перенос атмосферных и водных загрязняющих веществ, следует предпринять целенаправленные экологические меры, охватывающие все вредные и опасные источники загрязнений. Следует завершить паспортизацию технологического и экологического уровня промышленных предприятий региона, придать экологическому паспорту юридический статус и утвердить перечень предприятий подлежащих перепрофилированию или модернизации.

Важно в частности, как можно скорее определить судьбу Байкальского целлюлозно-бумажного комбината, расположенного непосредственно на берегу озера Байкал. Серьезной совместной экспертизы со стороны Республики Бурятия и Иркутской области заслуживает предложение о сопряженных мероприятиях по перепрофилированию Байкальского целлюлозно-бумажного и модернизации Селенгинского целлюлозно-картонного комбинатов. В случае, если по этому предложению будет достигнуто соглашение, в нем должны быть заложены жесткие экологические требования.

4. Энергетика

Очень важное значение для экономического развития региона имеет его топливно-энергетический комплекс, который, как известно, во многих случаях оказывает негативное воздействие на окружающую среду. Одна из проблем Байкальского региона — территориальные диспропорции в потенциале энергетических ресурсов. Например, Прибайкальская зона региона имеет определенный избыток их, а Забайкальская — недостаток. Поэтому необходима разработка федеральной программы по развитию региональной структуры производства и межрегионального транспорта электрической энергии и других видов энергетических ресурсов. Наряду с этим особое внимание необходимо уделить использованию энергосберегающих технологий, а также нетрадиционных источников производства энергии, опирающихся на возобновляемые ресурсы.

Вторая проблема связана с необходимостью технологической модернизации работающих на угле теплоэнергетических станций (ТЭЦ и ГРЭС) с целью уменьшения их негативного воздействия на окружающую среду. Целесообразно максимально использовать ресурсы природного газа Иркутской области. Соответствующие мероприятия по решению этой проблемы также должны быть включены в упомянутую программу.

5. Наилучшие ресурсосберегающие и природощадящие технологии

Представляется целесообразным просить ЮНЕСКО и ЮНИДО содействовать проведению в регионе специального рабочего совещания, посвященного возможностям доступа к мировым банкам данных по наилучшим ресурсосберегающим и экологичным технологиям, особенно в области топливно-энергетического комплекса, машиностроения и сельскохозяйственного производства.

6. Сельское хозяйство

Структурные изменения в сельскохозяйственном секторе могут быть начаты раньше, чем в промышленном. Они должны включать, в частности, улучшение существующей системы землепользования с переводом сильно эродированной пашни в луга и пастбища, а также сокращение численности овец и одновременное наращивание поголовья крупного рогатого скота и возрождение коневодства. Развитие сельского хозяйства должно исходить из природных возможностей региона в сочетании с установленными экологическими требованиями, понятными всему населению, такими как уменьшение использования химических удобрений и ядохимикатов путем применения традиционных и современных агротехнических методов сохранения плодородия почв и оптимизации питания растений.

Необходимо научить крестьянина как использовать землю в условиях устойчивого развития. Другой задачей должен быть региональный маркетинг производимой продукции.

Правительством Республики Бурятия и администрацией Читинской области поддержана "Комплексная программа политики землепользования для российской территории бассейна озера Байкал", разработанной группой российских и американских специалистов. В настоящее время желательно уделить должное внимание анализу и обсуждению эффективности реализуемых на основе этой программы модельных проектов.

7. Охрана природы и планирование ландшафтов

Планирование и реализацию соответствующих мероприятий можно рассматривать как важную основу развития Байкальского региона. Необходима согласованная региональная стратегия для всех важнейших видов землепользования (сельское хозяйство, лесное хозяйство, промышленность, населенные пункты, туризм). Кроме того, следует расширить перечень и размеры особо охраняемых территорий. Для каждой территории необходимо детальное зонирование по видам использования, что должно быть осуществлено с участием местного населения (открытое планирование) и с ориентацией на основные цели устойчивого развития.

Так как Байкальский регион занимает большую территорию, представляется целесообразным апробировать весь процесс, начиная с планирования и до реализации опытного проекта, в конкретном месте. Это необходимо сделать для каждого вида землепользования (см. выше).

8. Лесное хозяйство

Развитие лесного хозяйства может быть осуществлено только на базе экосистемного подхода. В лесопользовании необходим переход от сплошных рубок к ведению лесного хозяйства, исходя из принципов устойчивого развития. Это означает, что все виды хозяйствования осуществляются с полным использованием материалов леса и сохранением экологических функций лесов региона. Для этого необходимо:

- планирование рубок с учетом экологической роли лесов, их состояния и экономической и технологической доступности;
- использование новых экологически чистых технологий лесопользования;
- создание малых и экономически выгодных хозяйств на основе безотходной технологии;
- увеличение объема заготовок лиственных лесов до фактических пропорций в лесном фонде;
- увеличение объема рубок ухода, т.к. эти ресурсы могут покрыть значительную часть региональных рыночных потребностей.

Серьезное внимание необходимо уделить лесомелиоративным приемам борьбы с эрозией и дефляцией почв, созданию полочных и парковых лесов на эродированных землях.

9. Экотуризм

Экотуризм может стать важным элементом социально-экономического развития региона и инструментом защиты природы, но эта деятельность должна развиваться на базе новых природоохранных принципов, а также в рамках соответствующих организационных мероприятий, обеспечивающих эффективный контроль за этой сферой деятельности. Целесообразно на основе разработанной концепции развития экотуризма в Байкальском регионе на первой стадии выбрать 3-5 подходящих точек, в которых могли бы быть размещены туристические базы принципиально нового типа и объявить международный конкурс проектов на автономные байкальские эко-поселения, учитывающие этнокультурные традиции региона и основанные на использовании новейших достижений в области прогрессивных материалов, возобновляемых источников энергии, уничтожения и утилизации отходов и т.д.

10. Мониторинг и контроль

Следует расширить мониторинг природной среды в бассейне озера Байкал. Особое внимание целесообразно обратить на более широкое использование автоматизированных станций контроля, средств и методов космического мониторинга, а также биологических наблюдений, на сопряжение мониторинговых служб субъектов Федерации с формированием интегрированной сети и на создание современной интегрированной геоинформационной системы.

Существенное внимание должно быть уделено реализации мероприятий по предотвращению антропогенных катастроф (например, промышленных и железнодорожных аварий вблизи побережья озера) и снижению негативных последствий природных катастроф, учитывая, в частности, высокую сейсмичность Байкальского региона.

11. Местное население

Важным элементом устойчивого развития является сохранение культурных и природозащитных традиций народов региона. Необходимо поддерживать все усилия, предпринимаемые в этом направлении. В качестве примеров можно упомянуть издание "Атласа тибетской медицины" и многотомного издания "Памятники фольклора народов Сибири и Дальнего Востока".

12. Образование

Совещание поддерживает энтузиазм и инициативу жителей Байкальского региона и особенно Республики Бурятия в области экологического образования в целях повышения уровня понимания и поддержки мер по защите окружающей среды, включая культурные традиции, а также их развитие. Это необходимо для людей всех возрастов, но особое внимание следует обратить на подрастающее поколение и молодежь.

13. Здравоохранение

Одним из важных критериев устойчивого развития является состояние здоровья населения. В связи с этим необходимо всемерно поддерживать научные исследования и практические мероприятия по сохранению и укреплению здоровья человека.

Участники совещания надеются, что оно даст дополнительный импульс расширению международного сотрудничества в области науки и административного управления Байкальским регионом. Ряд приоритетных областей для такого сотрудничества был упомянут в приложении к Решению Международного рабочего совещания НАТО/СО РАН по научной политике, состоявшегося в ноябре 1993 года в Новосибирске, и в ходе дискуссий на Международном рабочем совещании ИНТАС/СО РАН "Байкал — природная лаборатория для исследования изменений окружающей среды и климата" в Иркутске в мае 1994 г. Кроме того, очень важно отметить острую необходимость сотрудничества в области экологической экономики и разработки интегрированных индикаторов устойчивого развития.

В сотрудничестве акценты должны перемещаться от обмена специалистами и организации научных встреч в направлении разработки и реализации демонстрационных проектов.

Движение к устойчивому развитию в Байкальском регионе предполагает тесное сотрудничество Российской Федерации и Республики Монголия в области экологических, культурных и экономических проблем. Необходимо расширять это сотрудничество с учетом рекомендаций одного из основных документов Конференции ООН в Рио-де-Жанейро "Повестка дня на XXI век".

Участники рабочего совещания надеются, что к данному решению присоединятся многочисленные научные организации и общественные движения и что все программы, планы и мероприятия, ориентированные на движение к устойчивому развитию, будут самым широким образом обсуждаться с общественностью.

List of Workshop Participants

Dr. Vladimir K. AGALOV
Narodny Khural of Republic Buryatia
Committee on Economics
Chairman
Soviet Square
670000 Ulan-Ude
Tel: 7 (301-22) 25-212
RUSSIA

Dr. Nikolai A. ALDOKHIN
Limnological Institute of SB RAS
3 Ulan-Batorskaya Str.
664033 Irkutsk
Tel: 7 (395-2) 460-405
RUSSIA

Yuri I. ALEKSEEV
State Committee of Land Resources and Management of Republic Buryatia
Chairman
11a Revolution 1905 Str.
670034 Ulan-Ude
Tel: 7 (301-22) 21-811
RUSSIA

Prof. Amirkhan M. AMIRKHANOV
Ministry of the Environment Protection and Natural Resources of Russia
Deputy Federal Minister
4/6 B.Grusinskaya Str.
123812 Moscow
Tel: 7 (095) 124-0471
RUSSIA

Zinaida N. ANTONTSEVA
Committee of State Duma of Russian Federation
Expert
2 Okhotny ryad Str.
103009 Moscow
Tel: 7 (095) 292-6009
RUSSIA

Prof. Dumaajav BAATAR
Mongolian Academy of Sciences
President
3 Sukhbaatar
Ulanbaatar 11
Tel: (976-1) 322-216
Fax: (976-1) 321-638
MONGOLIA

Dr. Alexander R. BATUEV
Institute of Geography of SB RAS
1 Ulan-Batorskaya Str.
664033 Irkutsk
E-mail: root@igs.irkutsk.su
RUSSIA

Dr. Aleksander D. BADLUEV
Ministry of Education
Soviet Square
670000 Ulan-Ude
Tel: 7 (301-22) 22-146
RUSSIA

Prof. Aleksey Ts. BALDUEV
Buryatsky Agricultural Institute
Rector
8 Pushkin Str.
670020 Ulan-Ude
Tel: 7 (301-22) 42-611
RUSSIA

Dr. Bair G. BALZHIROV
Ministry of Health Protection of Republic Buryatia
Minister
Soviet Square
670000 Ulan-Ude
Tel: 7 (301-22) 22-377
RUSSIA

Dr. Vladimir S. BATURIN
Irkutsk Computing Centre of SB RAS
Deputy Director
130 Lermontov Str.
664033 Irlutsk
E-mail: snvas@icc.ccsoan.irkutsk.su.
RUSSIA

Dr. Larisa D. BAZAROVA
Buryat Scientific Centre of SB RAS
Vice-Chairman
6 Sakhyanova Str.
670042 Ulan-Ude
Tel: 7 (301-22) 33-324
E-mail: root@bien.buriatia.su
RUSSIA

Prof. Aleksey V. BELOV
Irkutsk Scientific Center of SB RAS
Vice Chairman
134 Lermontov Str.
664033 Irkutsk
Tel: 7 (395-2) 465-613
Fax: 7 (395-2) 462-952
RUSSIA

Dr. I. BORISENKO
Buryat Geological Institute of SB RAS
6 Sakhyanova Str.
670042 Ulan-Ude
Tel: 7 (301-22) 35-035
Fax: 7 (301-22) 63-244
E-mail: root@bien.buriatia.su
RUSSIA

Valentin D. BROVCHAK
Ministry of the Environment Protection
and Natural Resources of Russia
4/6 B.Gruzinskaya Str.
123812 Moscow
Tel: 7 (095) 125-5216
RUSSIA

Tchoi-Dorzhy BUDAEV
Central Ecclesiastical Administration
of Buddhists of the Russian Federation
Hambo-Lama
Soviet Square
670000 Ulan-Ude
Tel: 7 (301-22) 27-048
RUSSIA

Luis Veiga DA CUNHA
NATO Scientific and Environmental
Affairs Division
1110 Brussels
Tel: (32-2) 728-4229
Fax: (32-2) 728-4232
E-mail: HAAAJ01@blekuo11.bitnet
BELGIUM

Prof. G.-Nimbu B. DANDARON
Buryat Branch of Novosibirsk State University
Rector
25 Pushkin Str.
670034 Ulan-Ude
Tel: 7 (301-22) 42 522
RUSSIA

Dr. George Donald DAVIS
Ecologically Sustainable Development, Inc.
President
2 Church str., P.O.Box 848
Elizabethtown, NY 12932
Tel: 1 (518) 873 3200
Fax: 1 (518) 873 2686
E-mail: esd@igc.apc.org
USA

Dr. G. Gordon DAVIS
Ecologically Sustainable Development, Inc.
Vice President
2 Church str., P.O.Box 848
Elizabethtown, NY 12932
Tel: 1 (518) 873 3200
Fax: 1 (518) 873 2686
E-mail: esd@igc.apc.org
USA

Larisa I. DEDOVA
Committee of the Environment Protection and
Natural Resources of Irkutskaya Oblast
16 Parkovaya Str.
664033 Irkutsk
Tel: 7 (395-2) 277 850
Fax: 7 (395-2) 276 573
RUSSIA

Prof. Nikolay L. DOBRETSOV
Institute of Geology of SB RAS
3 University Ave.
630090 Novosibirsk
Tel: 7 (383-2) 354 650
Fax: 7 (383-2) 352 692
RUSSIA

Dr. Harold K. EIDSVIK
UNESCO
World Heritage Centre
7 Place Fontenoy,
F- 75007 Paris
Tel: (331) 45 68 1443
Fax: (331) 40 56 9570
FRANCE

Dr. Valery D. ERMIKOV
Siberian Division
of Russian Academy of Sciences
Head of Research Management Department
17 Lavrentiev Ave.
630090 Novosibirsk
Tel: 7 (383-2) 356 619
Fax: 7 (383-2) 354 846
Telex: 133128 POISK SU
E-mail: vde@hq.soan.nsk.su
RUSSIA

Dr. Mikhail D. EVDOKIMENKO
Institute of Forest of SB RAS
Akademgorodok
660036 Krasnoyarsk
Fax: 7 (391-2) 433 686
E-mail: evag@ifor.krasnoyarsk.su
RUSSIA

Dr. Irina P. GLAZYRINA
Administration of the Chita Region
8 Tchaikovsky Str.
672021 Chita
Tel: 7 (302-22) 32 011, 7 (302-22) 66 505
Fax: 7 (302-22) 68 618
E-mail: chita@glas.apc.org
RUSSIA

Dr. Evgeny A. GOLUBEV
East Siberian Technological Institute
Politology and Sociology
20 Tereshkova Str.
670031 Ulan-Ude
Tel: 7 (301-22) 30 342
RUSSIA

Dr. Helmut GRIMM
National Park Office
Schleswig-Holstein Wadden Sea
Schlossgarten 1
D-25837 Toenning
Tel: 49 (4861) 61 630
Fax: 49 (4861) 459
GERMANY

Stanislav L. GROMOV
Commission of Russian Federation
for Land Resources and Land Use
Deputy Chairman
5/1 Leo Tolstoy Str.
119862 Moscow, G-21
Tel: 7 (095) 339-5804, 7 (095) 246-3598
Fax: 7 (095) 246-1007
Telex: 111455 RUNO
RUSSIA

Valery E. GULGONOV
State Committee of the Environment
Protection and Nature Resources of
Republic Buryatia
Chairman
24 Solnechnaya Str.
670071 Ulan-Ude
Tel: 7 (301-22) 31-716
RUSSIA

Dr. Christina von HAAREN
Institute of Landscape Management
and Nature Conservation of the
University of Hannover
Herrenhäuser Str. 2
D-30419 Hannover
Tel: 0511-762-3796
Fax: 0511-762-3791
GERMANY

Sergey P. ILYIN
Ministry of Russian Federation
for the Affairs of the Nationalities
19 Trubnikovsky per.
117916 Moscow
Tel: 7 (095) 248-8541
Fax: 7 (095) 291-5715
RUSSIA

Prof. Anatoly B. IMETKHENOV
Baikal Institute for Nature
Management of SB RAS
6 Sakhyanova Str.
670042 Ulan-Ude
Tel: 7 (301-22) 31-565
Fax: 7 (301-22) 63-244
E-mail: root@bien.buriatia.su
RUSSIA

Yuri I. KALASHNIKOV
State Committee of Fishery
of Republic Buryatia
Chairman
28 Kirova Str.
670000 Ulan-Ude
Tel: 7 (301-22) 24-573
Fax: 7 (301-22) 26-840
RUSSIA

Dr. Zb. KARPOWICZ
The World Conservation Union IUCN,
European Programme
219 Huntingdow Rd.
Cambridge CB2 0DL
Tel: ++44 0223 277 802
Fax: ++44 0223 277 175
E-mail: Zbig.Karpowicz@wemc.org.uk
UK

Dr. Lyubov K. KISHCHENKO
Agricultural Institute
Molodezhny Settlement
664038 Irkutsk
Tel: 7 (395-2) 469-440
Telex: 231522 rtgz SUNIVA
RUSSIA

Prof. Edward G. KONNIKOV
Buryat Institute of Geology of SB RAS
Director
2 Pavlov Str.
670015 Ulan-Ude
Tel: 7 (301-22) 30-955
Fax: 7 (301-22) 38-965
E-mail: root@bien.buriatia.su
RUSSIA

Dr. Azat O. KONURBAEV
Khazach Academy of Sciences
28 Schevchenko Str.
480021 Alma-Ata
Tel: 7 (327-2) 623-896
Fax: 7 (327-2) 696-116
KHAZACHSTAN

Prof. Valentin A. KOPTYUG
Chairman of SB RAS
17 Lavrentiev Ave.
630090 Novosibirsk
Tel: 7 (383-2) 356-619
Fax: 7 (383-2) 354-846
Telex: 133128 POISK SU
E-mail: vak@hg.soan.nsk.su
RUSSIA

Prof. Vladimir M. KORSUNOV
Buryat Institute of Biology of SB RAS
Director
6 Sakhyanova Str.
670042 Ulan-Ude
Tel: 7 (301-2) 31-211
E-mail: root@bien.buriatia.su
RUSSIA

Prof. Alexander L. KOVALEVSKY
Buryat Geological Institute of SB RAS
6 Sakhyanova Str.
670042 Ulan-Ude
Tel: 7 (301-22) 36-013
RUSSIA

Dr. Winfried KRAHL
Institute of Environmental Protection of the
Land Baden-Württemberg
Griesbachstrasse 3
D-76185 Karlsruhe
Tel: 0721/983-1424
Fax: 0721/983-1414
GERMANY

Prof. Valery V. KULESHOV
Institute of Economics and
Industrial Engineering of SB RAS
Director
17 Lavrentiev Ave.
630090 Novosibirsk
Tel: 7 (383-2) 350-536
Fax: 7 (383-2) 355-580
E-mail: uno-ie@sinor.nsk.su
RUSSIA

Dr. jur Michael KUSKE
European Union - Delegation of the
European Commission in Russia
Advisor
2/10 Astakhovsky per.
109028 Moscow
Tel: 7 (503) 956-3610 or 7 (502) 220-4649
Fax: 7 (503) 956-3939 or 7 (502) 220-4654
RUSSIA

Prof. Michael I. KUZMIN
Institute of Geochemistry of SB RAS,
Director
1a Favorsky Str.
664033 Irkutsk
Tel: 7 (395-2) 460-500
Fax: 7 (395-2) 462-559
E-mail: mikuzmin@igc.irkutsk.su
RUSSIA

Dr. Jon M. LONEY
Tennessee Valley Authority,
Resource Group
400 West Summit Hill Drive, WT 8C
Knoxville, Tennessee 37902 1499
Tel: 1 (615) 632 3012
Fax: 1 (615) 632 6855
USA

Prof. Askold A. MAIBORODA
State Medical Institute
Rector
1 Krasnogo Vosstania Str.
664003 Irkutsk
Tel: 7 (395-2) 333-597
RUSSIA

Prof. Igor M. MALKOVSKY
Institute of Geography
of the National Academy of Sciences
of Kazakhstan Republic
Vice Director
99 Pushkin Str.
480100 Almaty
Fax: 7 (327-2) 633-411
KAZAKHSTAN

Dr. Irina I. MAKSIMOVA
Limnological Institute of SB RAS
26 B.Dmitrovka Str., Moscow
Tel: 7 (095) 292-6821
Fax: 7 (095) 292-4653
RUSSIA

Prof. Vjatcheslav V. MANTATOV
Baikal Foundation
President
24/2 Solnechnaya Str.
670031 Ulan-Ude
Tel: 7 (301-22) 35-272
Fax: 7 (301-22) 63-244
E-mail: root@techn.buriatia.su
RUSSIA

Anne C.B. MEGIER
University of Bodenkultur
Hirschengasse 15/a
A-1060 Wien
Tel: +43 1 57 38 732
Austria

Prof. Anatoly P. MERENKOV
Siberian Energy Institute of SB RAS
Director
130 Lermontov Str.
664033 Irkutsk
Tel: 7 (395-2) 461-702
Fax: 7 (395-2) 462-796
Telex: 231510 MIRAJ SU
E-mail: root@sei.irkutsk.su
RUSSIA

Prof.Dr. Horst MIERHEIM
Federal Environmental Agency
Director
Mauerstrasse 52
D-10117 Berlin (Mitte)
Tel: 030/23 145-736
Fax: 030/23 156 38
GERMANY

Dr. Bedrich MOLDAN
Charles University,
Centre for Environmental Scholarship
Petrska 3, CZ-11000, Praha 1
Tel: (42-2) 231 5334
Fax: (42-2) 231 5324
E-mail: bedrich.moldan@ruk.cuni.cz
CZECH REPUBLIC

Prof.Dr. Bernhard MULLER
Dresden University of Technology
Mommsenstrasse 13
D-01062 Dresden
Dienstgeboude: Zeunerstrasse 1b
Tel: (0351) 463-5143 or (0351) 463-2994
Fax: (0351) 463-7064
GERMANY

Bavuu MYAGMARJAV
Ministry for Nature and the Environment
5 Hudaldaany Gudamj
Ulaanbaatar 11
Tel: (976-1) 329-619
Fax: (976-1) 321-401
Telex: 79343 MACNE MH
MONGOLIA

Prof. Bimba B.-M. NAMSALOV
Baikal Institute for
Nature Management of SB RAS
6 Sakhyanova Str.
670042 Ulan-Ude
Fax: 7 (301-22) 63-244
E-mail: root@bien.buriatia.su
RUSSIA

Dr. Sergey D. NAMSARAEV
Ministry of Education of Republic Buryatia
Minister
Soviet Square
670000 Ulan-Ude
RUSSIA

Prof. Kuzma A. NIKIFOROV
Buryat Institute of Natural Sciences of SB RAS
6 Sakhyanova Str.
670042 Ulan-Ude
Tel: 7 (301-22) 30-380
Fax: 7 (301-22) 63-244
E-mail: root@bien.buriatia.su
RUSSIA

Dr. Alexander M. NOVIKOV
Ministry of Sciences and Technical Policy
of the Russian Federation
11 Tverskaya Str.
103905 Moscow
Tel: 7 (095) 229-2302
Fax: 7 (095) 925-9609
Telex: 570224 TRIOD
RUSSIA

Michael OBERSTEINER
IIASA (International Institute for
Applied Systems Analysis)
A-2361 Laxenburg
Tel: +43 2236 71521 O
Telex: 079137 IIASA A
Telefax: +43 22236 71313
E-mail: oberstei@iiasa.ac.at.
AUSTRIA

Dr. I. OGORODNIKOV
International Project "ECODOM"
P.O.Box 728
630090 Novosibirsk
Tel: 7 (383-2) 357-392
E-mail: ecodom@glas.apc.org.
RUSSIA

Dr. Werner D'OLIERE-OLTMANNS
National Park Berchtesgaden
Doktorberg 6
D-83471 Berchtesgaden
Fax: 49 (4961) 459
GERMANY

Oleg I. ORLOV
Committee of the State Duma
for the Affairs of the Federation
Expert
36 Novy Arbat Str.
121205 Moscow
Tel: 7 (095) 290-8529
Fax: 7 (095) 290-8504
RUSSIA

Prof. Lyubov V. PETROVA
Ministry of the Russian Federation
for the Problems of the Human Activities
8 Stroiteley Str., stroenie 2
117916 Moscow
Tel: 7 (095) 930-5141
Fax: 7 (095) 251-5715
RUSSIA

Dr. Leonid V. POTAPOV
President of Republic Buryatia
Soviet Square
670000 Ulan-Ude
RUSSIA

Bair B. RALDIN
State Committee on Land Resources
and Management of Republic Buryatia
11a Revolution 1905 Str.
670034 Ulan-Ude
RUSSIA

Prof. Yury G. RANDALOV
Buryat Institute of Social Sciences of SB RAS
6 Sakhyanova Str.
670042 Ulan-Ude
Fax: 7 (301-22) 63-244
E-mail: root@bien.buriatia.su
RUSSIA

Vladimir B. SAGANOV
Narodny Khural of Republic Buryatia
Committee of Finance, Budget
Vice-Chairman
Soviet Square
670000 Ulan-Ude
RUSSIA

Prof. Ryurik K. SALYAEV
Siberian Institute of Plant Physiology
and Biochemistry of SB RAS
Director
132 Lermontov Str.
664033 Irkutsk
Tel: 7 (395-2) 460-721
E-mail: rksal@sifibr.irkutsk.su
RUSSIA

Dr. Boris G. SANEEV
Siberian Energy Institute of SB RAS
130 Lermontov Str.
664033 Irkutsk
Fax: 7 (395-2) 462-796
Telex: 231510 MIRAJ SU
E-mail: root@sei.irkutsk.su.
RUSSIA

Prof. Garmazhab L. SANZHIEV
Buryat Institute of Social Sciences of SB RAS
6 Sakhyanova Str.
670042 Ulan-Ude
Tel: 7 (301-22) 40-997
RUSSIA

Prof. Mikhail F. SAVTCHENKOV
Regional Centre of Medical Ecology
Director
16 Timiryazeva Str.
664003 Irkutsk
Tel: 7 (395-2) 274-816
Fax: 7 (395-2) 274-813
RUSSIA

Heinrich SCHMAUDER
Federal Agency for Nature Conservation
Konstantinstr. 110
D-53179 Bonn
Tel: (0228) 8491-181
Fax: (0228) 8491-200
GERMANY

Prof. Johann SCHREINER
Norddeutsche Naturschutzakademie
Hof Mohr,
D-29640 Schneverdingen
Tel: 49 (5199) 98-913
Fax: 49 (5199) 432
GERMANY

Prof. Igor M. SEDOV
Ministry of Russian Federation
for the Affairs of the Nationalities,
Association of Social-Economic Development
and Certification of Vitability of Man
Vice President
51 Tyulyaeva Str.
350088 Krasnodar
Tel: 7 (861-2) 360-827
RUSSIA

Mikhail I. SEMENOV
Narodny Khural of Republic Buryatia
Chairman
Soviet Square
670000 Ulan-Ude
Tel: 7 (301-22) 21-848
RUSSIA

Prof. Igor V. SEVERSKY
Institute of Geography
of the National Academy of Sciences
of Kazakhstan Republic
Director
99 Pushkin Str.
480100 Almaty
Tel: 7 (327-2) 618-129
Fax: 7 (327-2) 633-411
KAZAKHSTAN

Prof. Karl Sh. SHAGZHIEV
Baikal Institute
for Nature Management of SB RAS
6 Sakhyanova Str.
670042 Ulan-Ude
Tel: 7 (301-22) 30-080
Fax: 7 (301-22) 63-244
E-mail: root@bien.buriatia.su
RUSSIA

Dr. David A. SNEATH
University of Cambridge
Research Co-Ordinator
Newnham Road
Cambridge, C83 9EY
Tel: (44-223) 300 586
Fax: (44-223) 300 589
E-mail: cam.Phx.DS114
UK

Prof. Alexander B. SOKTOEV
Institute of Philology of SB RAS
Director
17 Lavrentiev Ave.
630090 Novosibirsk
Tel: 7 (383-2) 356-821
Fax: 7 (383-2) 357-791
RUSSIA

Dr. Michael A. SPASOV
Government of Republic Buryatia
First Vice-Chairman
Soviet Square
670000 Ulan-Ude
Tel: 7 (301-22) 22-877
RUSSIA

Prof. Izyaslav P. STABIN
Ministry of Russian Federation
for the Affairs of Nationalities,
National Scientific and Coordination Centre
for the Problems of the Northern Peoples
8 Stroitel Str., stroenie 2
117916 Moscow
Tel: 7 (095) 930-5141
Fax: 7 (095) 291-5715
RUSSIA

John Massey STEWART
International Union for the Conservation of Nature (IUCN)
20 Hillway, London N6 6QA
Tel: (44-81) 341-3544
Fax: (44-81) 341-5292
UK

Dr. Tatiana A. STRIZHOVA
Chita Institute of Natural Resources of SB RAS
16 Nedorezov Str.
672090 Chita
Tel: 7 (302-22) 12-498
Fax: 7 (302-22) 12-498
E-mail: chita@las.apc.org
RUSSIA

Dr. Alexander P. SUKHODOLOV
Correspondent of newspaper "Science in Siberia"
Irkutsk
Tel: 7 (395-2) 234-250
RUSSIA

Dr. Alexander N. SUTURIN
Limnological Institute of SB RAS
3 Ulan-Batorskaya Str.
664033 Irkutsk
Tel: 7 (395-2) 466-951
RUSSIA

Dr. Trofim T. TAISAEV
Buryat State Pedagogical Institute
Geography Faculty
24 Smolina Str.
670000 Ulan-Ude
Tel: 7 (301-22) 30-955
RUSSIA

Prof. Vladimir V. TETELMIN
Committee of State Duma on Ecology
Vice-Chairman
2 Okhotny Ryad Str.
103009 Moscow
Tel: 7 (095) 292-6909
RUSSIA

Prof. Arnold K. TULOKHONOV
Baikal Institute of Nature Management of SB RAS
Director
8 Sakhiyanova Str.
670042 Ulan-Ude
Tel: 7 (301-22) 38-035
Fax: 7 (301-22) 63-444
Telex: 219117 LOTOS SU
E-mail: root@bien.buriatia.su
RUSSIA

Yury H. UDODOV
Committee of the Environment Protection and Natural Resources of Irkutskaya Oblast
12 Parkovaya Str.
664012 Irkutsk
Tel: 7 (395-2) 277-850
Fax: 7 (395-2) 276-573
RUSSIA

Prof.Dr. Martin UPPENBRINK
Federal Agency for Nature Conservation
President
Konstantinstr. 110
D-53179 Bonn
Tel: (0228) 8491-210
Fax: (0228) 8491-200
GERMANY

Prof. Evgeny A. VAGANOV
Institute of Forest of SB RAS
Director
Akademgorodok
66036 Krasnoyarsk
Tel: 7 (391-2) 494-695
Telex: 288018 SOSNA SU
E-mail: dndr@ifor.krasnoyarsk.su
RUSSIA

Prof. A.A. VARLAMOV
State University on Land Management
Vice-Rector
15 Kazakova Str.
103064 Moscow
Tel: 7 (095) 261 6691
RUSSIA

Prof. Valerian E. VIKULOV
Buryat Scientific Center of SB RAS
Vice Chairman
8 Sakhyanova Str.
670042 Ulan-Ude
Tel: 7 (301-22) 31-891
Fax: 7 (301-22) 63-244
Telex: 219111 PTBSB
RUSSIA

Dr. Alexey A. VOINOV
Maryland International Institute
for Ecological Economics
MIIEE, CBL, P.O.Box 38,
Solomons, MD 20688
Tel: (1-410) 326-7207
Fax: (1-410) 326-7354
E-mail: voinov@cbl.umd.edu
USA

Prof. Vladimir V. VOROBIEV
Institute of Geography of SB RAS
Director
1 Ulan-Batorskaya Str.
664033 Irkutsk
Tel: 7 (395-2) 462-920
E-mail: root@irigs.irkutsk.su
RUSSIA

Anatoly M. VOZMILOV
Regional Committee for the Ecology
and Land Use
23 Bogomyakova Str.
672000 Chita
Tel: 7 (302-22) 37-403
Fax: 7 (302-22) 68-618
E-mail: chita@las.apc.org
RUSSIA

Prof. Dr. E. WENNIES
Universität Osnabrück
FB Biologiechemie
D-49069 Osnabrück
Telex: 94760 UNIOS D.
GERMANY

V. YAKOVENKO
Vice-Governor of Irkutskaya Oblast
Kirov Square, Dom Sovetov
664027 Irkutsk
Tel: 7 (395-2) 341-651
RUSSIA

Dr. Bair B. ZHAMSUEV
Committee of the State Duma for the
Affairs of the Nationalities
Chairman
1 Okhotny ryad Str., Moscow
Tel: 7 (095) 925-2958
RUSSIA

The Partnership Sub-Series incorporates activities undertaken in collaboration with NATO's Cooperation Partners, the countries of the CIS and Central and Eastern Europe, in Priority Areas of concern to those countries.

The volumes published as a result of these activities are:

Vol. 1: **Clean-up of Former Soviet Military Installations.** Edited by R. C. Herndon, P. I. Richter, J. E. Moerlins, J. M. Kuperberg, and I. L. Biczó. 1995

Vol. 2: **Cleaner Technologies and Cleaner Products for Sustainable Development.** Edited by H. M. Freeman, Z. Puskas, and R. Olbina. 1995

Vol. 3: **Remediation and Management of Degraded River Basins.** Edited by V. Novotny and L. Somlyódy. 1995

Vol. 4: **Earthquakes Induced by Underground Nuclear Explosions.** Edited by R. Console and A. Nikolaev. 1995

Vol. 5: **Transportation Infrastructure.** Edited by R. M. Gutkowski and J. Kmita. 1996

Vol. 6: **Sustainable Development of the Lake Baikal Region.** Edited by V. A. Koptyug and M. Uppenbrink. 1996